Functional Selectivity of G Protein-Coupled Receptor Ligands

The Receptors

Kim A. Neve, Series Editor

For other titles published in this series, go to
www.springer.com/series/7668

Functional Selectivity of G Protein-Coupled Receptor Ligands

New Opportunities for Drug Discovery

Edited by

Kim A. Neve

VA Medical Center, Portland, Oregon, USA

Humana Press

Editor
Kim A. Neve
VA Medical Center
3710 SW US Veterans Hospital Road
Portland, OR 97239-2999, USA
nevek@ohsu.edu

ISBN 978-1-4939-6116-0 ISBN 978-1-60327-335-0 (eBook)
DOI 10.1007/978-1-60327-335-0

Printed on acid-free paper

springer.com

Preface

Functional selectivity refers to the observation that different ligands acting at one subtype of receptor that couples to multiple signaling pathways can vary in their ability to activate the signaling pathways; that is, one drug can be an agonist at pathway A and an antagonist or partial agonist at pathway B, and another drug can have the reverse profile. As discussed by Bryan L. Roth in his introductory chapter, this is not a new notion, with evidence for functional selectivity accumulating over the past 20 years. During the 1990s, as molecular cloning of G protein-coupled receptors (GPCRs) facilitated the unequivocal demonstration that coupling of one type of receptor to multiple signaling pathways is a general characteristic of GPCRs, a number of investigators also demonstrated that ligands differed in their ability to activate those signaling pathways. In other words, the ligands were functionally selective.

During the late 1990s and the early part of this decade, Terry Kenakin and other investigators placed functional selectivity within a theoretical framework, sophisticated structural studies of GPCRs provided a mechanistic basis for the phenomenon by identifying ligand-specific and signaling pathway-specific receptor conformations, the concept of ligand-selective signaling was expanded to include other responses to receptor activation such as phosphorylation and internalization, and functionally selective ligands were identified for many more classes of receptors. The purpose of this book is to review that work.

This phenomenon has many names, including agonist-directed stimulus trafficking, ligand-biased signaling, and ligand-induced differential signaling. The authors and I debated the best name to use in this book, with Ligand-Induced Bias in Downstream Outcome (LIBIDO) being a brief front-runner, and eventually compromised on the use of *functional selectivity*, despite the concerns of some that it is too broad and could include both cell-specific and ligand-specific aspects of differential signaling. In spite of the cell-specific factors that influence selective responses to ligands, I sought to focus this book on the aspect of ligand-selective signaling that can best be controlled for drug development – the ability of ligands to stabilize receptor conformations with distinct functional properties.

The attentive reader will note that the authors are not in complete agreement concerning what functional selectivity is or what its significance is for the basic tenets of pharmacology, with some followers of Hume believing that there are no ligand-specific characteristics, that everything depends on the cellular context in which the receptor is expressed, and, therefore, that there is no basis for ascribing any property such as intrinsic efficacy to a ligand. My own view is more Kantian; I believe that intrinsic efficacy is an invariant characteristic of a ligand–receptor pair that results from the receptor conformation(s) stabilized by that ligand, but that when referring to the intrinsic efficacy of a ligand we must also specify a particular functional response. Cell-specific factors have always been important because, for example, binding of isoproterenol to a β-adrenergic receptor will not stimulate cyclic AMP accumulation if the cell lacks Gαs or adenylate cyclase, but whereas previously one would refer to isoproterenol as a full agonist at the β-adrenergic receptor despite its seeming lack of efficacy in cells lacking the necessary components, now it is also necessary to add "for stimulation of adenylate cyclase." Despite the differences of opinion, the authors agreed not to dispose of the concept of intrinsic efficacy. (Some of the authors may be mouthing "yet" as they read the end of that sentence.)

This book is organized into two parts, with Part I containing six chapters that focus on theoretical or mechanistic aspects of functional selectivity and that cut across subfamilies of GPCRs, and Part II being composed of seven chapters that focus on subfamilies of therapeutically relevant receptors where there is considerable evidence of ligand functional selectivity. This format intentionally produces considerable overlap between the two parts, so that each chapter in one part collects information that is scattered throughout several chapters of the other part. Although, ideally, the reader would begin with the introductory chapter, my goal was to produce a book in which any of the chapters would be an appropriate point of entry into the topic.

Portland, OR Kim A. Neve

Contents

Contributors

Jean-Martin Beaulieu • *CRULRG, Research Institute, Université Laval, Quebec City, QC, Canada*

Kelly A. Berg • *Department of Pharmacology, University of Texas Health Science Center, San Antonio, TX, USA*

Anushree Bhatnagar • *Department of Molecular Cardiology, Lerner Research Institute, The Cleveland Clinic Research Foundation, Cleveland, OH, USA*

Laura M. Bohn • *Departments of Pharmacology and Psychiatry, The Ohio State University College of Medicine, Columbus, OH, USA*

Barbara Bosier • *Laboratoire de Pharmacologie Expérimentale, Université Catholique de Louvain, 1200 Brussels, Belgium*

Justin Brown • *Department of Pharmacology, University of North Carolina School of Medicine, Chapel Hill, NC, USA*

Marc G. Caron • *Departments of Cell Biology, Medicine, and Neurobiology, Duke University Medical Center, Durham, NC, USA*

R.A. John Challiss • *Department of Cell Physiology and Pharmacology, University of Leicester, Leicester, UK*

William P. Clarke • *Department of Pharmacology, University of Texas Health Science Center, San Antonio, TX, USA*

Emmanuel Hermans • *Laboratoire de Pharmacologie Expérimentale, Université Catholique de Louvain, 1200 Brussels, Belgium*

Allyn C. Howlett • *Department of Physiology and Pharmacology, Wake Forest University Health Sciences, Winston-Salem, NC, USA*

Andrew Kant • *Curriculum in Toxicology, University of North Carolina School of Medicine, Chapel Hill, NC, USA*

Sadashiva Karnik • *Department of Molecular Cardiology, Lerner Research Institute, The Cleveland Clinic Research Foundation, Cleveland, OH, USA*

Terry Kenakin • *Biological Reagents and Assay Development, GlaxoSmithKline Research and Development, Research Triangle Park, NC, USA*

Richard B. Mailman • *Departments of Pharmacology and Neurology, Penn State University College of Medicine–Milton S. Hershey Medical Center, Hershey, PA, USA*

Richard R. Neubig • *Department of Pharmacology, The University of Michigan Medical School, Ann Arbor, MI, USA*

Kim A. Neve • *Portland VA Medical Center and Department of Behavioral Neuroscience, Oregon Health and Science University, Portland, OR, USA*

Graciela Piñeyro • *Département de Psychiatrie, Université de Montréal, Montréal, QC, Canada*

Bryan L. Roth • *Pharmacology and Medicinal Chemistry, University of North Carolina at Chapel Hill Medical School, Chapel Hill, NC, USA*

Rachel L. Thomas • *Department of Cell Physiology and Pharmacology, University of Leicester, Leicester, UK*

Yan-Min Wang • *Department of Psychiatry, University of North Carolina School of Medicine, Chapel Hill, NC, USA*

Jennifer L. Whistler • *Department of Neurology, Ernest Gallo Clinic and Research Center, University of California, San Francisco, Emeryville, CA, USA*

Part I
Theoretical and Mechanistic Aspects of Functional Selectivity

Chapter 1
Historical Overview of the Concept of Functional Selectivity

Bryan L. Roth

Abstract A historical overview of the concept of drug action and functional selectivity is provided. The earliest notions of drug-receptor interactions could not account for the diversity of effects induced by ligand-binding to a receptor. Indeed, it has been clear at least since 1965 that the simple notions of "intrinsic activity" cannot not account for agonist and partial agonist actions at receptors. More recent findings have unequivocally demonstrated that functional selectivity is a fact. In this chapter, the therapeutic implications of this conceptual revolution are emphasized.

Keywords Collateral agonism, Antagonist, Agonist, Serotonin receptor

1.1 Introduction

It is now an accepted tenant of molecular pharmacology that many drugs exert their actions in vivo via interactions with *receptors* – an idea once considered so novel *(1,2)*, that this concept was "despised" by many pharmacologists *(3)*. The recent cloning and annotation of the human genome *(4,5)* has revealed the *receptorome* (the entire complement of receptors in the human genome *(6)*) to be one of the largest functionally annotated classes of genes in the genome. The receptorome, which constitutes in excess of 5% of the human genome *(7)*, is dominated by the G-protein coupled receptor (GPCR) superfamily, which represent more than 50% of the open reading frames identified as "receptors" *(6,7)*. GPCRs also represent the single largest target class of proteins exploited for therapeutic drug discovery efforts (i.e., "druggable genome" *(8,9)*). In this book, we will focus entirely on GPCRs.

Since at least the 1950s, it has been an accepted tenant of quantitative pharmacology that agonists are drugs that activate receptors *(3,10)* and that a full agonist will always fully activate a receptor. This concept was dubbed "efficacy" and defined as

B.L. Roth
Departments of Pharmacology and Medicinal Chemistry, University of North Carolina at Chapel Hill Medical School, Chapel Hill, NC 27599-7365
e-mail: bryan_roth@med.unc.edu

K. Neve (ed.), *Functional Selectivity of G Protein-Coupled Receptor Ligands*,
DOI: 10.1007/978-1-60327-335-0_1,

"a new factor, called ***efficacy***, to be postulated which, together with affinity for the receptors, determines the potency of an agonist drug…" *(3)*. Ariens proposed that each agonist had an "intrinsic activity" *(10)* that was "specific" to each agonist, antagonist, and partial agonist. This concept was further codified by Furchgott as the concept of "relative agonist efficacy" *(11)* and experimental approaches to estimate relative efficacies among various partial agonists devised. It has been clear for many years, however, that the notions of "intrinsic activity" or "relative efficacy" cannot fully describe the complexities of drug actions at GPCRs. In this book the emerging notion of ***functional selectivity*** *(12)*, also known as biased agonism *(13)*, differential receptor-linked effector actions *(14)*, agonist-directed trafficking of receptor stimulus *(15)*, or tissue-specific intrinsic efficacy *(16)*, will be highlighted.

1.2 Early History of the Concept of Functional Selectivity

The first inklings of the notion of functional selectivity came from Portoghese *(17)* when he suggested that if the binding modes of drugs of a common class are different one might expect differential pharmacological responses. He proposed a new notion of "configurational selectivity" to account for these differences in physiological responses induced by chemically distinct partial agonists. Later, in studies of isolated blood vessels, Matthews and colleagues identified a series of α-adrenergic agonists that differentially induced contractile responses and intra and extracellular Ca++ mobilization *(8–20)*. They proposed that additional selective adrenergic agonists might be developed, which differentially activated these pathways – an idea that we exploited as a novel way to treat circulatory collapse *(21)*. Around the same time, we proposed a similar schema for 5-HT_2-serotonin receptors *(14,22)* and predicted that "each class of serotonergic receptor may be linked to one or more distinct biochemical transduction mechanisms. The possibility is raised that selective agonists and antagonists might be developed which have specific effects on a particular receptor-linked effector system…" *(14)*. These early studies were criticized at the time (and later) for failing to take into account the likely hetereogeneity of receptors *(23)*.

1.3 Agonist Functional Selectivity in the Post-Genomic Age

With the cloning of GPCR subtypes, which differentially coupled to signaling cascades, it became clear that receptor pharmacology was considerably more complicated than previously anticipated. Early on, however, it was evident that individual cloned human GPCRs could activate multiple G proteins. One of the first unequivocal demonstrations of this was with the 5-HT_{1A} receptor, which was demonstrated to activate multiple G proteins *(24)*. Fairly soon thereafter, it was demonstrated for the 5-HT1A receptor that the intrinsic activity for various signaling pathways showed tremendous variation depending upon the agonist employed *(25–28)*. In my estimation, these studies by the Raymond group represent the cleanest demonstration of functional selectivity. Later studies by Berg et al. *(15)* confirmed that in the same cell type, different agonists displayed different "intrinsic efficacies" for various signaling pathways at 5-HT$_{2A}$ receptors.

1.4 Functionally Selective Antagonists

Although it has now been obvious for more than 20 years that agonist-mediated functional selectivity is a feature of nearly every GPCR-effector-linked signaling pathway, it has been evident for an even longer period of time that the same pharmacological property applied to antagonists as well. The first demonstration of this were the observations by many investigators *(29–31)* that various 5-HT$_{2A}$ receptor antagonists had a differential ability to induce receptor "down-regulation" in vivo. Again, in the pre-genomic era of GPCR research, these studies were widely criticized, and it was not until some years later that we conclusively demonstrated that 5-HT$_{2A}$ receptors were internalized *(32)* and down-regulated *(33)* in vitro and in vivo. Others have similarly reported antagonist-induced internalization (*(34)*, see also *(12)* for review). The differential ability of 5-HT$_{2A}$ antagonists to induce receptor internalization and down-regulation has been dubbed an example of collateral efficacy/functional selectivity *(12)*. It is likely that the differential ability of antagonists to induce receptor down and up-regulation will have therapeutic implications for the treatment of a variety of diseases *(35)*.

1.5 The Future

As this brief historical overview suggests, the notion of functional selectivity is not a particularly new one, although it is currently trendy. The theoretical legitimacy of this notion was held up for many years because of quaint notions of drug action, which were based on a limited number of empirical observations and a lack of appreciation of the complexities of the interactions of drugs with receptors. As we and others have recently suggested *(35)*, functional selectivity has profound therapeutic implications *(12,36,37)*, and it is likely that in the future drugs targeting GPCRs will be developed, which activate distinct effector pathways and have enhanced therapeutic efficacies with fewer side effects.

References

1. Langley J. On the reaction of cells and of nerve-endings to certain poisons, chiefly as regards the reaction of striated muscle to nicotine and to curare. J Physiol 1905;33:374–413.
2. Ehrlich P. Chemotherapeutics: scientific principles, methods and results. Lancet 1913;2:445–51.

3. Stephenson RP. A modification of receptor theory. Br J Pharmacol Chemother 1956;11:379–93.
4. Venter JC, Adams MD, Myers EW, et al. The sequence of the human genome. Science 2001;291:1304–51.
5. Lander ES, Linton LM, Birren B, et al. Initial sequencing and analysis of the human genome. Nature 2001;409:860–921.
6. Armbruster BN, Roth BL. Mining the receptorome. J Biol Chem 2005;280:5129–32.
7. Strachan RT, Ferrara G, Roth BL. Screening the receptorome: an efficient approach for drug discovery and target validation. Drug Discov Today 2006;11:708–16.
8. Hopkins AL, Groom CR. The druggable genome. Nat Rev Drug Discov 2002;1:727–30.
9. Russ AP, Lampel S. The druggable genome: an update. Drug Discov Today 2005;10: 1607–10.
10. Ariens EJ. Affinity and intrinsic activity in the theory of competitive inhibition. I. Problems and theory. Arch Int Pharmacodyn Ther 1954;99:32–49.
11. Furchgott RF. The use of haloalkylamines in the differentiation of receptors and in the determination of dissociation constants of agonist-receptor complexes. In: Harper N, Simmonds A, eds. Advances in Drug Research. New York: Academic Press; 1966:21–55.
12. Urban JD, Clarke WP, von Zastrow M, et al. Functional selectivity and classical concepts of quantitative pharmacology. J Pharmacol Exp Ther 2007;320:1–13.
13. Jarpe MB, Knall C, Mitchell FM, Buhl AM, Duzic E, Johnson GL. [D-Arg1,D-Phe5,D-Trp7,9,Leu11]Substance P acts as a biased agonist toward neuropeptide and chemokine receptors. J Biol Chem 1998;273:3097–104.
14. Roth BL, Chuang D-M. Minireview: multiple mechanisms of serotonergic signal transduction. Life Sci 1987;41:1051–64.
15. Berg KA, Maayani S, Goldfarb J, Scaramellini C, Leff P, Clarke WP. Effector pathway-dependent relative efficacy at serotonin type 2A and 2C receptors: evidence for agonist-directed trafficking of receptor stimulus. Mol Pharmacol 1998;54:94–104.
16. Kenakin T. Are receptors promiscuous? Intrinsic efficacy as a transduction phenomenon. Life Sci 1988;43:1095–101.
17. Portoghese PS. A new concept on the mode of interaction of narcotic analgesics with receptors. J Med Chem 1965;8:609–16.
18. Jim KF, Macia RA, Matthews WD. An evaluation of the ability of a series of full alpha-1 adrenoceptor agonists to release internal calcium in venous smooth muscle. J Pharmacol Exp Ther 1985;235:377–81.
19. Matthews WD, Macia RA, Beckeringh JJ, et al. Calcium utilization in the vasoconstriction to enantiomers of SK&F 89748-A. J Pharmacol Exp Ther 1985;232:330–6.
20. Matthews WD, Forman DL. Vascular alpha adrenergic receptors and signal transduction. Prog Clin Biol Res 1989;286:11–32.
21. Chernow B, Roth BL. Pharmacologic manipulation of the peripheral vasculature in shock: clinical and experimental approaches. Circ Shock 1986;18:141–55.
22. Nakaki T, Roth BL, Chuang DM, Costa E. Phasic and tonic components in 5-HT2 receptor-mediated rat aorta contraction: participation of Ca + + channels and phospholipase C. J Pharmacol Exp Ther 1985;234:442–6.
23. Hieble JP, DeMarinis RM, Matthews WD. Evidence for and against heterogeneity of alpha 1-adrenoceptors. Life Sci 1986;38:1339–50.
24. Fargin A, Raymond JR, Regan JW, Cotecchia S, Lefkowitz RJ, Caron MG. Effector coupling mechanisms of the cloned 5-HT1A receptor. J Biol Chem 1989;264:14848–52.
25. Boddeke HW, Fargin A, Raymond JR, Schoeffter P, Hoyer D. Agonist/antagonist interactions with cloned human 5-HT1A receptors: variations in intrinsic activity studied in transfected HeLa cells. Naunyn Schmiedebergs Arch Pharmacol 1992;345:257–63.
26. Mulheron JG, Casanas SJ, Arthur JM, Garnovskaya MN, Gettys TW, Raymond JR. Human 5-HT1A receptor expressed in insect cells activates endogenous G(o)-like G protein(s). J Biol Chem 1994;269:12954–62.

27. Raymond JR. Multiple mechanisms of receptor-G protein signaling specificity [editorial]. Am J Physiol 1995;269:F141–58.
28. Raymond JR, Mukhin YV, Gelasco A, et al. Multiplicity of mechanisms of serotonin receptor signal transduction. Pharmacol Ther 2001;92:179–212.
29. Peroutka SJ, Snyder SH. Regulation of serotonin2 (5-HT2) receptors labeled with [3H] spiroperidol by chronic treatment with the antidepressant amitriptyline. J Pharmacol Exp Ther 1980;215:582–7.
30. Andree TH, Mikuni M, Tong CY, Koenig JI, Meltzer HY. Differential effect of subchronic treatment with various neuroleptic agents on serotonin2 receptors in rat cerebral cortex. J Neurochem 1986;46:191–7.
31. Bergstrom DA, Kellar KJ. Adrenergic and serotonergic receptor binding in rat brain after chronic desmethylimipramine treatment. J Pharmacol Exp Ther 1979;209:256–61.
32. Berry S, Shah M, Khan N, Roth B. Rapid agonist-induced internalization of the 5-hydroxytryptamine2A receptor occurs via the endosome pathway in vitro. Mol Pharmacol 1996;50:306–13.
33. Willins D, Berry S, Alsayegh L, Backstrom J, Sanders-Bush E, Roth B. Clozapine and other 5-hydroxytryptamine2A receptor antagonists alter the subcellular distribution of 5-hydroxytryptamine2A receptors in vitro and in vivo. Neurosci 1999;91:599–606.
34. Roettger B, Ghanekar D, Rao R, et al. Antagonist-stimulated internalization of the G protein-coupled cholecystokinin receptor. Mol Pharmacol 1997;51:357–62.
35. O'Connor KA, Roth BL. Finding new tricks for old drugs: an efficient route for public-sector drug discovery. Nat Rev Drug Discov 2005;4:1005–14.
36. Urban JD, Vargas GA, von Zastrow M, Mailman RB. Aripiprazole has functionally selective actions at dopamine D2 receptor-mediated signaling pathways. Neuropsychopharmacology 2007;32:67–77.
37. Mailman RB. GPCR functional selectivity has therapeutic impact. Trends Pharmacol Sci 2007;28:390–6.

Chapter 2
Functional Selectivity: Theoretical Considerations and Future Directions

Terry Kenakin

Abstract The selective activation of receptors by some agonists to emphasize some but not all aspects of the receptor signaling capability was proposed on theoretical grounds in 1995 because of data showing reversal of relative orders of potency for different stimulus pathways linked to a single receptor. These data precluded the notion that all agonists produce a single receptor active state. Since that time, a number of different lines of evidence indicate that ligands can bias receptor toward different pathways in cells. Conformational selection within the ensemble of conformation receptors formed during normal function theoretically is capable of producing functional selectivity; this chapter discusses the thermodynamic nature of this effect. Finally, although functional selectivity is a well-documented pharmacological phenomenom duplicated in many laboratories, it is still unclear whether it can be harnessed to produce therapeutically unique effect; it is hoped that studies in translational medicine with functionally selective ligands will furnish the link to therapy.

Keywords Receptor theory, Functional selectivity, Receptor signaling

2.1 Introduction: Linear View of Efficacy

The concept of agonist efficacy was required when it was observed that agonist occupancy curves and functional response curves did not coincide with respect to location along the concentration axis (functional curves are shifted to the left of occupancy curves). To accommodate this, it is necessary to invoke some property of the agonist operative in the production of tissue response, specifically a property of the agonist variously referred to as "intrinsic activity" *(1)*, efficacy *(2)* and intrinsic efficacy *(3)*. These were not terms rooted in physiology but rather were mathematical

T. Kenakin
Department of Biological Reagents and Assay Development
GlaxoSmithKline Research and Development Research Triangle Park, NC 27709
e-mail: Terry.P.Kenakin@gsk.com

K. Neve (ed.), *Functional Selectivity of G Protein-Coupled Receptor Ligands*,
DOI: 10.1007/978-1-60327-335-0_2,

terms inserted to make experimental results coincide with theory. A more physiological approach yielded the standard that largely has replaced these early theories, namely Operational Theory where efficacy and tissue sensitivity is quantified by a term τ *(4)*. Although the settings for these efficacies have changed over the years, the basic assumption driving all of them has not, namely that efficacy stems from a single activated receptor state. This assumption has furnished the basis for receptor and drug classification such as agonist potency ratios. The necessity for invoking functional selectivity as described in this volume stems from violations of these predictions of classical receptor theory.

When considering equiactive concentrations of agonist under null conditions, it is assumed that the ability of each agonist to produce response is subject to identical cellular constraints leaving the difference in potency to be solely due to molecular features controlling ligand affinity and efficacy. If this premise is violated, then the resulting potency ratio loses this immutable property of being dependent on chemical features of the agonists and takes on tissue-specific characteristics. This can be illustrated by analyzing potency ratios for agonists at different stages of stimulus–response coupling. Figure 2.1 shows the effects of β-adrenoceptor agonists on two cardiac functions in rat atria that are both mediated by elevation of cytosolic cyclic AMP through the same receptor activation process *(5)*. At some point in the cell, the two processes utilize the cyclic AMP to yield positive inotropy (increased strength of contraction) and the other to elicit lusitropy (increased rate of relaxation). It also is observed that the lusitropic response is more sensitive to agonist stimulation probably because of a more efficient coupling of the relaxation mechanism to elevated cyclic AMP. It can be seen in Fig. 2.1 that relative potency ratios of two β-adrenoceptor agonists, isoproterenol and pirbuterol, yield the same relative potency when each is compared within the same stimulus–response pathway. This is consistent with equal ratios of τ since this term reflects the efficacy of the agonists (a constant molecular term), the receptor density (constant for both agonists since testing is done in the same tissue), and K_E (reflecting the efficiency of coupling of receptors to the response cascade, which is common for the two agonists when the same pathway is compared).

When relative potencies are compared across two different stimulus–response coupling cascades, then different K_E values for the response are operative. It can be seen that when this prerequisite is violated (compare relative potency for inotropy vs. lusitropy), the potency ratio is much different and not reflective of only agonist efficacy. However, within a given pathway (inotropy or lusitropy), potency ratios agree, a finding consistent with each agonist producing a single receptor active state. In general, such potency ratios have been very consistent for receptor and agonist classification over the years, since the definition of efficacy by Stephenson *(2,)* up to approximately 10 years ago. What has changed since that time is that the number of vantage points to view receptor activation has greatly increased with improving technology. Now there are many methods available to measure agonist interaction with the receptor and the resulting change in receptor behavior beyond simple organ response (i.e., guinea pig ileal contraction as was available to Stephenson), and these increased vantage points have shown an astounding

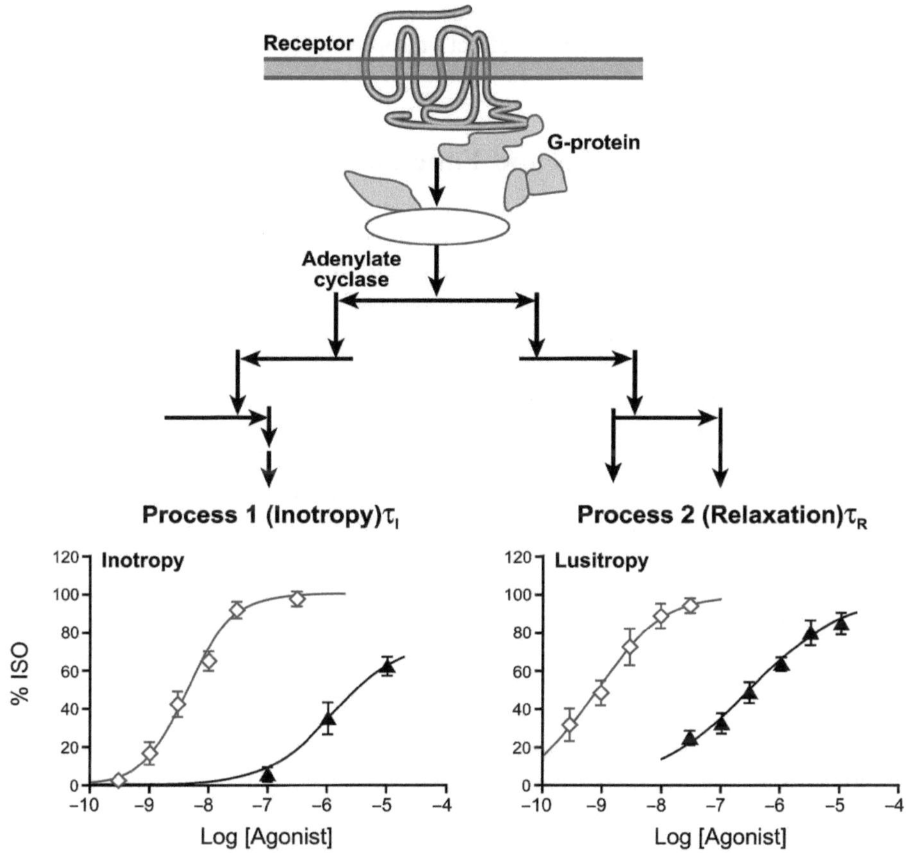

Isoproterenol-Pirbuterol Potency Ratios	
Inotropy-Inotropy	304
Lusitropy-Lusitropy	329
Inotropy-Lusitropy	1444

Fig. 2.1 The coupling of cardiac β-adrenoceptors to adenylate cyclase which, in turn, activates stimulus–response systems to increase cardiac inotropy and myocardial relaxation rate. The coupling is more efficient for relaxation than it is for inotropy, therefore agonists such as isoproterenol (*diamonds*) and pirbuterol (*triangles*) are more potent for lusitropy (relaxation). Within a given stimulus–response cascade, potency ratios are preserved in agreement with receptor theory predictions (tissue effects cancel). However, if this is violated and potencies are compared across stimulus–response coupling cascades the potency ratio is different

heterogeneity in receptor behavior with activation by different ligands. The other important observation is the fact that various receptors have been shown to be pleiotropic with respect to the number of G-proteins with which they interact (particularly with Family B (secretin) receptors). This phenomenon allowed some of the first early opportunities to quantitatively measure more than one consequence of receptor activation by an agonist, namely the effect on separate G-protein pathways.

2.2 Multiple G-Proteins and Functional Selectivity

Although the most simple model dictates that a receptor couples monotonically to a single signaling pathway, there is no a priori reason for this to be the case for all receptors. Moreover, the demonstration that receptors pleiotropically couple to multiple G-proteins suggests that multiple coupling with differential signaling is possible. An early formal model depicting such multi receptor coupling behavior is based on two G-proteins interacting with one receptor *(6)* – see linkage model schematic Fig. 2.2. Intrinsic to such models is the fact that receptor species bound to ligand and/or G-proteins are energetically different than those that are not. Therefore, given two G-proteins, G_1 and G_2, the energy required to form the two ternary species ARG_1 and ARG_2 will fundamentally be different. This furnishes thermodynamic reasons for a given ligand to not be equally adept at producing two such ternary species. The same argument applies to different spontaneous receptor/G-protein complexes, an idea supported by the fact that receptors have different intrinsic affinities for different G-proteins biochemically. The model also fits with the notion that proteins adopt a variety of conformations in accordance to variations in thermal energy *(7–12)*. For example, mutation data, such as that reported for the α_2-adrenoceptor, indicate that multiple receptor conformations are able to activate G-proteins, i.e., there can be multiple receptor active conformations. The model shown in Fig. 2.2 accommodates

$$\frac{[ARG_1]}{[R_{total}]} = \frac{\alpha\gamma L([A]/K_A)([G_1]/K_1)}{[A]/K_A\left(\alpha L(1+\gamma[G_1]/K_1+\beta[G_2]/K_2)+1\right) + L\left(1+[G_1]/K_1) + [G_2]/K_2\right) + 1}$$

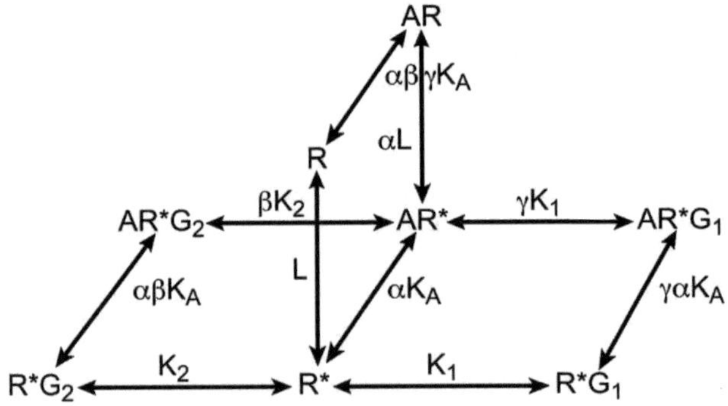

Fig. 2.2 Model of one receptor interacting with two G-proteins (G_1 and G_2). The affinity of the receptor differs for each G-protein (K_1 and K_2) as does that of the ligand-bound receptor (γK_1 βK_2). The receptor forms the active state of the receptor through selective affinity (α). The equation for production of one of the ternary complex species can be used to demonstrate how different agonists can traffic stimulus to different G-proteins and thus how reversals in potency ratios for agonists can occur through selective values of β and γ. This model was presented on theoretical grounds six years before the first experimental evidence to show such behavior was described in the literature *(6)*

this by the presence of specific parameters γ and β, which denote the possibility of differing affinities of the ligand-bound receptor for each G-protein. Interestingly, the model also predicts that different ligands have the capability of actually reversing their relative potency for different G-proteins *(6)*.

The link between differential G-protein coupling and receptor conformation comes from mutation study data indicating that different regions of the receptor interact with different G-proteins. Under these circumstances, it would be highly unlikely that different receptor conformations would expose different regions of the receptor protein in identical ways. The corollary to this idea then is that different receptor conformations in systems that couple to multiple G-proteins would lead to differential activation of the G-protein pathways. The only theoretical piece missing to link these ideas to ligand functional selectivity is the ability of ligands to stabilize different receptor conformations.

A model of efficacy proposed by Burgen *(13)*, namely conformational selection, is useful in imagining the interaction of a ligand with multiple receptor conformations. Burgen's view is that ligands stabilize various conformations by having selectively higher affinities for them (these will preferentially be stabilized). In turn, the preferential stabilization of some conformations in a system of freely interchangeable conformations necessitates that favored conformations will be formed at the expense of other conformations (Le Chatelier's principle, '… *If a dynamic equilibrium is disturbed by changing the conditions, the position of the equilibrium moves to counteract the change…*'). Therefore, when a ligand enters a collection of conformations (to be referred to as an ensemble), it could, by selective micro-affinities, create a new preferred ensemble (Fig. 2.3). Interestingly, since

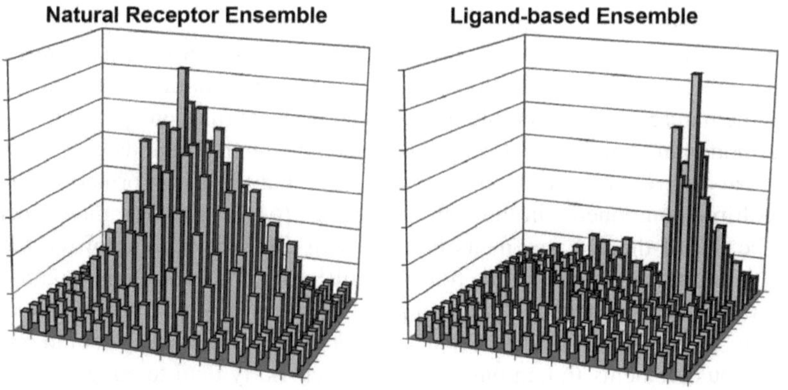

Fig. 2.3 Histograms depicting the relative abundance of receptor conformations for a receptor at rest (*left panel*) and in the presence of a ligand that has different affinities for the various states (*right panel*). In the latter case, the conformations for which the ligand has high affinity are stabilized and therefore enriched at the expense of other conformations. The composition of the new collection of conformations depends upon the molecular structure of the agonist; therefore, there is no a priori reason to suppose that the same ligand-bias will be formed by every agonist

multiple conformations are involved, it need not be that each type of ligand would stabilize an identical ensemble of conformations. In fact, since affinity is specific to chemical structure, it might be postulated that different ligands would *not* form identical ensembles, i.e., that ligands would produce different bias in the receptor conformation that subsequently interacts with the cell *(14)*.

It can be shown that a ligand with varying affinities for a range of receptor conformations necessarily will change the composition of the conformational collection through binding. Assume an ensemble of receptor conformations R (denoted as the root "inactive" state) to R_i. It can be shown that the fraction of receptors not in the R state in the absence of a ligand is given by *(15)*:

$$\rho_{\text{nonR}} = \frac{\sum_{i=1}^{n} L_i}{\left(1+\sum_{i=1}^{n} L_i\right)} \tag{2.1}$$

Where L_1 to L_i are the allosteric constants for the various states ($L_i = [R_i]/[R]$). In the presence of a ligand A having an affinity of K for R and $\psi_1 K_1$ to $\psi_i K_i$ for each of the other states, this expression changes to:

$$\rho_{\text{nonR}} = \frac{\sum_{i=1}^{n} L_i + [A]/K \sum_{i=1}^{n} \psi_i L_i}{[A]/K\left(1+\sum_{i=1}^{n} \psi_i L_i\right) + \left(1+\sum_{i=1}^{n} L_i\right)} \tag{2.2}$$

It can be seen that (2.2) reduces to (2.1) (i.e., there will be no change in the make-up of the conformational ensemble) *only* if Ψ_1 to $\Psi_i = 1$, i.e., only if the affinity of the ligand for every single conformation is *identical*. If this is not the case, then the fraction of conformations different from R in the absence and presence of a ligand will change. By definition, this means that the binding of the ligand will change the nature of the ensemble of the receptor conformations present.

Thermodynamic and theoretical predictions indicate that ligands have the ability to stabilize different receptor conformations, and that these conformations interact with multiple components in the cell membrane *(16)*. In addition, the expectation would be that if these components interact with different regions of the receptor protein, then heterogenous interaction with differing conformations would result. This puts all of the theoretical pieces in place to describe ligand-specific functional selectivity. At this point in time, it remained for an experimental system to combine these various elements to demonstrate this effect. Early data to suggest this came from studies on the PACAP receptor, and this led to the first formal mechanistic model of ligand-specific functional selectivity *(17)*; this model is shown in Fig. 2.4a. The model, formally identical to the one shown in Fig. 2.2, was invoked to describe a particularly striking experimental phenomenon seen in the literature with PACAP receptors, a pleiotropic receptor that activates pathways to elevate cyclic AMP and IP_3. Specifically, it was seen that two PACAP peptide fragments

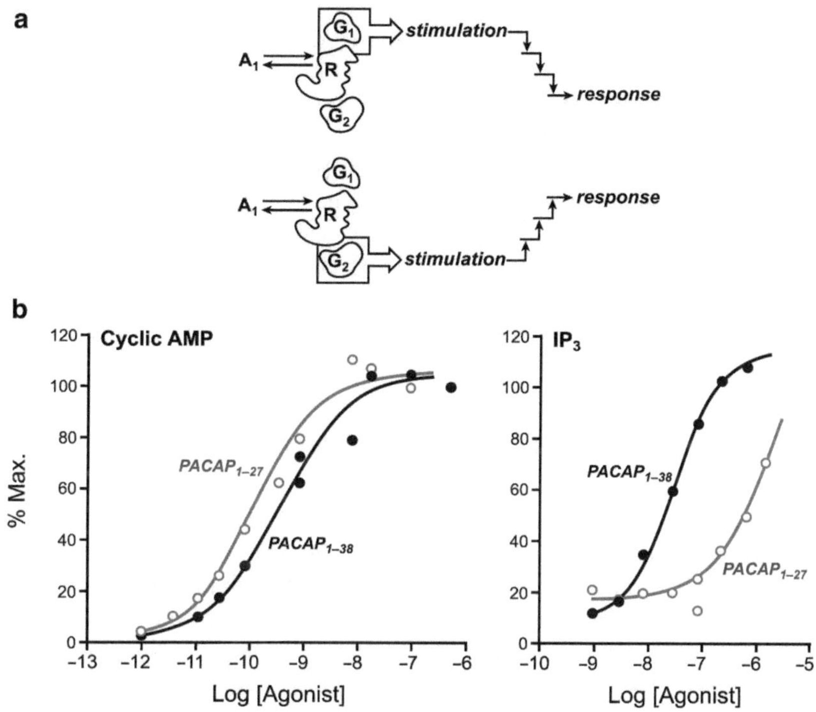

Fig. 2.4 Theoretical model of biased agonism *(17)* based on a one-receptor/two G-protein model. The first data to support this model were reported for PACAP receptors where reversed relative potencies of $PACAP_{1-27}$ and $PACAP_{1-38}$ are clearly inconsistent with a single receptor state produced by these two agonists (model from *(17)*; data from *(18)*)

($PACAP_{1-27}$ and $PACAP_{1-38}$) produced elevated cyclic AMP and IP_3 in cells but the relative potency of these two agonists for these pathways was *reversed (18)*. Thus, the relative efficacy of $PACAP_{1-27}$ for cyclic AMP elevation is higher than that for $PACAP_{1-38}$ but lower for elevation of IP_3. This phenomenon is not compatible with these agonists producing a single active state of the receptor that goes on to activate these two pathways. In contrast, it suggests that $PACAP_{1-27}$ produces an active state with higher efficacy for cyclic AMP stimulus components (relative to $PACAP_{1-38}$) and that $PACAP_{1-38}$ produces an active state with higher efficacy for IP_3 stimulus components.

The model depicted in Fig. 2.4 is sufficient to describe the differential signaling properties of $PACAP_{1-27}$ and $PACAP_{1-38}$ but no doubt other models are capable of doing this. The more important outcome of the analysis of the PACAP data is the demonstration of total inconsistency of such behavior with a single receptor active state model of agonist function. These data provided a serious question to the assumption that agonists form only one receptor active state to induce response but it should be noted that the conceptual thread described here is not the only one questioning the linear concept of agonist efficacy (*vide infra*).

This phenomenon originally was labeled as "stimulus trafficking" when first described *(17)* but subsequently has been referred to in the literature by a number of labels including "biased agonism," "collateral efficacy," "receptor-based functional selectivity," "conformation-based functional selectivity," and simply "functional selectivity." In subsequent years, versions of this phenomenon, namely differential signaling by different agonists acting on the same receptor, also have been described in a variety of settings beyond multiple G-protein activation including desensitization, phosphorylation, receptor internalization, and, recently and notably in β-arrestin/receptor interactions *(19–25)*. Also over the past decade, advances in technology have led to independent data to support the notion that different ligands stabilize different conformations of the same receptor *(26–30)*.

2.3 Links to Established Allosteric Mechanisms

The previous discussion is concerned with ligands that bind in special ways to the receptor to produce an active effect. However, there is no mechanistic difference between this effect and long established models describing allosteric effects of molecules, i.e., standard allosteric molecules such as muscarinic modulators that bind to receptors to stabilize certain conformations that have special properties with respect to their interaction with natural ligands. The operational differences between these established mechanisms and functionally active ligands may involve the geography of binding (i.e., functional antagonists may or may not bind to the natural orthosteric endogenous agonist binding site), and the fact that the allosteric effect is expressed through an active receptor property (functional agonism) as opposed to modification of the effects of other ligands (allosteric modulation). However, the lines become blurred in these distinctions with allosteric agonists such as alcuronium where effects on endogenous agonists are mixed with a direct agonism by the allosteric ligand *(31)*.

Functional selectivity is remarkably similar to classical allosteric modulation. Thus the binding of an allosteric modulator can impose functional selectivity on endogenous agonism through bias of the conformations possible with the binding of the endogeonous agonist. For example, in systems containing CRTH2 receptors, the allosteric modulator *N*α-tosyltryptophan causes the natural agonist prostaglandin D2 to change from Gi and β-arrestin activation to solely Gi-activation (with no concomitant β-arrestin interaction; *(32)*). Similarly, the allosteric modulator AMD3100 blocks natural agonist (SDF-1α)-mediated chemotaxis via CXCR4 receptor but not the effects of peptide fragments RSVM and ASLW *(33)*. Also, the natural agonist neurokinin A acts via NK2 receptors to activate Gs and Gq, while the allosteric modulator LP1805 changes this pattern to one of enhanced Gq activation and blockade of Gs activation *(34)*. The key to these effects is the fact that an allosteric mechanism allows the receptor to be permissive and edit the effects of other ligands by cobinding with them. The relationship between functional agonism and classical allosteric mechanisms is illustrated schematically in Fig. 2.5.

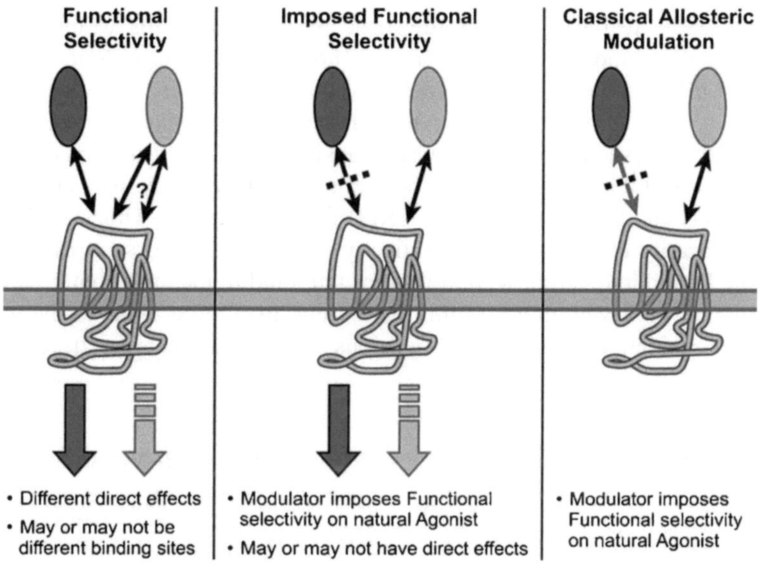

Fig. 2.5 Relationship of functional selectivity (stimulus trafficking) to conventional ligand-induced allosterism. The production of different receptor active states (*left panel*) may or may not involve a binding site for the selective agonism separate from the endogenous binding site. Conventional models of allosterism (*right panel*) describe a modulator binding to a site separate from the endogenous site to modify the interaction of the receptor with the endogenous agonist. There is no formal difference between this model and one that describes direct consequences of modulator binding (i.e., agonism) or modification of endogenous signaling by a modulator (*middle panel*)

2.4 Beyond G-Proteins and Application to Therapeutics

At this point in time, there is little doubt that ligands are able to exhibit functional selectivity and the question now becomes, is it physiologically relevant and can pharmacology harness such a potentially powerful mechanism to therapeutic advantage? Some of the earliest work in this area, originating from work closely associated with therapeutics (namely dopamine treatment of CNS disorders; for review see *(35)*), suggests that direct therapeutic advantages may be derived from functional selectivity. Functional selectivity also has been associated with CNS behavior patterns for serotonin ligands through the 5-HT_{2A} receptor providing further links to the molecular mechanism and therapeutic events *(36)*. For future work in this field, two general ideas may be relevant. The first is that the efficacy of any given ligand is defined by the assay used to detect it, e.g., the ERK stimulating activity of propranolol was not detected for 40 years until propranolol was tested in an ERK assay *(37–39)*. A related idea is that binding of ligands to receptors is an active, not passive, process and that ensembles of receptor conformations are changed by the binding of ligands (i.e., see *(40)*). Therefore, the most generic screening assay available may be the most efficient since this would detect all compounds that bind to the receptor with no reference to predefined efficacy. Functional efficacies then could be detected in various other assays on a smaller scale. Thus, a generic screen (i.e., bioluminescence resonance energy transfer (BRET) and fluorescence resonace energy transfer (FRET)) of a million compounds might detect 300 that bind and then these could be tested in 5–10 therapeutically oriented assays to determine possible useful activity (Fig. 2.6). This may be preferable to arbitrarily choosing a therapeutically oriented assay to start with and risk not seeing ligand interaction with molecules that bind to the target but do not elicit that particular observed effect. In this sense, any ligand that binds to the receptor should be considered a potentially efficacious drug in a variety of settings.

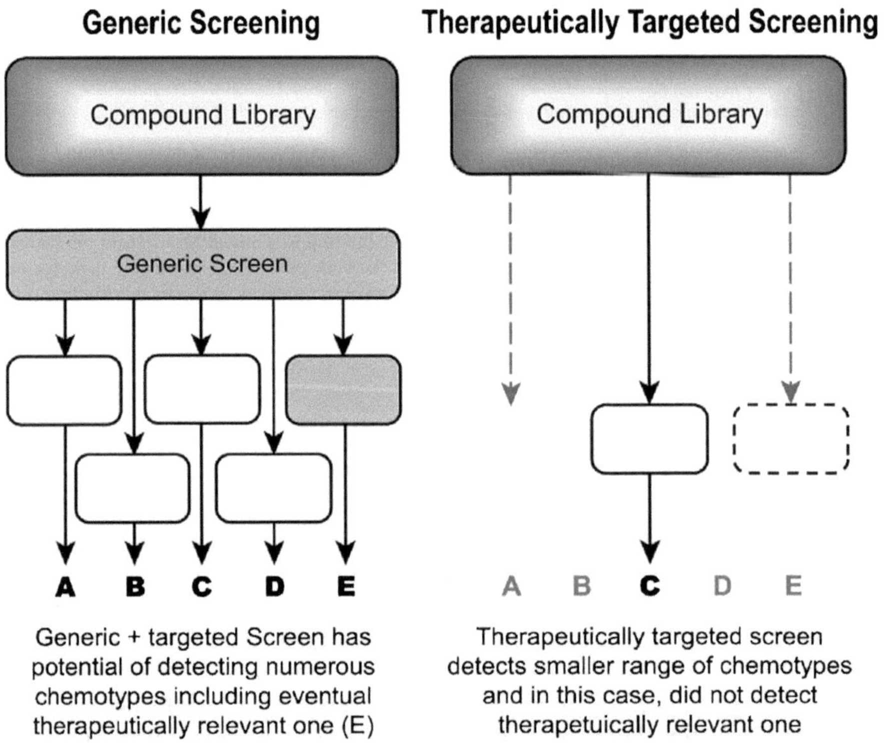

Fig. 2.6 Two modes of screening. In generic screening (i.e., BRET or FRET detection of ligand-receptor interaction), the fact that the ligand-bound receptor is thermodynamically different from the unliganded receptor predicts that all molecules that bind to the receptor will be detected. Secondary testing of the subset of binding molecules (much smaller set than the original library) can then bind compounds with respect to function. On the *right* is shown a therapeutically relevant screen where a specific receptor coupling pathway is chosen for detection. This may shorten the process if evidence is strong that the pathway is all that is required for therapeutic activity. On the other hand, ligands with unknown potential will be missed and the approach will not work if the chosen pathway is the incorrect one

It may be useful to speculate on where functional selectivity of natural system might be a useful physiological control. One such area might be in systems with pleiotropic receptor coupling. For example, associations between selective coupling and physiology have been made for the thyrotropin receptor, which couples to Gs and Gq protein; the Gs protein coupling may be associated with thyroid growth and differentiation, while the Gq coupling may be more associated with thyroid hormone synthesis *(41)*. Another case may be the orexin receptor, where selective agonism may have significance with respect to differences in adrenal steroid production and release (Gs protein) and catchecholamine release (Gq protein) *(42,43)*.

A second area where functional selectivity may be important is in redundant systems, and it raises the question whether or not natural systems make use of this potentially powerful mechanism. There are suggestions that this may be the case. Thus, studies show that ligand-bound receptor active states (some with natural ligands such as catecholamines, dopamine, and natural enkephalins) differ from spontaneously formed constitutive active states *(44–46)*. This would be a way to achieve fine control of signaling through the same receptor in response to hormonal input vs cellular constitutive setpoints.

Another setting where functional selectivity may be important is systems where the chemical input to the receptor is redundant. Perhaps the most redundant and pleiotropic receptor system of all is the chemokine system, where multiple natural agonists are known to activate a range of receptors (Fig. 2.7). It might be expected that this redundancy could naturally be exploited to yield subtle differences in signaling for physiological benefit. Evidence of such functional selectivity is emerging;

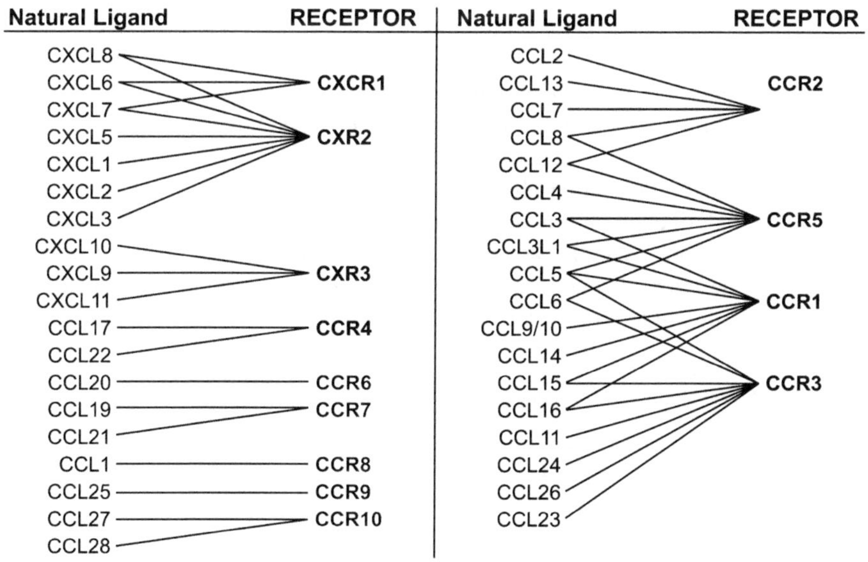

Fig. 2.7 Redundancy of chemical input to chemokine receptors. Taken from *(47)*

for example, two natural agonists for the CCR7 chemokine receptor, namely CCL19 and CCL21, differ in the type of pathway stimulation they elicit through this receptor. Specifically, only CCL19, not CCL21, causes the receptor to undergo agonist-dependent phosphorylation and recruitment of β-arrestin as a means of terminating the G-protein stimulus *(48)*.

A therapeutic application of functional selectivity theoretically can be found in the treatment of AIDS through CCR5-mediated blockade of HIV-1 entry. Specifically, a number of allosteric CCR5 HIV-1 entry inhibitors have been described (see *49,50* for review) and these function through prevention of the virus binding to and utilizing CCR5 for infection. Interestingly, separate data indicate that the natural chemokine system can be beneficial in the delay of AIDS from HIV-1 infection *(51–56)*. Although suggestive, these studies are difficult to interpret since measurement of elevated chemokines is technically difficult as chemokines are produced and utilized at the site of action. A novel way around this limitation has been reported in a large clinical trial (1,064 patients) where the gene copy number for a variable chemokine ligand for CCR5 (CCL3L1) is strongly correlated with AIDS survival *(57)*. Specifically, patients with high gene copy numbers for CCL3L1 have a much greater rate of survival and slowed progression to AIDS than patients with a low gene copy number for this chemokine. At present, on the one hand, all clinically tested CCR5 HIV-1 inhibitors block chemokine function as well as HIV-1 entry; theoretically, an allosteric modulator that prevents the utilization of CCR5 by HIV-1 but otherwise allows the natural chemokine system to function through this receptor (i.e., exhibits functional selectivity) could increase the efficacy for treatment of AIDS. On the other hand, it may be simplistic to suggest that preservation of chemokine function could uniformly be beneficial since CCR5 receptor activation through natural chemokines is known to produce a variety of effects, some not conducive to protection against AIDS (Table 2.1). Thus, although activation of AKT and increased neuronal survival in AIDS dementia have been suggested to be useful effects of chemokine receptor stimulation in HIV-infected patients, other signals such as activation of P38 leading to immunocompetent cell death, Gi-protein-mediated increased replication of HIV virus, and nonspecific inflammation have negative ramifications *(58–61)*. This makes it incumbent upon pharmacologists to understand the pathophysiology of the target to define which functions mediated by the receptor would be therapeutically beneficial. One protective action of CCR5 activation that has been identified as beneficial is the internalization of the CCR5 receptor since this would block HIV-1 entry and also preclude viral resistance through mutation *(51–56)*.

Table 2.1 Physiological consequences of CCR5 receptor activation

Adverse effects	Beneficial effects
• Immunocompetent cell death (↑P38)	• Immune preservation (↑AKT)
• ↑HIV replication	• ↑CCR5 internalizaiton
• ↑Inflammation	• ↑Neuronal survival (AIDS Dementia)

Data from *(58–61)*

2.5 Conclusions and Perspective

The ability of different ligands to direct signals to different cellular interactants to produce texture in signaling (a general phenomenon referred to as "functional selectivity") is well established in mechanistic, theoretical, thermodynamic, and experimental terms. It would be tempting to believe that such a versatile physiological system would not be redundant in normal biology but rather would be employed for fine control of chemical signaling to cells. However, to date, sound data to associate this pharmacological mechanism with normal physiology or pathophysiology is still lacking. However, this gap should be eliminated in the near future as new classified functional ligands are introduced into clinical therapy; it would be hoped that translational medicine will connect functional selectivity with therapeutic phenotypic behavior. The supply end of this process still requires functionally selective ligands and the key to finding these is the appropriate assay system (both in screening and lead optimization of new chemical entities). At the very least, the concept of receptors as "on-off" switches (as first described by John Newport Langley (1852–1926) – *(62)*) is laid to rest with the appreciation of the complexity of seven transmembrane signal processing capability.

References

1. Ariens EJ. Affinity and intrinsic activity in the theory of competitive inhibition. Arch Int Pharmacodyn Ther 1954; 99:32–49.
2. Stephenson RP. A modification of receptor theory. Br J Pharmacol 1956; 11:379–93.
3. Furchgott RF. The use of β-haloalkylamines in the differentiation of receptors and in the determination of dissociation constants of receptor-agonist complexes. In: Harper NJ, Simmonds AB, eds. Advances in Drug Research, Vol. 3, Academic Press, New York; 1966: 21–55
4. Black JW, Leff P. Operational models of pharmacological agonism. Proc R Soc Lond [Biol] 1983; 220:141–62.
5. Kenakin TP, Ambrose JR, Irving PE. The relative efficiency of β-adrenoceptor coupling to myocardial inotropy and diastolic relaxation: Organ selective treatment for diastolic dysfunction. J Pharmacol Exp Ther 1991: 257;1189–97.
6. Kenakin TP, Morgan PH. The theoretical effects of single and multiple transducer receptor coupling proteins on estimates of the relative potency of agonists. Mol Pharmacol 1989; 35:214–22.
7. Fraunfelder H, Parak F, Young RD. Conformational substrates in proteins. Annu Rev Biophys Biophys Chem 1988; 17; 451–79.
8. Fraunfelder H, Sligar SG, Wolynes PG. The energy landscapes and motions of proteins. Science 1991; 254;1598–603.
9. Hilser J, Freire E. Predicting the equilibrium protein folding pathway: structure-based analysis of staphylococcal nuclease. Protein Struct Funct Genet 1997; 27:171–83.
10. Hilser J, Dowdy D, Oas TG, Freire E. The structural distribution of cooperative interactions in proteins: Analysis of the native state ensemble. Proc Natl Acad Sci USA 1998; 95:9903–8.
11. Onaran HO, Costa T. Agonist efficacy and allosteric models of receptor action. Ann NY Acad Sci 1997; 812:98–115.

12. Onaran HO, Scheer A, Cotecchia S, Costa T. A look at receptor efficacy. From the signaling network of the cell to the intramolecular motion of the receptor. In: Kenakin TP, Angus JA, eds. The Pharmacology of Functional, Biochemical, and Recombinant Systems Handbook of Experimental Pharmacology, Vol. 148, Springer, Heidelberg; 2000; 217–80.
13. Burgen ASV. Conformational changes and drug action. Fed Proc 1966: 40;2723–8.
14. Kenakin TP. Efficacy at G protein coupled receptors. Nat Rev (Drug Discovery) 2002; 1:103–9.
15. Kenakin, TP. Collateral efficacy as pharmacological problem applied to new drug discovery. Expert Opin Drug Disc 2006; 1:635–52.
16. Kenakin TP. Collateral efficacy in drug discovery: taking advantage of the good (allosteric) nature of 7TM receptors. Trends Pharmacol Sci 2007; 28:407–15.
17. Kenakin TP. Agonist-receptor efficacy II: Agonist-trafficking of receptor signals. Trends Pharmacol Sci 1995; 16:232–8.
18. Spengler D, Waeber C, Pantaloni C, Holsboer F, Bockaertet J, Seeburgt PH et al. Differential signal transduction by five splice variants of the PACAP receptor. Nature (London) 1993; 365:170–5.
19. Kenakin TP. Efficacy at G Protein Coupled Receptors. Annu Rev Pharmacol Toxicol 2002; 42:349–79.
20. Kenakin TP. Efficacy in drug receptor theory: Outdated concept or under-valued tool? Trends Pharmacol Sci 1999; 20:400–5.
21. Urban JD, Clarke WP, von Zastrow M, Nichols DE, Kobilka B, Weinstein H et al. Functional selectivity and classical concepts of quantitative pharmacology. J Pharmacol Exp Ther 2007; 320:1–13.
22. Hermans E. Biochemical and pharmacological control of the multiplicity of coupling at G-protein receptors. Pharmacol Ther 2003; 99:25–44.
23. Perez DM, Karnick SS. Multiple signaling states of G-protein coupled receptors Pharmacol Rev 2005; 57:147–61.
24. Kukkonen JP. Regulation of receptor-coupling to (multiple) G proteins: A challenge for basic research and drug discovery. Recept Channels 2004; 10:167–83.
25. Kenakin TP. Collateral efficacy in drug discovery: Taking advantage of the good (allosteric) nature of 7TM receptors (Special Issue on Allosterism and Collateral Efficacy in Drug Discovery). Trends Pharmacol Sci 2007; 28:407–15.
26. Gether U, Lin S, Kobilka BK. Fluorescent labeling of purified β2-adrenergic receptor: evidence for ligand specific conformational changes. J Biol Chem 1995; 270:28268–75.
27. Ghanouni P, Gryczynski Z, Steenhuis JJ, Lee TW, Farrens DL, Lakowicz JR et al. Functionally different agonists produce distinct conformations in G-protein coupling domains of the β2-adrenergic receptor. J Biol Chem 2001; 276:24433–6.
28. Palanche T, Ilien B, Zoffmann S,Reck MP, Nucher B, Edelstein SJ et al. The neurokinin A receptor activates calcium and cAMP responses through distinct conformational states. J Biol Chem 2001; 276:34853–61
29. Swaminath G, Xiang Y, Lee TW,Steenhuis J, Parnot C, Kobilka BK et al. Sequential binding of agonists to the β2 adrenoceptor: Kinetic evidence for intermediate conformational states. J Biol Chem 2004; 279:686–91.
30. Hruby VJ, Tollin G. Plasmon-waveguide resonance (PWR) spectroscopy for directly viewing rates of GPCR/G-protein interactions and quantifying affinities Curr Opin Pharmacol 2007; 7:507–14.
31. Jakubic J, Bacakova L, Lisá V, El-Fakahany EE. Tucek of muscarinic acetylcholine receptors via their allosteric binding sites Proc Natl Acad Sci USA 1996; 93:8705–9
32. Mathiesen JM, Ulven T, Martini L, Gerlach LO, Heineman A, Kostenis E. Identification of indole derivatives exclusively interfering with a G protein-independent signaling pathway of the prostaglandin D2 receptor CRTH2. Mol Pharmacol 2005; 68:393–402.
33. Sachpatzidis A, Benton BK, Manfredi JP, Wang H, Hamilton A, Dohlman HG, Loliset E. Identification of allosteric peptide agonists. J Biol Chem 2003; 278:896–907.

34. Maillet EL, Pellegrini N, Valant C, Bucher B, Hibert M, Bourguignon J-J, Galzi J-L. A novel, conformation-specific allosteric inhibitor of the tachykinin NK2 receptor (NK2R) with functionally selective properties. FASEB J 2007; 21:2124–34.
35. Mailman RB, GPCR functional selectivity has therapeutic impact. Trends Pharmacol Sci 2007; 28:390–7.
36. Schmid CL, Raehal KM, Bohn LM. Agonist-directed signaling of the serotonin 2A receptor depends on β-arrestin2 interactions in vivo. Proc Natl Acad Sci USA 2008; 105:1079–84.
37. Galandrin S, Bouvier M. Distinct signaling profiles of β1 and β2 adrenergic receptor ligands toward adenylyl cyclase and mitogen-activated protein kinase reveals the pluridimensionality of efficacy. Mol Pharmacol 2006; 70:1575–84.
38. Azzi M, Charest PG, Angers S, Rousseau G, Kohout M. β-arrestin-mediated activation of MAPK by inverse agonists reveals distinct active conformations for G-protein-coupled receptors. Proc Natl Acad Sci USA 2003; 100:11406–11
39. Baker JG, Hall IP, Hill SJ. Agonist and inverse agonist actions of β-blockers at the human β2-adrenoceptor provide evidence for agonist-directed signaling. Mol Pharmacol 2003; 64:1357–69
40. Kenakin TP, Onaran O. The ligand paradox between affinity and efficacy: Can you be there and not make a difference? Trends Pharmacol Sci 2002; 23:275–80.
41. Vassart G, Dumont D. The thyrotropin receptor and the regulation of thyrocyte function and growth Endocr Rev 1992; 13:596–611.
42. Mazzocchi G, Malendowicz LK, Aragona F,Nussdorfer GG. Human pheochromocytomas express orexin receptor type 2 gene and display an in vitro secretory response to orexins A and B. J Clin Endocrinol Metab 2001; 86:4818–21.
43. Mazzocchi G, Malendowicz LK, Gottardo AF, Nussdorfer GG. Orexin A stimulates cortisol secretion from human adrenocortical cells through activation of the adenylate cyclase-dependent signaling sascade J Clin Endocrinol Metab 2001; 86:778–82.
44. Zhou Y-Y, Cheng H, Song L-S, Wang D, Lakatta EG, Xiao R-P. Spontaneous β2-adrenergic signaling fails to modulate L-type Ca2+ current in mouse ventricular myocytes. Mol Pharmacol 1999; 56:485–93.
45. Wiens BL, Nelson CS, Neve KA. Contribution of serine residues to constitutive and agonist-induced signaling via the D2s dopamine receptor: evidence for multiple, agonist-specific active conformations. Mol Pharmacol 1998; 54:435–44.
46. Liu JG, Ruckle MB, Prather PL. Constitutively active μ-opioid receptors inhibit adenylyl cyclase activity in intact cells and active G-proteins differently than the agonist [D-Ala2,N-MePhe4, Gly-ol5]enkephalin. J Biol Chem 2001; 276:37779–86.
47. Wells TNC, Power CA, Shaw JP,Proudfoot AEI. Chemokine blockers – therapeutics in the making? Trends Pharmacol Sci 2004; 27:41–47
48. Kohout TA, Nicholas SL, Perry SJ, Reinhart G, Junger S, Struthers RS. Differential desensitization, receptor phosphorylation, β-arrestin recruitment, and ERK1/2 activation by the two endogenous ligands for the CC chemokine receptor 7. J Biol Chem 2004; 279:23214–22
49. Kazmierski W, Peckham JP, Duan M, Kenakin TP, Jenkinson S, Gudmundsson KS et al. Recent progress in discovery of new CCR5 and CXCR4 chemokine receptor antagonists as inhibitors of HIV-1 entry. Part 2. Curr Med Chem Anti-Infective Agents 2005; 4:2456–72.
50. Kazmierski WM, Kenakin TP, Gudmundsson KS. Peptide, peptidomimetic and small-molecule drug discovery targeting HIV-1 host-cell attachment and entry through gp120, gp41, CCR5 and CXCR4. Chem Biol Drug Dis 2006; 67:13–26.
51. Xiang J, George SL, Wünschmann S, Chang Q, Klinzman D, Stapletonet JT. Inhibition of HIV-1 replication by GB virus C infection through increases in RANTES, MIP-lα, MIP-1β, and SDF-1 Lancet 2004; 363:2040–6.
52. Garzino-Demo A, Moss RB, Margolick JB, Cleghorn F, Sill A, Blattner WA et al. Spontaneous and antigen-induced production of HIV-inhibitory β-chemokines are associated with AIDS-free status. Proc Natl Acad Sci USA 1999; 96:11986–91.

53. Heredia A, Davis C, Amoroso A, Dominique JK, Le N, Klingebiel E et al. Induction of G1 cycle arrest in T lymphocytes results in increased extracellular levels of β-chemokines: A strategy to inhibit R5 HIV-1. Proc Natl Acad Sci 2003; 100:4179–84
54. Ullum H, Cozzi A, Victor J, Aladdin H, Phillips AN, Gerstoft J et al. Production of β-chemokines in human immunodeficiency virus (HIV) infection: Evidence that high levels of macrophage inflammatory protein-1β are associated with a decreased risk of HIV disease progression. J Infect Dis 1998; 177:331–7.
55. Lori F, Jessen H, Foli A,Lisziewicz J, Matteo PS. Long-term suppression of HIV-1 by hydroxyurea and didanosine. JAMA 1997; 277:1437–8
56. Rogez C, Martin M, Dereuddre-Bosquet N, Martal J, Dormont D, Clayette P. Human immunodeficiency virus activity of tau Interferon in human macrophages: involvement of cellular factors and β-chemokines. J Virol 2003; 77:12914–20
57. Gonzalez E, Kulkarni H, Bolivar H, Mangano A, Sanchez R, Catano G et al. The influence of *CCL3L1* gene-containing segmental duplications on HIV-1/AIDS susceptibility. Science 2005; 307:1434–40.
58. Choi W-T, Kaul M, Kumar S, Wang J, Kumar IMK, Dong CZ et al. Neuronal apoptotic signaling pathways probed and intervened by synthetically and modularly modified (SMM) chemokines. J Biol Chem 2007; 282:7154–63.
59. Kaul M, Lipton SA. Chemokines and activated macrophages in HIV gp120-induced neuronal apoptosis. Proc Natl Acad Sci USA 1999; 96:8212–6.
60. Tyner JW, Uchida O, Kajiwara N,Kim EY, Patel AC, O'Sullivan MP et al. CCL5-CCR5 interaction provides antiapoptotic signals for macrophage survival during viral infection. Nat Med 2005; 11:1180–7.
61. Vhlahakis SR, Villasis-Keever A, Gomez T,Vanegas M, Vlahakis N, Payaet CV. G protein-coupled chemokine receptors induce both survival and apoptotic singaling pathways. J Immunol 2002; 169:5546–54.
62. Holmstedt B, Liljestrand G. Readings in Pharmacology, Raven Press, New York, 1981.

Chapter 3
Agonist-Selective Coupling of G Protein-Coupled Receptors

Barbara Bosier and Emmanuel Hermans

Abstract During the last 20 years, molecular and biochemical data concerning G protein-coupled receptors (GPCRs) have accumulated, providing a detailed characterization of the structure and function of this large family of receptors. Initially viewed as simple tran sducing proteins interacting with intracellular adapters that confer signaling specificity and amplification, the last decade has revealed the extreme complexity and flexibility offered by these membrane receptors. Indeed, the capacity to interact with several unrelated G proteins, which was originally considered as a peculiar property of some recombinant receptors, is now demonstrated for the vast majority of GPCRs. The mechanisms governing and regulating this multiplicity of coupling have been deeply investigated, highlighting the physiological and pharmacological consequences, which are herein reviewed. Of particular importance is the emerging concept of functional selectivity, which explains the capacity of a ligand to selectively orientate the coupling of a receptor with a subset of G proteins. Obviously, future studies should help to transpose functional selective ligands into functional selective drugs showing enhanced clinical efficacy with lower unwanted side effects. In addition, ligands endowed with functional selectivity constitute relevant tools for exploring the GPCR functions in physiological and pathological processes.

Keywords Functional selectivity, G protein-coupled receptor, Multiplicity of coupling, Intracellular signaling, Receptor conformation, Signal transduction

3.1 Introduction

The evolution of pluricellular organisms has been accompanied by the apparition and enlargement of several receptor families widely distributed in all tissues. Responding to a limited set of key endogenous transmitters, a complex system of

B. Bosier and E. Hermans
Laboratoire de Pharmacologie Expérimentale, Université catholique de Louvain, 1200 Brussels, Belgium
e-mail: emmanuel.hermans@uclouvain.be

K. Neve (ed.), *Functional Selectivity of G Protein-Coupled Receptor Ligands*,
DOI: 10.1007/978-1-60327-335-0_3,

receptor subtypes allows the generation of multifaceted but integrated responses. Ultimate functions of receptor activation may be divided in two main goals. First, both ionotropic receptors, eliciting immediate and transient responses, and metabotropic receptors, mediating slower responses, are able to modify cell-to-cell communication. Second, alterations in metabolic activities as well as cell differentiation and/or proliferation are commonly achieved through interactions of transmitters with both tyrosine kinase-coupled receptors and G protein-coupled receptors (GPCR). Considering the importance of these activities during development, as well as in the control of tissue function and dysfunction, these receptor systems constitute major targets in pharmacology. Among these, GPCRs have received the largest attention, and today these membrane molecules are probably the best characterized proteins in mammalian cells.

GPCRs are involved in a wide array of physiological functions after activation by various extracellular stimuli including neurotransmitters, hormones, chemokines, Ca^{2+} ions, and even light. Though several studies have demonstrated the existence of G protein-independent signalings elicited after activation of some receptors, the common biochemical feature of GPCRs – as indicated by their name – is their interaction with heterotrimeric G proteins ensuring the coupling with intracellular signaling partners. Molecular assemblies of Gα, Gβ, and Gγ subunits specifically transduce signals from the receptors to diverse effectors, which include enzymes and ion channels. The binding of endogenous transmitters or exogenous ligands stabilizes defined conformations of critical domains of the seven transmembrane helix pocket and intracellular domains of the receptor. When interacting with appropriate G proteins, the active conformation of the receptor facilitates the exchange of a molecule of GDP by a molecule of GTP within the active site of the Gα subunit. The binding of GTP causes the dissociation of the heterotrimeric complex and both the GTP-bound Gα subunit and the released Gβγ complex are able to interact with intracellular or membrane effectors. The signaling pathways that are specifically activated by the receptor are essentially dictated by the nature of the interacting G protein. Although the diversity in Gα subunits and related intracellular signaling partners is well documented, more recent studies also revealed the diversity of signals associated with the Gβγ complexes. The intrinsic GTPase activity of the Gα subunit rapidly hydrolyses GTP into GDP, restoring its initial inactive conformation as well as its affinity for the βγ complex. For many receptors, the molecular determinants involved in the coupling and activation of defined G proteins have been characterized. With some exceptions, these studies have highlighted the role of membrane proximal regions of the second and third intracellular loops and of the COOH-terminus of the receptor in driving the coupling.

A key characteristic of the multistep signaling system associated with GPCRs is its complexity, which results from the intrinsic property of response amplification throughout the signaling cascade and from the diversity of regulatory mechanisms that control the nature, amplitude, and duration of the cellular responses. In particular, this complexity and specificity of GPCR signaling partly relies on the existence of numerous closely-related molecular species of the G protein subunits. Depending on the nature of the Gα subunit, a simplistic classification gathers G proteins into three major families, $G_{i/o}$, $G_{q/11}$, and G_s, and each family shows a specific influence on intracellular effectors. At least 23 Gα subunits derived from 17 different genes have been identified, whereas no less than 6 and 12 different molecular species of Gβ and Gγ subunits, respectively, have been described. Although all combinations do not necessarily exist in nature, the diversity of heterotrimeric assemblies is particularly large, contributing to the complexity, but also selectivity of intracellular signals activated by GPCRs through either defined Gα or the Gβ/Gγ subunits. In comparison to the variety of receptors and G proteins, the number of downstream effectors appears rather limited, and many closely-related G proteins (e.g., G_{i-1}, G_{i-2}, G_{i-3}) ensure the coupling with the same intracellular effector. It is also likely that a given receptor has the possibility to interact independently with many G proteins from a single class (e.g., *(1–3)*). However, the exact nature of the particular G proteins involved in the coupling is probably critical in modulating the efficacy, potency, and duration of the cell signaling. Within one class of G proteins, the subtype involved in the transduction from the receptor to the effectors will depend on its availability at the vicinity of the receptor and may thus differ from one cell/tissue to another *(4–7)*.

3.2 Diversity of Signaling Through GPCR

As from the discovery of conventional GPCR signaling systems, it has been evidenced that a given neurotransmitter can trigger the production of a variety of second messengers in distinct experimental models. Indeed, these early observations inevitably raised questions regarding the specificity of coupling of these receptors. The complexity of intracellular responses to a single neurotransmitter or activating ligand is the consequence of the diversity of molecular mechanisms involved in the signal handling at multiple steps of the amplification cascade. The detailed knowledge of intracellular signaling pathways and their functional interconnection allows to dissect these mechanisms.

3.2.1 Multiplicity of Receptor Subtypes

For the majority of transmitters interacting with GPCRs, a set of closely-related receptor subtypes has been identified (e.g., the multitude of receptors for catecholamines, acetylcholine, or glutamate). These receptors belonging to the same family and possessing a high degree of sequence homology may show similar or totally distinct G protein-coupling specificities. This explains how a given neurotransmitter can activate a variety of signaling cascades. For instance, type 1 and 5 metabotropic glutamate receptors (mGluR) preferentially activate $G_{q/11}$-type G proteins, whereas other members of the same family (mGluR2-4 and mGluR6-8) efficiently interact with $G_{i/o}$. Receptor subtypes from a given family may also differ in their coupling efficiency, leading to different intensities of the cellular responses (e.g., bradykinin and angiotensin). Since these receptor subtypes frequently show structural differences in their extracellular and transmembrane domains, selective ligands have been successfully developed, permitting the detailed characterization of their functional properties. For several decades, the design of such subtype-selective ligands has constituted the main goal in the development of new drugs interacting with GPCRs. Indeed, the enhanced specificity of a drug for a defined receptor subtype is commonly associated with improved pharmacotherapeutic outcomes with fewer unwanted side effects.

3.2.2 Multiple Signaling by a Single G Protein

A second level of signaling divergence is related to the ability of a single G protein subtype to independently elicit the activation or inhibition of multiple intracellular cascades. Thus, upon activation, both Gα and Gβγ subunits may contribute to the modulation (in a synergistic or antagonistic manner) of either the same or unrelated effectors. Best described is the dual signaling through $G_{i/o}$, involving inhibition of adenylyl cyclase by the Gα subunit and concomitant stimulation of phospholipase C isoforms by Gβγ subunits *(8–12)*. The involvement of $G_{i/o}$-type G proteins in both responses is commonly addressed using pertussis toxin, which inactivates these G proteins and thereby impairs adenylyl cyclase function and secondarily related phospholipase C activation *(13,14)*. In addition, the complexity in cell signaling may also result from the indirect modulation of distal intracellular effectors and from the intricacy of signaling pathways. Thus, activation of phospholipase A2 is commonly initiated by increases in intracellular calcium concentration after opening of membrane channels or after release from intracellular stores *(15–17)*. Indirect (G_s-independent) activation of adenylyl cyclase by protein kinase C and calcium is also documented, providing a link between $G_{q/11}$-type G proteins and the production of cyclic AMP *(18–21)*. Considering that this complexity of signaling patterns initiates at a distal step of the amplification cascade, it is unlikely that synthetic ligands acting on the cell surface receptors could provide adequate tools for manipulating the cellular responses that are intrinsically linked.

3.2.3 Multiplicity of G Protein Coupling

Finally, as detailed below, it is now generally accepted that further complexity in cell signaling initiated through GPCRs arises from the simultaneous (or consecutive) coupling of a given receptor with distinct types/subtypes of G proteins. Such multiple coupling was initially proposed to explain that some $G_{i/o}$-coupled receptors mediating inhibition of adenylyl cyclase concomitantly promote phosphoinositides hydrolysis through a pertussis toxin-insensitive pathway *(19,22–25)*. Likewise, examples for the simultaneous activation and inhibition of adenylyl cyclase by a single GPCR subtype were also published *(26,27)*. The ability of a single receptor subtype to independently interact with a set of distinct G proteins was demonstrated using a variety of biochemical approaches, but also by the use of synthetic ligands capable of selectively orienting the signaling cascade (see later).

3.3 Multiplicity of G Protein Coupling

3.3.1 Biochemical Relevance

As summarized in Table 3.1, several methods have been developed to specifically examine the type (and subtype) of G proteins involved in the responses associated with receptor activation. During the last decade, experimental data evidencing multiplicity of coupling with G proteins belonging to totally distinct families (notably G_s, G_i, and G_q) have been reported for almost all cloned GPCRs. Thus, in contrast to the simple paradigm that a given receptor always interacts with a particular G protein, it became generally accepted that simultaneous functional coupling with distinct unrelated G proteins can be observed. However, the physiological relevance of these experimental observations remains a question of debate. Indeed, as outlined here above, the concomitant activation of several signaling pathways by a neurotransmitter or a synthetic ligand could actually result from a variety of mechanisms. It is common that closely-related receptor subtypes or receptor splice variants are constitutively coexpressed in preparation of tissues containing a large variety of cell types. In such models, it is therefore difficult to assign the responses induced by nondiscriminating ligands to the activation of distinct receptor subtypes or from the complexity of signaling through a single receptor. Models of transfected cells where a single receptor subtype is specifically expressed have also been frequently used *(58,62,84,86,104–114)*. However, it is demonstrated that the high level of receptor expression achieved in recombinant or reconstituted systems would disrupt the specificity of coupling and favour the promiscuous/unfaithful coupling with the most abundant G proteins *(115)*. Finally, the signal complexity is likely to be influenced by the cell type used as both the specificity and efficacy of receptor coupling were shown to largely depends on the expression level of the receptor and signaling partners *(6,78,116–124)*.

3.3.2 Physiological Relevance

As single GPCRs would functionally interact with several G proteins, the intracellular response induced upon activation would reflect the simultaneous involvement of multiple signaling cascades triggered with distinct efficacies and/or potencies.

Table 3.1 Example of validated methods used to evidence the functional coupling of GPCRs with distinct G proteins

Selective G protein inactivation	Cell treatment with pertussis toxin	Adrenoceptor α2 *(26)* Neurotensin *(28)* Adrenoceptor β2 *(29,30)* Thrombin *(31)* Cholecystokinin B *(32)* Platelet-activating factor *(33)* Dopamine D2 *(34)* Endothelin B *(35)*
	Cell treatment with cholera toxin	Adrenoceptor α2 *(26)* Adrenoceptor β2 *(30)* GnRH *(36)* GLP1 *(37)* Cannabinoid CB1 *(38)* Dopamine D5 *(39)*
	Use of synthetic inhibitors (e.g., YM-254890 for inhibiting Gq)	Endothelin B *(35)* PAR-2 *(40)*
	Neutralization with antibodies raised against G protein subtypes	Vasopressin V1a *(41)* Serotonin 5HT2C *(42)* Dopamine D1 *(43)* Adrenoceptor β *(44)* VIP *(44)* Dopamine D3 *(45)* Platelet-activating factor *(33)*
	Antisense/knock out strategies	Adrenoceptor α2 *(46)* Vasopressin V1a *(41)* Muscarinic M1 *(47)* Adrenoceptor β *(44)* VIP *(44)* See also *(48)* for review
Interaction with recombinant G proteins	Receptor/G protein fusion	Adenosine A1 *(49)* Prostanoid IP *(50)* Neurotensin NTS-1 *(51)* Adrenoceptor β2 *(52,53)* Opioid mu *(54,55)* Dopamine D2 *(56)* See also *(57)* for review
	G proteins expressed in simplified cells models (e.g., Sf9)	Dopamine D2 *(58–60)* Chemokin CXCR4 *(61)* Serotonin 5HT1A *(62,63)* Glutamate metabotropic mGluR (Group I, IIet III) *(64)* Leukotriene B4 *(65)*
	G proteins expressed in artificial lipid vesicles	Endothelin A/B *(66)* Purinergic P2Y12 *(67)* Adenosine A1 *(68)*
	G proteins expressed in G proteins depleted eukaryotic cells	Somatostatin *(69)* Dopamine D2 *(56,70)* Glutamate metabotropic mGluR6 *(71)* Opioid mu *(72)* Cannabinoid CB1 *(73)*

Table 3.1 (continue)

Selective detection of activated G proteins	Measurement of gunaylyl nucleotide bound to G protein subtypes distinguished after G protein subunits immunoprecipitation using selective antibodies	Serotonin 5HT2C *(42)* Dopamine D2 *(74)* Thyrotropin *(75)* Melatonin Mel1a *(76)* Opioid mu *(2,3,77)* Metabotropic glutamate mGluR1 *(78,79)* Luteinizing hormone *(80)* Histamine H2 *(81)* Prostacyclin *(82)* Muscarinic M1, M2, M3 *(83)* Sphingosine 1-phosphate *(84)* Dopamine D3 *(85)* Dopamine D1 *(43)* GnRH *(36)*
	Scintillation proximity assay with G protein subunits coated beads	Serotonin 5HT2C *(86)* Serotonin 5HT1A *(87,88)* Muscarinic M1 *(89)* Dopamine D1 *(90)* GABA B *(91)*
Selective detection of interacting G proteins	Coimmunoprecipitation of receptor and G protein subunits	Cannabinoid CB1 *(73,92)* Orexin OX1R *(93)* Melatonin MT1 *(94)* Opioid mu *(95)*
Measure of GPCR conformational changes	Fluorescence and bioluminescence energy transfer	Adrenoceptor α2 *(96,97)* Opioid mu *(98)* Glutamate metabotropic mGluR1 *(99)*
	Plasmon waveguide resonance spectroscopy	Opioid delta *(100,101)* Cannabinoid CB1 *(102)* For review, see *(103)*

In several studies, combinations of these techniques were proven useful to dissect the complexity associated with GPCR function

This model is in total contradiction with classical views supporting that the nature of signaling is precisely achieved through the unique and efficient coupling of each receptor subtype with a defined G protein. As mentioned earlier, multiplicity in coupling was initially evidenced in expression systems where high densities of receptor are generated. Though these artificial models allow to specifically focus on a single receptor subtype, this approach raises the question of whether such complex signaling reveals artifactual coupling or whether it is a genuine property of GPCRs. Indeed, the complexity in signaling was initially found to depend on the receptor expression level *(26,105,125,126)* suggesting that the measure of pharmacologic properties of drugs depends on the relative stoechiometry of receptors and other cellular components *(127,128)*. Alternatively, these models could also facilitate the detection of discrete but physiologically relevant couplings with distinct G proteins, which physiologically contribute to the fine-tuning of the cell responsiveness. Indeed, differences in agonist efficacy and potency are frequently observed when considering the predominant and alternative couplings. Though several experimental biases could result in the detection of unfaithful or artifactual promiscuous couplings, there is accumulating data obtained in physiological systems and using validated technical approaches demonstrating that multiplicity of coupling is an intrinsic property of the vast majority of GPCRs. Of importance are those studies evidencing the existence of dual or multiple G protein couplings in tissues or cultured cell models where GPCRs are constitutively expressed at rather low densities *(33,43,90,109,129–136)*.

The diversity in G protein couplings could also be regarded as an adaptation mechanism that supports modifications in receptor signaling associated with physiological up or down regulation of receptor expression in target cells. Thus, in addition to the well-described desensitization of G protein-dependent signaling after prolonged or repeated exposure to agonists, recent studies have evidenced that a molecular switch in the coupling of a receptor with distinct G proteins could participate in the regulation of cellular signaling. The first evidence for distinct dynamics of desensitization between two signaling cascades was provided by Daaka and colleagues showing that protein kinase A-mediated phosphorylation of the β2-adrenergic receptor resulted in the uncoupling from G_s-dependent signaling, whereas activation of G_i was simultaneously induced *(90)*. Similar findings have been reported for a variety of other GPCRs, and it has been suggested that the dual or multiple G protein coupling provides a sophisticated mechanism that permits a temporal resolution of the cellular signaling elicited during the sustained stimulation of a receptor *(137–139)*.

Obviously, the possibility for a receptor to activate a wide range of intracellular signaling pathways could inevitably lead to a loss of signaling specificity. Therefore, a major question that arises concerns the identification of physiological mechanisms ensuring a regulation of the respective contribution of the different couplings and related cells signals. Molecular, pharmacological, and biochemical studies have shown that functional interaction with distinct type of G proteins is influenced by a variety of factors, explaining the large variability of responses patterns observed between different models. Likely, the couplings are subjected to dynamic controls, involving the regulation of the expression, compartmentalization, and activity of both the receptor and other signaling partners. Thus, the expression level of the different G proteins or other adapting proteins and their presence at the vicinity of the target receptor certainly accounts for determination of signaling specificity in a given cell type. Likewise, cell type-dependent alterations in the receptor structure (by RNA editing, or posttranslational modification) have been shown to influence the interaction with intracellular G proteins. Together, these mechanisms could support the variety of functional responses obtained with rather specific drugs in distinct tissues *(43,140)*. A better knowledge of this complexity of cell signaling through GPCR could help to explain the sometimes unexpected difference in clinical efficacy of drugs that have been characterized with similar profiles in in-vitro models.

3.4 Consequences of the Multiplicity of G Protein Coupling

3.4.1 Revisiting the Receptor Conformational Model

At the molecular level, the mechanism involved in the activation of GPCRs upon interaction with appropriate ligands has received considerable attention during the last decades. Several models have been proposed, amongst which, the extended ternary complex model *(141,142)* appears as the most suitable one for explaining the pharmacodynamic activities of the majority of interacting ligands. This model predicts that receptors exist in either an inactive state/conformation (R) or an active state/conformation (R*) that efficiently couples to a defined G protein, leading to the functional species (R*G). In a given environment, an equilibrium spontaneously establishes between the inactive (usually predominant) and active conformations. The existence of the active conformation in the absence of ligand justifies the constitutive activation of related signaling cascades. Ligands endowed with an agonist profile show higher affinity for the active state. This conformation is therefore stabilized, explaining the ability of agonists to enhance the functional response (receptor activation). On the opposite, ligands acting as inverse agonists preferentially interact and stabilize the inactive state, decreasing the constitutive activity of the system. Some ligands indifferently interact with the active and inactive conformations. On their own, these ligands do not alter the spontaneous equilibrium, but are recognized as antagonists as they impair the binding of other ligands. In summary, these two conformation models provide a molecular basis for classical concepts of pharmacology and help to explain the properties of drugs acting as agonist, antagonist, partial agonist, and inverse agonist *(143–145)*. Thus, in this model, the value of intrinsic efficacy, which traditionally describes the activating property of a receptor-interacting ligand, is viewed as a geometric parameter corresponding to the ratio of the affinities for the active and inactive conformations of the receptor *(146)*.

The concept of multiplicity of coupling adds another degree of complexity when trying to define the model of GPCR activation by ligands. The ability of a given receptor to independently activate distinct G proteins is best explained by considering the existence of several active conformations of the receptor differing in their coupling efficiencies with these distinct G proteins *(147–149)* (Fig. 3.1). Thus, in a simplistic view, the receptor would be either in an inactive conformation R or would adopt two independent active conformations (R' and R''), which show different abilities to functionally interact with distinct G proteins. Thus, these active conformations would differentially expose critical intracellular domains of the receptor, enabling a preferential recognition by a subset of G proteins. Although recent biophysical studies have validated this multiple active conformation states model, early experimental evidence arose from mutagenesis studies aimed at depicting the molecular determinants of the receptors involved in their functional coupling. Though details concerning the localization of critical coupling domains have been obtained, conserved sequences driving the coupling have not been strictly identified (for reviews, see *(115,150–153)*).

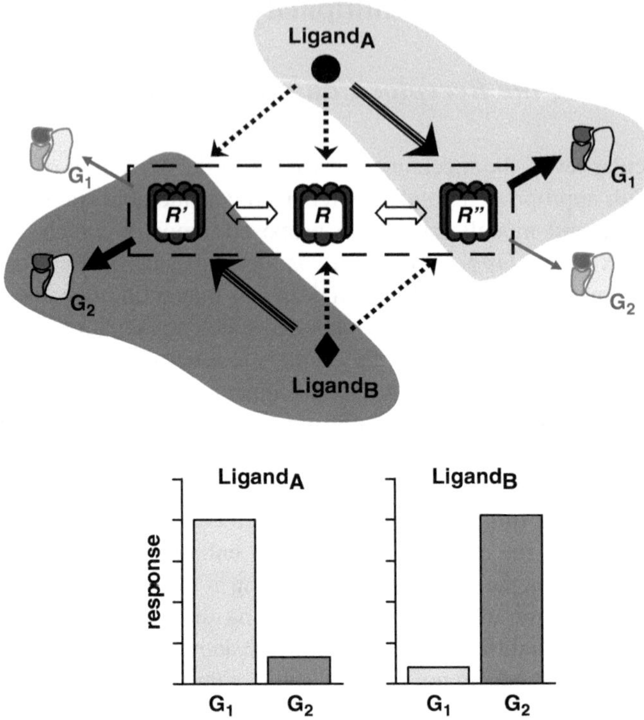

Fig. 3.1 The multiple active conformations model and the properties of functional selective ligands. *Upper panel:* the complexity of the responses associated with several drugs acting on GPCRs is best explained by considering a model in which the receptor can adopt multiple conformations (indicated as R, R' and R"). These conformations differ in their functional coupling with G proteins (generic G proteins indicated as G_1 and G_2). Thus, conformation R' efficiently interacts with G_2 while conformation R" favours G_1 activation. The pharmacodynamic properties of interacting ligands reflect their affinity for these conformation states. Functionally selective ligands distinctly stabilize these conformations and thereby promote the activation of a subset of G proteins while reducing the activity of others. In the example shown, ligand A shows a high affinity for conformation R" and a modest affinity for R and R', resulting in a predominant activation of G_1, as illustrated in the *lower panel.* On the opposite, ligand B preferentially activates G_2 as it shows a higher affinity for conformation R'. Ligands showing modest or no functional selectivity are thought to interact with both active conformations, leading to complex responses related to concomitant G_1 and G_2 activation (not illustrated)

Nevertheless, several studies have highlighted the existence of critical regions of the receptor whose alteration differentially affect the intracellular signaling cascade triggered by the agonist *(104,116,154–158)*. More recently, specific domains ensuring selectivity in coupling were identified in the intracellular loops *(159–166)* and in the membrane proximal region of the COOH-terminus *(66,167–171)*. Rather convincing are those studies showing that distinct peptides corresponding to sequences present within intracellular regions of the receptor are able to differentially activate intracellular effectors *(172)*. Consistently, these peptides were found effective in disrupting the molecular interaction between GPCR and selected G proteins. Nevertheless, it remains difficult, in most cases, to definitely assign a particular domain of the receptor to the coupling with G_s, $G_{i/o}$, and $G_{q/11}$ type G proteins. In addition, it is likely that the multiple couplings of a given receptor could involve common intracellular domains, limiting the interpretation of mutagenesis studies. Thus, deletion or substitution of these domains could severely impair the less efficient coupling, while the preferential coupling would appear partially affected *(104,157)*. Yet, these trivial criticisms should not obscure convincing data highlighting the existence of distinct molecular determinants of a single receptor that ensure the coupling with multiple signaling cascades. Thus, some key mutations were found to alter a particular coupling while leaving the other responses quantitatively unaffected or even enhanced *(161,162,168,173)*. Finally, it is noteworthy that naturally occurring splice variants of many GPCRs have been shown to differentially couple to multiple intracellular effectors *(174–176)*. Consistent with mutagenesis studies, alternative splicings associated with alteration of critical sequences within intracellular regions of the receptor were shown to modify the functional selectivity of the receptor (e.g., type 1 mGluR, 5HT4 receptor or prostaglandin E2 receptors) *(174,175,177–182)*. Indeed, the demonstrations that distinct molecular determinants of the receptor contribute to independent G protein couplings are consistent with the working model suggesting the existence of two or even more active conformations *(148,183)*. Of importance, this model suggests that the nature of the functional coupling is not solely influenced by the availability of the signaling partners, but is also dictated by intrinsic properties of the receptor. Consistent with this theoretical model, several sophisticated biochemical studies involving mutagenesis of the receptors or biophysical analysis have accumulated, allowing to better characterize the different receptor conformations and the equilibriums that establish in the presence or absence of appropriate ligands. Indeed, exciting data have been obtained using dynamic fluorescence or bioluminescence measures, allowing to delineate the conformational changes induced by diverse agonists of the receptor, which correlate with biological responses *(99,184)*. Also particularly relevant are those studies identifying mutations of the receptors that cause robust receptor-mediated constitutive coupling that is restricted to the activation of a single signaling pathway among those ordinarily activated by the agonist *(183,185,186)*. Finally, some modifications of the receptor structure were found to selectively change the pattern of cellular signaling triggered by a given agonist while leaving the properties of other ligands unaffected *(73,187–189)*. Several detailed reviews with respect to the multistate model and its extensions have been recently published, see *(190–194)*.

3.4.2 Functional Selectivity

The initial discovery that a single receptor subtype may independently couple to distinct and unrelated G proteins implies that the agonist tested in this model is able to equally stabilize the different conformations of the receptor supporting these couplings. Alternatively, this agonist could also stabilize a conformation that enables activation of two distinct signaling pathways. More importantly, the multiple active states model could also explain the differences in the coupling efficiency leading to distinct signaling cascades observed with some ligands. Thus, a preferential coupling measured experimentally could reflect the peculiar property of the ligand, which preferentially interacts with one of the active conformations of the receptor *(195)*. This model suggests that each ligand could exhibit different affinities for the distinct active states of the receptor, and thereby could orientate the coupling toward a defined signaling cascade. Through these assumptions, this model established the basis for the concept of signaling selective agonism *(196)* also referred to as "agonist specific trafficking of receptor signaling" largely developed by Kenakin *(197)* and more recently termed "functional selectivity" *(198)*, suggestive of its potential importance for future drug development.

At the pharmacodynamic level, an agonist showing robust functional selectivity should be regarded as having a largely predominant affinity for the active conformation that supports the coupling with one type of G proteins, while minimally stabilizing other possible active conformations. Alternatively, agonists showing poor functional selectivity should present similar affinities for the diverse active conformations of the receptors. However, as the active conformations probably differ in the structure of the ligand recognition domains, one should predict that it is unlikely that a given ligand would equally recognize distinct active conformations. This would explain that for the majority of ligands, a preferential signaling cascade should be identified, as mentioned earlier. Furthermore, it is worth noting that a ligand showing strong functional selectivity will behave as a powerful agonist when considering the signaling cascade supported by the conformation that is stabilized, but will also behave as an antagonist when considering other signalings possibly associated with the receptor (Fig. 3.2). Thus, assuming an independent coupling with distinct G proteins, a given ligand would possess distinct intrinsic efficacies as it may simultaneously act as full agonist, partial agonist, antagonist, or inverse agonist at the same receptor subtype when considering the different signaling cascades.

Initially, this model explained how modest structural modifications of some peptide agonists of GPCRs showing coupling promiscuity caused marked changes in the relative activation of the diverse signaling cascades *(77,199–202)*. As summarised in Table 3.2, this property of signaling-selective agonism is not restricted to peptide ligands of GPCRs, as functionally selective ligands were also identified for the several other GPCRs, including those for monoamines, purines, amino-acids, and cannabinoids. Worth mentioning, endogenous ligands such as endocannabinoids acting at CB_2 receptors were found to show functional selectivity *(214)*. This is also exemplified by the naturally occuring glycosylated form of the follicle-stimulating hormone, which differentially affects adenylyl cyclase and phospholipase C when compared with the non-glycosylated peptide *(241)*. Likewise, the binding of neurokinin A to its cognate NK2 tachykinin receptor was shown to elicit modulation of both adenylyl cyclase and phospholipase C, whereas its natural analogue neurokinin *(4–10)* only mediates phosphoinositide hydrolysis *(138)*.

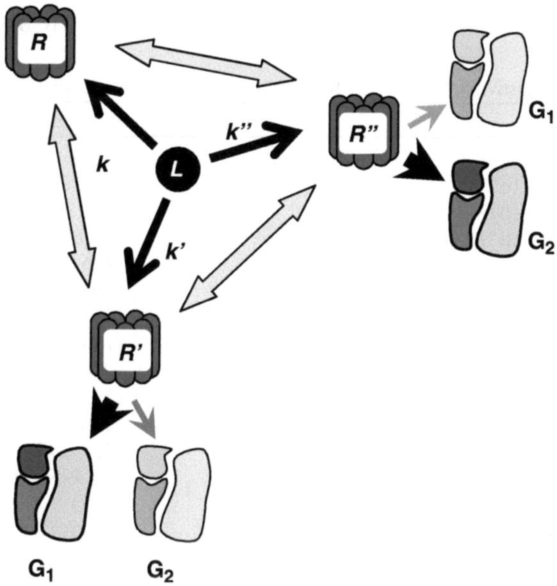

k' = k = k"	L is a *neutral antagonist*
k' >> k = k"	L is a *functional selective agonist.* It acts as an agonist when considering G_1-dependent signals, but acts as an antagonist when considering G_2-dependent signals.
k' ≈ k" >> k	L is an agonist devoid of functional selectivity
k' > k" >> k	L shows some *functional selectivity,* activating both G_1-and G_2-dependent signals, but with a predominant activity through G_1.

Fig. 3.2 Functional selectivity and intrinsic efficacy. In a simplistic view, the pharmacodynamic profile (agonist, partial agonist, antagonist, or inverse agonist) of a ligand is defined by its intrinsic efficacy, which can be viewed as the ratio of affinities for the active and the inactive conformation of the receptor (k and k', respectively). However, as most GPCRs are thought to adopt several active or inactive conformations, interacting ligands are likely to show distinct affinities for these conformations (k' and k"). Therefore, it is not possible to define the ligand activity by a single value of intrinsic efficacy. Assuming an independent coupling with distinct G proteins, any ligand would possess distinct intrinsic efficacies as it may simultaneously act as full agonist, partial agonist, antagonist, or inverse agonist at the same receptor subtype when considering the different signaling cascades. This model helps to explain how a ligand showing strong functional selectivity will behave as an agonist when considering the signaling cascade supported by the conformation that is stabilized, but will also behave as an antagonist when considering other signalings associated with the receptor through other couplings (see text for details and examples)

Table 3.2 Nonexhaustive list of GPCRs for which functional selective ligands have been identified

Receptor	References
Adenosine receptor (A1)	*(114)*
Adrenergic receptor (α2)	*(203–206)*
Adrenergic receptor (β1)	*(148,207)*
Adrenergic receptor (β2)	*(53,119,208–210)*
Adrenergic receptor (β3)	*(126)*
Bombesin receptor	*(199)*
Calcitonin receptor	*(211)*
Cannabinoids receptors (CB1)	*(133,212,213)*
Cannabinoids receptors (CB2)	*(214)*
Chemokine receptor (U51)	*(215)*
Corticotropin-releasing factor receptor (type1)	*(216)*
Dopamine receptor (D1)	*(217–219)*
Dopamine receptor (D2)	*(58,59,90,90,134,158,220–222)*
Histamine (H1)	*(223)*
Muscarinic receptors (m1)	*(224)*
Muscarinic receptors (m2)	*(225)*
Muscarinic receptors (m4)	*(225)*
Melanocortin receptor (MC4)	*(226)*
Octopamine/tyramine receptor	*(200)*
Opioid receptors (delta)	*(227,228)*
Opioid receptors (mu)	*(72,77,113)*
Oxytocin receptor	*(229)*
Parathyroid hormone receptor	*(202)*
Pituitary adenylyl cyclase-activating polypeptide receptor (PACAP1)	*(201)*
Prostaglandin	*(230)*
Serotonin receptor (5HT1A)	*(231,232)*
Serotonin receptor (5HT2A)	*(233)*
Serotonin receptor (5HT2C)	*(86,178,234–237)*
Somatostatin receptor (SST3)	*(238)*
Substance P receptor	*(199)*
Thrombin receptor (PAR1)	*(239)*
Vasopressin (V2)	*(240)*

3.4.3 Physiological and Pharmacological Implications

This exciting behavior of GPCR ligands that instruct the nature of the receptor coupling is now frequently suggested to participate in the complex clinical responses observed with atypical drugs. Thus, recent in vitro data indicate that the functional selectivity of the partial agonist aripiprazole at the dopamine D2 receptor could support its therapeutic efficacy as a new generation of antipsychotic *(220)*. Similarly, it is likely that the differential activation of subsets of signaling cascade by selected psychoactive substances interacting with serotonin receptors could explain their hallucinogen action *(242)*. The role of 5HT receptors in the control of appetite and food intake is largely described, and it was recently suggested that efficient anorectic effects could be improved by selecting ligands showing functional selectivity at 5HT2C receptors, favoring phospholipase C over phospholipase A2 activation *(234)*.

As discussed earlier, ligands showing a strong functional selectivity are expected to show opposite pharmacodynamic profiles when examining different signaling cascades. Accordingly, examples of ligands with mixed agonist and antagonist or inverse agonist properties at the same receptor subtype have been described *(212,216,243)*, opening new avenues for the design of pharmacological tools with enhanced selectivity. Thus, it is expected that the use of functionally selective ligands could contribute to optimize the therapeutic outcome and minimize putative side effects possibly associated with the complexity of signals triggered by a single receptor subtype. An interesting example is provided by the high affinity histamine H1 ligand Trans-PAT, which was shown to behave as competitive antagonist or full agonist when examining phospholipase C or adenylyl cyclase activation, respectively *(223)*. As these signaling cascades have been associated with histamine-mediated allergic responses or with modulatory influences on psychiatric functions, the possibility to specifically promote adenylyl cyclase activation should be privileged for the development of psychoactive histamine H1 receptor ligands devoid of peripheral side effects.

The demonstration that a receptor can interact with distinct G proteins and that selective ligands are able to orientate the coupling has stimulated a renewed interest for studies related to the control of receptor desensitization. Agonist-induced GPCR regulation has been extensively characterized, and it is well established that receptor phosphorylation by a variety of kinases promotes its interaction with intracellular proteins that interfere with G protein coupling. As distinct regions of the receptor are involved in the interaction of a given receptor with multiple G proteins, one may assume that phosphorylation of the receptor may differentially affect the efficiency of coupling with unrelated G proteins. Hence, as mentioned earlier (molecular switch), there are several studies evidencing a phosphorylation-mediated receptor uncoupling from a given G protein, which leaves other couplings unaffected or even promoted *(244)*. It was also documented that when a receptor interacts with multiple G proteins, a subset of the downstream signaling pathways could be responsible for triggering the receptor desensitization process *(79)*. Therefore, it is likely that functional selectivity will be associated with an agonist-selective pattern of receptor desensitization *(245,246)*. Accordingly, Stout and colleagues recently reported on differences in the kinetic and agonist profile of desensitization of distinct signaling activated by the 5-HT2C receptor, suggesting that GPCR desensitization is both agonist and effector pathway-dependent *(247)*. For many transmission systems, the possibility to pharmacologically manipulate receptor regulation independently of the functional response expected from common agonists constitutes a promising challenge in therapy. For instance, different profiles of μ opioid receptor phosphorylation, desensitization, and internalization are observed after exposure to either morphine or DAMGO, which are suggested to stabilize distinct conformations of the same receptor *(121)*. Considering the implication of receptor regulation in the development of tolerance, a better control of opioid receptor trafficking by selected agonists could certainly contribute to enhance the pharmacological efficacy and safety of opioid analgesics. Conversely, some ligands could efficiently induce receptor internalization without producing the expected functional response *(248,249)*. These ligands could constitute appropriate pharmacological tools when aiming at decreasing tissue responsiveness without causing substantial activation of the receptor *(250,251)*.

Finally, the better knowledge of the concept of functional selectivity has stimulated a considerable interest for allosteric modulators of GPCRs. Contrasting with common orthosteric ligands, allosteric modulators recognize an auxiliary site of the receptor in a noncompetitive manner. Thereby, these ligands are thought to impose structural changes in the receptor conformations, not only influencing the recognition by orthosteric ligands, but also orientating the possibilities of couplings to G proteins *(252,253)*. Thus, allosteric modulators that offer the possibility to achieve a fine tuning of the pharmacological responses to endogenous or exogenous ligands constitute promising tools in future development of drugs acting on GPCRs.

3.5 Conclusions

During the last decade, accumulating experimental data have demonstrated the validity of the concept of multiplicity of coupling at GPCRs, which reinforces the complexity of signaling associated with this family of membrane proteins. From a pharmacological perspective, a better understanding of the molecular functioning of GPCRs undoubtedly opens attractive perspectives for the identification or development of innovative drugs. Indeed, the possibility to pharmacologically manipulate the coupling of the targeted receptors using defined functionally selective ligands should be considered as a valuable approach to improve the selectivity of drugs. Besides, allosteric ligands, which could subtly modulate the subset of signaling cascades and biochemical responses triggered by endogenous transmitters, certainly hold promises for future pharmacology. Hence, these recently described concepts appear to raise a renewed interest for drugs acting at GPCRs with the expectation to obtain enhanced selectivity and thereby higher clinical efficacy. Yet, apart from a few convincing examples, the physiological consequences of this complexity remain largely unresolved. Moreover, little is presently known regarding the importance to orientate the coupling of a given receptor in one or the other direction to obtain a beneficial outcome. Therefore, it is likely that these new ligands should first be used as research tools to better define the pharmacological relevance that could later be exploited in therapeutic drugs.

Acknowledgments This study was supported by grants from the National Fund for Scientific Research. EH is Research Director of the F.N.R.S.

References

1. Albert PR, Robillard L. G protein specificity: Traffic direction required. Cell Signal 2002;14:407–18

2. Burford NT, Wang DX, Sadee W. G-protein coupling of mu-opioid receptors (OP3): elevated basal signalling activity. Biochem J 2000;348:531–7
3. Chalecka-Franaszek E, Weems HB, Crowder AT, Cox BM, Cote TE. Immunoprecipitation of high-affinity, guanine nucleotide- sensitive, solubilized mu-opioid receptors from rat brain: coimmunoprecipitation of the G proteins G(alpha o), G(alpha i1), and G(alpha i3). J Neurochem 2000;74:1068–78
4. Milligan G. The stoichiometry of expression of protein components of the stimulatory adenylyl cyclase cascade and the regulation of information transfer. Cell Signal 1996;8:87–95
5. Neubig RR. Membrane organization in G-protein mechanisms. FASEB J 1994;8:939–46
6. Ostrom RS, Post SR, Insel PA. Stoichiometry and compartmentation in G protein-coupled receptor signaling: implications for therapeutic interventions involving G(s). J Pharmacol Exp Ther 2000;294:407–12
7. Ostrom RS, Insel PA. The evolving role of lipid rafts and caveolae in G protein-coupled receptor signaling: implications for molecular pharmacology. Br J Pharmacol 2004;143:235–45
8. Blank JL, Brattain KA, Exton JH. Activation of cytosolic phosphoinositide phospholipase C by G-protein beta gamma subunits. J Biol Chem 1992;267:23069–75
9. Boyer JL, Waldo GL, Harden TK. Beta gamma-subunit activation of G-protein-regulated phospholipase C. J Biol Chem 1992;267:25451–6
10. Exton JH. Regulation of phosphoinositide phospholipases by hormones, neurotransmitters, and other agonists linked to G proteins. Annu Rev Pharmacol Toxicol 1996;36:481–509
11. Morris AJ, Scarlata S. Regulation of effectors by G-protein alpha- and beta gamma-subunits. Recent insights from studies of the phospholipase c-beta isoenzymes. Biochem Pharmacol 1997;54:429–35
12. Rhee SG, Bae YS. Regulation of phosphoinositide-specific phospholipase C isozymes. J Biol Chem 1997;272:15045–8
13 Fargin A, Yamamoto K, Cotecchia S et al. Dual coupling of the cloned 5-HT1A receptor to both adenylyl cyclase and phospholipase C is mediated via the same Gi protein. Cell Signal 1991;3:547–57
14. Zgombick JM, Borden LA, Cochran TL, Kucharewicz SA, Weinshank RL, Branchek TA. Dual coupling of cloned human 5-hydroxytryptamine1D alpha and 5-hydroxytryptamine1D beta receptors stably expressed in murine fibroblasts: inhibition of adenylate cyclase and elevation of intracellular calcium concentrations via pertussis toxin-sensitive G protein(s). Mol Pharmacol 1993;44:575–82
15. Clark JD, Lin LL, Kriz RW et al. A novel arachidonic acid-selective cytosolic PLA2 contains a Ca(2+)-dependent translocation domain with homology to PKC and GAP. Cell 1991;65:1043–51
16. Hirabayashi T, Kume K, Hirose K et al. Critical duration of intracellular Ca2+ response required for continuous translocation and activation of cytosolic phospholipase A2. J Biol Chem 1999;274:5163–9
17. Schievella AR, Regier MK, Smith WL, Lin LL. Calcium-mediated translocation of cytosolic phospholipase A2 to the nuclear envelope and endoplasmic reticulum. J Biol Chem 1995;270:30749–54
18. Choi EJ, Wong ST, Hinds TR, Storm DR. Calcium and muscarinic agonist stimulation of type I adenylylcyclase in whole cells. J Biol Chem 1992;267:12440–2
19. Felder CC, Jose PA, Axelrod J. The dopamine-1 agonist, SKF 82526, stimulates phospholipase-C activity independent of adenylate cyclase. J Pharmacol Exp Ther 1989;248:171–5
20. Sunahara RK, Dessauer CW, Gilman AG. Complexity and diversity of mammalian adenylyl cyclases. Annu Rev Pharmacol Toxicol 1996;36:461–80
21. Taussig R, Gilman AG. Mammalian membrane-bound adenylyl cyclases. J Biol Chem 1995;270:1–4
22. Crawford KW, Frey EA, Cote TE. Angiotensin II receptor recognized by DuP753 regulates two distinct guanine nucleotide-binding protein signaling pathways. Mol Pharmacol 1992;41:154–62

23. Felder CC, Kanterman RY, Ma AL, Axelrod J. A transfected m1 muscarinic acetylcholine receptor stimulates adenylate cyclase via phosphatidylinositol hydrolysis. J Biol Chem 1989;264:20356–62
24. Jones SB, Halenda SP, Bylund DB. Alpha 2-adrenergic receptor stimulation of phospholipase A2 and of adenylate cyclase in transfected Chinese hamster ovary cells is mediated by different mechanisms. Mol Pharmacol 1991;39:239–45
25. Wolsing DH, Rosenbaum JS. The mechanism for the rapid desensitization in bradykinin-stimulated inositol monophosphate production in NG108-15 cells involves interaction of a single receptor with multiple signaling pathways. J Pharmacol Exp Ther 1993;266:253–61
26. Eason MG, Kurose H, Holt BD, Raymond JR, Liggett SB. Simultaneous coupling of alpha 2-adrenergic receptors to two G-proteins with opposing effects. Subtype-selective coupling of alpha 2C10, alpha 2C4, and alpha 2C2 adrenergic receptors to Gi and Gs. J Biol Chem 1992;267:15795–801
27. Fraser CM, Arakawa S, McCombie WR, Venter JC. Cloning, sequence analysis, and permanent expression of a human alpha 2-adrenergic receptor in Chinese hamster ovary cells. Evidence for independent pathways of receptor coupling to adenylate cyclase attenuation and activation. J Biol Chem 1989;264:11754–61
28. Gailly P, Najimi M, Hermans E. Evidence for the dual coupling of the rat neurotensin receptor with pertussis toxin-sensitive and insensitive G-proteins. FEBS Lett 2000;483:109–13
29. Katz A, Wu D, Simon MI. Subunits beta gamma of heterotrimeric G protein activate beta 2 isoform of phospholipase C. Nature 1992;360:686–9
30. Vasquez C, Lewis DL. The beta2-adrenergic receptor specifically sequesters Gs but signals through both Gs and Gi/o in rat sympathetic neurons. Neuroscience 2003;118:603–10
31. Ogino Y, Tanaka K, Shimizu N. Direct evidence for two distinct G proteins coupling with thrombin receptors in human neuroblastoma SH-EP cells. Eur J Pharmacol 1996;316:105–9
32. Pommier B, Da-Nascimento S, Dumont S et al. The cholecystokininB receptor is coupled to two effector pathways through pertussis toxin-sensitive and -insensitive G proteins. J Neurochem 1999;73:281–8
33. Shi LC, Wang HY, Horwitz J, Friedman E. Guanine nucleotide regulatory proteins, Gq and Gi1/2, mediate platelet-activating factor-stimulated phosphoinositide metabolism in immortalized hippocampal cells. J Neurochem 1996;67:1478–84
34. Vallar L, Muca C, Magni M et al. Differential coupling of dopaminergic D2 receptors expressed in different cell types. Stimulation of phosphatidylinositol 4,5-bisphosphate hydrolysis in LtK- fibroblasts, hyperpolarization, and cytosolic-free Ca2+ concentration decrease in GH4C1 cells. J Biol Chem 1990;265:10320–6
35. Morishita R, Ueda H, Ito H, Takasaki J, Nagata K, Asano T. Involvement of Gq/11 in both integrin signal-dependent and -independent pathways regulating endothelin-induced neural progenitor proliferation. Neurosci Res 2007;59:205–14
36. Krsmanovic LZ, Mores N, Navarro CE, Arora KK, Catt KJ. An agonist-induced switch in G protein coupling of the gonadotropin-releasing hormone receptor regulates pulsatile neuropeptide secretion. Proc Natl Acad Sci USA 2003;100:2969–74
37. Hallbrink M, Holmqvist T, Olsson M, Ostenson CG, Efendic S, Langel U. Different domains in the third intracellular loop of the GLP-1 receptor are responsible for Galpha(s) and Galpha(i)/Galpha(o) activation. Biochim Biophys Acta 2001;1546:79–86
38. Hampson RE, Mu J, Deadwyler SA. Cannabinoid and kappa opioid receptors reduce potassium K current via activation of G(s) proteins in cultured hippocampal neurons. J Neurophysiol 2000;84:2356–64
39. Sidhu A, Kimura K, Uh M, White BH, Patel S. Multiple coupling of human D5 dopamine receptors to guanine nucleotide binding proteins Gs and Gz. J Neurochem 1998;70:2459–67
40. Goon GF, Sloss CM, Cunningham MR, Nilsson M, Cadalbert L, Plevin R. G-protein-dependent and -independent pathways regulate proteinase-activated receptor-2 mediated p65 NFkappaB serine 536 phosphorylation in human keratinocytes. Cell Signal 2008;20:1267–74

41. Abel A, Wittau N, Wieland T, Schultz G, Kalkbrenner F. Cell cycle-dependent coupling of the vasopressin V1a receptor to different G proteins. J Biol Chem 2000;275:32543–51
42. Alberts GL, Pregenzer JF, Im WB, Zaworski PG, Gill GS. Agonist-induced GTP gamma S-35 binding mediated by human 5-HT2C receptors expressed in human embryonic kidney 293 cells. Eur J Pharmacol 1999;383:311–9
43. Jin LQ, Wang HY, Friedman E. Stimulated D-1 dopamine receptors couple to multiple G alpha proteins in different brain regions. J Neurochem 2001;78:981–90
44. Luo X, Zeng WZ, Xu X et al. Alternate coupling of receptors to G(s) and G(i) in pancreatic and submandibular gland cells. J Biol Chem 1999;274:17684–90
45. Newman-Tancredi A, Cussac D, Audinot V, Pasteau V, Gavaudan S, Millan MJ. G protein activation by human dopamine D3 receptors in high-expressing Chinese hamster ovary cells: A guanosine-5'-O-(3-[35S]thio)- triphosphate binding and antibody study. Mol Pharmacol 1999;55:564–74
46. Delmas P, Abogadie FC, Milligan G, Buckley NJ, Brown DA. betagamma dimers derived from Go and Gi proteins contribute different components of adrenergic inhibition of Ca2+ channels in rat sympathetic neurones. J Physiol 1999;518 (Pt 1):23–36
47. Kasahara J, Sugiyama H. Inositol phospholipid metabolism in Xenopus oocytes mediated by endogenous G(o) and Gi proteins. FEBS Lett 1994;355:41–4
48. Koller E, Gaarde WA, Monia BP. Elucidating cell signaling mechanisms using antisense technology. Trends Pharmacol Sci 2000;21:142–8
49. Bevan N, Palmer T, Drmota T et al. Functional analysis of a human A(1) adenosine receptor/green fluorescent protein/G(il)alpha fusion protein following stable expression in CHO cells. FEBS Lett 1999;462:61–5
50. Fong CW, Milligan G. Analysis of agonist function at fusion proteins between the IP prostanoid receptor and cognate, unnatural and chimaeric G- proteins. Biochem J 1999;342:457–63
51. Grisshammer R, Hermans E. Functional coupling with G alpha(q) and G alpha(il) protein subunits promotes high-affinity agonist binding to the neurotensin receptor NTS-1 expressed in *Escherichia coli*. FEBS Lett 2001;493:101–5
52. Seifert R, Wenzel SK, Kobilka BK. GPCR-Galpha fusion proteins: molecular analysis of receptor-G-protein coupling. Trends Pharmacol Sci 1999;20:383–9
53. Wenzel-Seifert K, Seifert R. Molecular analysis of beta(2)-adrenoceptor coupling to G(s)-, G(i)-, and G(q)-proteins. Mol Pharmacol 2000;58:954–66
54. Stanasila L, Lim WK, Neubig RR, Pattus F. Coupling efficacy and selectivity of the human mu-opioid receptor expressed as receptor-G alpha fusion proteins in Escherichia coli. J Neurochem 2000;75:1190–9
55. Massotte D, Brillet K, Kieffer B, Milligan G. Agonists activate Gi1 alpha or Gi2 alpha fused to the human mu opioid receptor differently. J Neurochem 2002;81:1372–82
56. Lane JR, Powney B, Wise A, Rees S, Milligan G. Protean agonism at the dopamine D2 receptor: (S)-3-(3-hydroxyphenyl)-N-propylpiperidine is an agonist for activation of Go1 but an antagonist/inverse agonist for Gi1,Gi2, and Gi3. Mol Pharmacol 2007;71:1349–59
57. Milligan G. Insights into ligand pharmacology using receptor-G-protein fusion proteins. Trends Pharmacol Sci 2000;21:24–8
58. Cordeaux Y, Nickolls SA, Flood LA, Graber SG, Strange PG. Agonist regulation of d2 dopamine receptor/g protein interaction. evidence for agonist selection of g protein subtype. J Biol Chem 2001;276:28667–75
59. Gazi L, Nickolls SA, Strange PG. Functional coupling of the human dopamine D2 receptor with G alpha i1, G alpha i2, G alpha i3 and G alpha o G proteins: evidence for agonist regulation of G protein selectivity. Br J Pharmacol 2003;138:775–86
60. Nickolls SA, Strange PG. The influence of G protein subtype on agonist action at D2 dopamine receptors. Neuropharmacology 2004;47:860–72
61. Kleemann P, Papa D, Vigil-Cruz S, Seifert R. Functional reconstitution of the human chemokine receptor CXCR4 with G(i)/G (o)-proteins in Sf9 insect cells. Naunyn Schmiedebergs Arch Pharmacol 2008

62. Barr AJ, Brass LF, Manning DR. Reconstitution of receptors and GTP-binding regulatory proteins (G proteins) in Sf9 cells. A direct evaluation of selectivity in receptor.G protein coupling. J Biol Chem 1997;272:2223–9
63. Okada M, Goldman D, Linnoila M, Iwata N, Ozaki N, Northup JK. Comparison of G-protein selectivity of human 5-HT2C and 5-HT1A receptors. Ann N Y Acad Sci 2004; 1025:570–7
64. Parmentier ML, Joly C, Restituito S, Bockaert J, Grau Y, Pin JP. The G protein-coupling profile of metabotropic glutamate receptors, as determined with exogenous G proteins, is independent of their ligand recognition domain. Mol Pharmacol 1998;53:778–86
65. Masuda K, Itoh H, Sakihama T et al. A combinatorial G protein–coupled receptor reconstitution system on budded baculovirus. Evidence for Galpha and Galphao coupling to a human leukotriene B4 receptor. J Biol Chem 2003;278:24552–62
66. Doi T, Sugimoto H, Arimoto N, Hiroaki Y, Fujiyoshi Y. Interactions of endothelin receptor subtypes A and B with G(i), G(o), and G(q) in reconstituted phospholipid vesicles. Biochemistry 1999;38:3090–9
67. Bodor ET, Waldo GL, Hooks SB, Corbitt J, Boyer JL, Harden TK. Purification and functional reconstitution of the human P2Y12 receptor. Mol Pharmacol 2003;64:1210–6
68. Figler RA, Lindorfer MA, Graber SG, Garrison JC, Linden J. Reconstitution of bovine A1 adenosine receptors and G proteins in phospholipid vesicles: betagamma-subunit composition influences guanine nucleotide exchange and agonist binding. Biochemistry 1997; 36:16288–99
69. Kozasa T, Kaziro Y, Ohtsuka T, Grigg JJ, Nakajima S, Nakajima Y. G protein specificity of the muscarine-induced increase in an inward rectifier potassium current in AtT-20 cells. Neurosci Res 1996;26:289–97
70. Banihashemi B, Albert PR. Dopamine-D2S receptor inhibition of calcium influx, adenylyl cyclase, and mitogen-activated protein kinase in pituitary cells: distinct Galpha and Gbetagamma requirements. Mol Endocrinol 2002;16:2393–404
71. Tian L, Kammermeier PJ. G protein coupling profile of mGluR6 and expression of G alpha proteins in retinal ON bipolar cells. Vis Neurosci 2006;23:909–16
72. Clark MJ, Furman CA, Gilson TD, Traynor JR. Comparison of the relative efficacy and potency of mu-opioid agonists to activate Galpha(i/o) proteins containing a pertussis toxin-insensitive mutation. J Pharmacol Exp Ther 2006;317:858–64
73. Anavi-Goffer S, Fleischer D, Hurst DP et al. Helix 8 Leu in the CB1 cannabinoid receptor contributes to selective signal transduction mechanisms. J Biol Chem 2007;282:25100–13
74. Alberts GL, Pregenzer JF, Im WB. Advantages of heterologous expression of human D2long dopamine receptors in human neuroblastoma SH-SY5Y over human embryonic kidney 293 cells. Br J Pharmacol 2000;131:514–20
75. Allgeier A, Offermanns S, Van-Sande J, Spicher K, Schultz G, Dumont JE. The human thyrotropin receptor activates G-proteins Gs and Gq/11. J Biol Chem 1994;269:13733–5
76. Brydon L, Roka F, Petit L et al. Dual signaling of human Mel1a melatonin receptors via G(i2), G(i3), and G(q/11) proteins. Mol Endocrinol 1999;13:2025–38
77. Chakrabarti S, Prather PL, Yu L, Law PY, Loh HH. Expression of the mu-opioid receptor in CHO cells: ability of mu-opioid ligands to promote alpha-azidoanilido[32P]GTP labeling of multiple G protein alpha subunits. J Neurochem 1995;64:2534–43
78. Selkirk JV, Price GW, Nahorski SR, Challiss RAJ. Cell type-specific differences in the coupling of recombinant mGlu1 alpha receptors to endogenous G protein sub-populations. Neuropharmacology 2001;40:645–56
79. Hermans E, Saunders R, Selkirk JV, Mistry R, Nahorski SR, Challiss RAJ. Complex involvement of pertussis toxin-sensitive G proteins in the regulation of type 1 alpha metabotropic glutamate receptor signaling in baby hamster kidney cells. Mol Pharmacol 2000;58:352–60
80. Herrlich A, Kuhn B, Grosse R, Schmid A, Schultz G, Gudermann T. Involvement of Gs and Gi proteins in dual coupling of the luteinizing hormone receptor to adenylyl cyclase and phospholipase C. J Biol Chem 1996;271:16764–72

81. Kuhn B, Schmid A, Harteneck C, Gudermann T, Schultz G. G proteins of the Gq family couple the H2 histamine receptor to phospholipase C. Mol Endocrinol 1996;10:1697–707
82. Lawler OA, Miggin SM, Kinsella BT. Protein kinase A-mediated phosphorylation of serine 357 of the mouse prostacyclin receptor regulates its coupling to G(s)-, to G(i)-, and to G(q)-coupled effector signaling. J Biol Chem 2001;276:33596–607
83. Offermanns S, Wieland T, Homann D et al. Transfected muscarinic acetylcholine receptors selectively couple to Gi-type G proteins and Gq/11. Mol Pharmacol 1994;45:890–8
84. Windh RT, Lee MJ, Hla T, An SZ, Barr AJ, Manning DR. Differential coupling of the sphingosine 1-phosphate receptors Edg-1, Edg-3, and H218/Edg-5 to the G(i), G(q), and G(12) families of heterotrimeric G proteins. J Biol Chem 1999;274:27351–8
85. Zaworski PG, Alberts GL, Pregenzer JF, Bin I, Slightom JL, Gill GS. Efficient functional coupling of the human D3 dopamine receptor to G(o) subtype of G proteins in SH-SY5Y cells. Br J Pharmacol 1999;128:1181–8
86. Cussac D, Newman -Tancredi A, Duqueyroix D, Pasteau V, Millan MJ. Differential activation of Gq/1 and Gi(3) proteins at 5- hydroxytryptamine(2C) receptors revealed by antibody capture assays: influence of receptor reserve and relationship to agonist-directed trafficking. Mol Pharmacol 2002;62:578–89
87. Mannoury LC, Elmestikawy S, Hanoun N, Hamon M, Lanfumey L. Regional differences in the coupling of 5-hydroxytryptamine-1A receptors to G proteins in the rat brain. Mol Pharmacol 2006;70:1013–21
88. Newman-Tancredi A, Cussac D, Marini L, Millan MJ. Antibody capture assay reveals bell-shaped concentration- response isotherms for h5-HT1A receptor-mediated G alpha(i3) activation: Conformational selection by high-efficacy agonists, and relationship to trafficking of receptor signaling. Mol Pharmacol 2002;62:590–601
89. DeLapp NW, McKinzie JH, Sawyer BD et al. Determination of [35S]guanosine-5'-O-(3-thio) triphosphate binding mediated by cholinergic muscarinic receptors in membranes from Chinese hamster ovary cells and rat striatum using an anti-G protein scintillation proximity assay. J Pharmacol Exp Ther 1999;289:946–55
90. Mannoury la Cour C, Vidal S, Pasteau V, Cussac D, Millan MJ. Dopamine D1 receptor coupling to Gs/olf and Gq in rat striatum and cortex: a scintillation proximity assay (SPA)/antibody-capture characterization of benzazepine agonists. Neuropharmacology 2007;52:1003–14
91. Mannoury la Cour C, Herbelles C, Pasteau V, de Nanteuil G, Millan MJ. Influence of ositive allosteric modulators on GABA(B) receptor coupling in rat brain: a scintillation proximity assay characterisation of G protein subtypes. J Neurochem 2008;105:308–23
92. Mukhopadhyay S, Howlett AC. Chemically distinct ligands promote differential CB1 cannabinoid receptor-Gi protein interactions. Mol Pharmacol 2005;67:2016–24
93. Magga J, Bart G, Oker-Blom C, Kukkonen JP, Akerman KE, Nasman J. Agonist potency differentiates G protein activation and Ca2+ signalling by the orexin receptor type 1. Biochem Pharmacol 2006;71:827–36
94. Bondi CD, McKeon RM, Bennett JM et al. MT1 melatonin receptor internalization underlies melatonin-induced morphologic changes in Chinese hamster ovary cells and these processes are dependent on Gi proteins, MEK 1/2 and microtubule modulation. J Pineal Res 2008;44:288–98
95. Chakrabarti S, Regec A, Gintzler AR. Biochemical demonstration of mu-opioid receptor association with Gsalpha: enhancement following morphine exposure. Brain Res Mol Brain Res 2005;135:217–24
96. Gibson SK, Gilman AG. Gialpha and Gbeta subunits both define selectivity of G protein activation by alpha2-adrenergic receptors. Proc Natl Acad Sci USA 2006;103:212–7
97. Nobles M, Benians A, Tinker A. Heterotrimeric G proteins precouple with G protein-coupled receptors in living cells. Proc Natl Acad Sci USA 2005;102:18706–11
98. Hasbi A, Nguyen T, Fan T et al. Trafficking of preassembled opioid mu-delta heterooligomer-Gz signaling complexes to the plasma membrane: coregulation by agonists. Biochemistry 2007;46:12997–3009
99. Tateyama M, Kubo Y. Dual signaling is differentially activated by different active states of the metabotropic glutamate receptor 1alpha. Proc Natl Acad Sci USA 2006;103:1124–8

100. Alves ID, Salamon Z, Varga E, Yamamura HI, Tollin G, Hruby VJ. Direct observation of G-protein binding to the human delta-opioid receptor using plasmon-waveguide resonance spectroscopy. J Biol Chem 2003;278:48890–7
101. Alves ID, Cowell SM, Salamon Z, Devanathan S, Tollin G, Hruby VJ. Different structural states of the proteolipid membrane are produced by ligand binding to the human delta-opioid receptor as shown by plasmon-waveguide resonance spectroscopy. Mol Pharmacol 2004;65:1248–57
102. Georgieva T, Devanathan S, Stropova D et al. Unique agonist-bound cannabinoid CB1 receptor conformations indicate agonist specificity in signaling. Eur J Pharmacol 2008;581:19–29
103. Hruby VJ, Tollin G. Plasmon-waveguide resonance (PWR) spectroscopy for directly viewing rates of GPCR/G-protein interactions and quantifying affinities. Curr Opin Pharmacol 2007;7:507–14
104. Chabre O, Conklin BR, Brandon S, Bourne HR, Limbird LE. Coupling of the alpha 2A-adrenergic receptor to multiple G-proteins. A simple approach for estimating receptor-G-protein coupling efficiency in a transient expression system. J Biol Chem 1994;269:5730–4
105. Cordeaux Y, Briddon SJ, Megson AE, McDonnell J, Dickenson JM, Hill SJ. Influence of receptor number on functional responses elicited by agonists acting at the human adenosine A(1) receptor: evidence for signaling pathway-dependent changes in agonist potency and relative intrinsic activity. Mol Pharmacol 2000;58:1075–84
106. Mullaney I, Carr IC, Milligan G. Overexpression of G(s)alpha in NG108-15, neuroblastomaXglioma cells: effects on receptor regulation of the stimulatory adenylyl cyclase cascade. FEBS Lett 1996;397:325–30
107. Offermanns S, Iida-Klein A, Segre GV, Simon MI. G alpha q family members couple parathyroid hormone (PTH)/PTH-related peptide and calcitonin receptors to phospholipase C in COS-7 cells. Mol Endocrinol 1996;10:566–74
108. Palmer TM, Gettys TW, Stiles GL. Differential interaction with and regulation of multiple G-proteins by the rat A3 adenosine receptor. J Biol Chem 1995;270:16895–902
109. Stanislaus D, Ponder S, Ji TH, Conn PM. Gonadotropin-releasing hormone receptor couples to multiple G proteins in rat gonadotrophs and in GGH3 cells: evidence from palmitoylation and overexpression of G proteins. Biol Reprod 1998;59:579–86
110. Nasman J, Kukkonen JP, Ammoun S, Akerman KEO. Role of G-protein availability in differential signaling by alpha 2-adrenoceptors. Biochem Pharmacol 2001;62:913–22
111. Sidhu A, sullivan M, Kohout T, Balen P, Fishman P. D1 dopamine receptors can interact with both stimulatory and inhibitory guanine nucleotide binding proteins. J Neurochem 1991;57:1445–51
112. Lauckner JE, Hille B, Mackie K. The cannabinoid agonist WIN55,212-2 increases intracellular calcium via CB1 receptor coupling to Gq/11 G proteins. Proc Natl Acad Sci USA 2005;102:19144–9
113. Saidak Z, Blake-Palmer K, Hay DL, Northup JK, Glass M. Differential activation of G-proteins by mu-opioid receptor agonists. Br J Pharmacol 2006;147:671–80
114. Cordeaux Y, Ijzerman AP, Hill SJ. Coupling of the human A1 adenosine receptor to different heterotrimeric G proteins: evidence for agonist-specific G protein activation. Br J Pharmacol 2004;143:705–14
115. Gudermann T, Schoneberg T, Schultz G. Functional and structural complexity of signal transduction via G-protein-coupled receptors. Annu Rev Neurosci 1997;20:399–427
116. Schneider H, Feyen JH, Seuwen K. A C-terminally truncated human parathyroid hormone receptor is functional and activates multiple G proteins. FEBS Lett 1994;351:281–5
117. Burt AR, Sautel M, Wilson MA, Rees S, Wise A, Milligan G. Agonist occupation of an alpha(2A)-adrenoreceptor-G(i1)alpha fusion protein results in activation of both receptor-linked and endogenous G(i) proteins - Comparisons of their contributions to GTPase activity and signal transduction and analysis of receptor-G protein activation stoichiometry. J Biol Chem 1998;273:10367–75

118. Esbenshade TA, Wang X, Williams NG, Minneman KP. Inducible expression of alpha 1B-adrenoceptors in DDT1 MF-2 cells: comparison of receptor density and response. Eur J Pharmacol 1995;289:305–10
119. Kukkonen JP, Nasman J, Akerman KEO. Modelling of promiscuous receptor-G(i)/G(s) – protein coupling and effector response. Trends Pharmacol Sci 2001;22:616–22
120. Theroux TL, Esbenshade TA, Peavy RD, Minneman KP. Coupling efficiencies of human alpha 1-adrenergic receptor subtypes: titration of receptor density and responsiveness with inducible and repressible expression vectors. Mol Pharmacol 1996;50:1376–87
121. Johnson EA, Oldfield S, Braksator E et al. Agonist-selective mechanisms of mu-opioid receptor desensitization in human embryonic kidney 293 cells. Mol Pharmacol 2006;70:676–85
122. Roy AA, Nunn C, Ming H et al. Up-regulation of endogenous RGS2 mediates cross-desensitization between Gs and Gq signaling in osteoblasts. J Biol Chem 2006;281:32684–93
123. Kudlacek O, Just H, Korkhov VM et al. The human D2 dopamine receptor synergizes with the A2A adenosine receptor to stimulate adenylyl cyclase in PC12 cells. Neuropsychopharmacology 2003;28:1317–27
124. Sato M, Blumer JB, Simon V, Lanier SM. Accessory proteins for G proteins: partners in signaling. Annu Rev Pharmacol Toxicol 2006;46:151–87
125. Zhu X, Gilbert S, Birnbaumer M, Birnbaumer L. Dual signaling potential is common among Gs-coupled receptors and dependent on receptor density. Mol Pharmacol 1994;56:460–9
126. Sato M, Horinouchi T, Hutchinson DS, Evans BA, Summers RJ. Ligand-directed signaling at the beta3-adrenoceptor produced by 3-(2-ethylphenoxy)-1-[(1,S)-1,2,3,4-tetrahydronapth-1-ylamino]-2S-2-propan ol oxalate (SR59230A) relative to receptor agonists. Mol Pharmacol 2007;72:1359–68
127. Nelson CP, Nahorski SR, Challiss RA. Constitutive activity and inverse agonism at the M2 muscarinic acetylcholine receptor. J Pharmacol Exp Ther 2006;316:279–88
128. Kenakin T. Collateral efficacy in drug discovery: taking advantage of the good (allosteric) nature of 7TM receptors. Trends Pharmacol Sci 2007;28:407–15
129. Kilts JD, Gerhardt MA, Richardson MD et al. Beta(2)-adrenergic and several other G protein-coupled receptors in human atrial membranes activate both G(s) and G(i). Circ Res 2000;87:705–9
130. Klein J, Reymann KG, Riedel G. Activation of phospholipases C and D by the novel metabotropic glutamate receptor agonist tADA. Neuropharmacology 1997;36:261–3
131. Laugwitz KL, Allgeier A, Offermanns S et al. The human thyrotropin receptor: a heptahelical receptor capable of stimulating members of all four G protein families. Proc Natl Acad Sci USA 1996;93:116–20
132. Santos-Alvarez J, Sanchez-Margalet V. G protein G alpha(q/11) and G alpha(i1,2) are activated by pancreastatin receptors in rat liver: studies with GTP-gamma S- 35 and azido-GTP-alpha-P-32. J Cell Biochem 1999;73:469–77
133. Bosier B, Tilleux S, Najimi M, Lambert DM, Hermans E. Agonist selective modulation of tyrosine hydroxylase expression by cannabinoid ligands in a murine neuroblastoma cell line. J Neurochem 2007;102:1996–2007
134. Mottola DM, Kilts JD, Lewis MM et al. Functional selectivity of dopamine receptor agonists. I. Selective activation of postsynaptic dopamine D2 receptors linked to adenylate cyclase. J Pharmacol Exp Ther 2002;301:1166–78
135. Sneddon WB, Yang Y, Ba J, Harinstein LM, Friedman PA. Extracellular signal-regulated kinase activation by parathyroid hormone in distal tubule cells. Am J Physiol Renal Physiol 2007;292:F1028–F1034
136. Magocsi M, Vizi ES, Selmeczy Z, Brozik A, Szelenyi J. Multiple G-protein-coupling specificity of beta-adrenoceptor in macrophages. Immunology 2007;122:503–13
137. Herrero I, Miras PM, Sanchez-Prieto J. Functional switch from facilitation to inhibition in the control of glutamate release by metabotropic glutamate receptors. J Biol Chem 1998;273:1951–8

138. Palanche T, Ilien B, Zoffmann S et al. The neurokinin A receptor activates calcium and cAMP responses through distinct conformational states. J Biol Chem 2001;276:34853–61
139. Paquette JJ, Wang HY, Bakshi K, Olmstead MC. Cannabinoid-induced tolerance is associated with a CB1 receptor G protein coupling switch that is prevented by ultra-low dose rimonabant. Behav Pharmacol 2007;18:767–76
140. Mannoury LC, Elmestikawy S, Hanoun N, Hamon M, Lanfumey L. Regional differences in the coupling of 5-hydroxytryptamine-1A receptors to G proteins in the rat brain. Mol Pharmacol 2006;70:1013–21
141. Lefkowitz RJ, Cotecchia S, Samama P, Costa T. Constitutive activity of receptors coupled to guanine nucleotide regulatory proteins. Trends Pharmacol Sci 1993;14:303–7
142. Kenakin T. Agonist-specific receptor conformations. Trends Pharmacol Sci 1997;18:416–7
143. Daeffler L, Landry Y. Inverse agonism at heptahelical receptors: concept, experimental approach and therapeutic potential. Fund Clin Pharmacol 2000;14:73–87
144. Kenakin T. Inverse, protean, and ligand-selective agonism: matters of receptor conformation. FASEB J 2001;15:598–611
145. Samama P, Cotecchia S, Costa T, Lefkowitz RJ. A mutation-induced activated state of the beta 2-adrenergic receptor. Extending the ternary complex model. J Biol Chem 1993;268:4625–36
146. Kenakin T. Drug efficacy at G protein-coupled receptors. Annu Rev Pharmacol Toxicol 2002;42:349–79
147. Kenakin T. Ligand-selective receptor conformations revisited: the promise and the problem. Trends Pharmacol Sci 2003;24:346–54
148. Krumins AM, Barber R. The stability of the agonist beta2-adrenergic receptor-Gs complex: evidence for agonist-specific states. Mol Pharmacol 1997;52:144–54
149. Scaramellini C, Leff P. A three-state receptor model: Predictions of multiple agonist pharmacology for the same receptor type. Ann N Y Acad Sci 1998;861:97–103
150. Gether U. Uncovering molecular mechanisms involved in activation of G protein-coupled receptors. Endocr Rev 2000;21:90–113
151. Gudermann T, Kalkbrenner F, Schultz G. Diversity and selectivity of receptor-G protein interaction. Annu Rev Pharmacol Toxicol 1996;36429:59-59
152. Wess J. G-protein-coupled receptors: Molecular mechanisms involved in receptor activation and selectivity of G-protein recognition. FASEB J 1997;11:346–54
153. Wess J. Molecular basis of receptor/G-protein-coupling selectivity. Pharmacol Ther 1998;80:231–64
154. Iida-Klein A, Guo J, Xie LY et al. Truncation of the carboxyl-terminal region of the rat parathyroid hormone (PTH)/PTH-related peptide receptor enhances PTH stimulation of adenylyl cyclase but not phospholipase C. J Biol Chem 1995;270:8458–65
155. Kosugi S, Mori T. The intracellular region adjacent to plasma membrane (residues 684-692) of the thyrotropin receptor is important for phosphoinositide signaling but not for agonist-induced adenylate cyclase activation. Biochem Biophys Res Commun 1994;199:1497–503
156. Liggett SB, Caron MG, Lefkowitz RJ, Hnatowich M. Coupling of a mutated form of the human beta 2-adrenergic receptor to Gi and Gs. Requirement for multiple cytoplasmic domains in the coupling process. J Biol Chem 1991;266:4816–21
157. Nussenzveig DR, Thaw CN, Gershengorn MC. Inhibition of inositol phosphate second messenger formation by intracellular loop one of a human calcitonin receptor. Expression and mutational analysis of synthetic receptor genes. J Biol Chem 1994;269:28123–9
158. Wiens BL, Nelson CS, Neve KA. Contribution of serine residues to constitutive and agonist-induced signaling via the D2S dopamine receptor: evidence for multiple, agonist-specific active conformations. Mol Pharmacol 1998;54:435–44
159. Conchon S, Barrault MB, Miserey S, Corvol P, Clauser E. The C-terminal third intracellular loop of the rat AT1A angiotensin receptor plays a key role in G protein coupling specificity and transduction of the mitogenic signal. J Biol Chem 1997;272:25566–72
160. Eason MG, Liggett SB. Identification of a Gs coupling domain in the amino terminus of the third intracellular loop of the alpha 2A-adrenergic receptor. Evidence for distinct structural determinants that confer Gs versus Gi coupling. J Biol Chem 1995;270:24753–60

161. Eason MG, Liggett SB. Chimeric mutagenesis of putative G-protein coupling domains of the alpha2A-adrenergic receptor. Localization of two redundant and fully competent gi coupling domains. J Biol Chem 1996;271:12826–32
162. Francesconi A, Duvoisin RM. Role of the second and third intracellular loops of metabotropic glutamate receptors in mediating dual signal transduction activation. J Biol Chem 1998;273:5615–24
163. Gilchrist RL, Ryu KS, Ji I, Ji TH. The luteinizing hormone/chorionic gonadotropin receptor has distinct transmembrane conductors for cAMP and inositol phosphate signals. J Biol Chem 1996;271:19283–7
164. Wade SM, Lim WK, Lan KL, Chung DA, Nanamori M, Neubig RR. G(i) activator region of alpha(2A)-adrenergic receptors: Distinct basic residues mediate G(i) versus G(s) activation. Mol Pharmacol 1999;56:1005–13
165. Lane JR, Powney B, Wise A, Rees S, Milligan G. G protein coupling and ligand selectivity of the D2L and D3 dopamine receptors. J Pharmacol Exp Ther 2008;325:319–30
166. Kushwaha N, Harwood SC, Wilson AM et al. Molecular determinants in the second intracellular loop of the 5-hydroxytryptamine-1A receptor for G-protein coupling. Mol Pharmacol 2006;69:1518–26
167. Hoare S, Copland JA, Strakova Z et al. The proximal portion of the COOH terminus of the oxytocin receptor is required for coupling to G(q), but not G(i) - Independent mechanisms for elevating intracellular calcium concentrations from intracellular stores. J Biol Chem 1999;274:28682–9
168. Najimi M, Gailly P, Maloteaux JM, Hermans E. Distinct regions of the high affinity neurotensin receptor mediate the functional coupling with PTx sensitive and insensitive G proteins. FEBS Lett 2002;512:329–33
169. Okamoto Y, Ninomiya H, Tanioka M, Sakamoto A, Miwa S, Masaki T. Cysteine residues in the carboxyl terminal domain of the endothelin-B receptor are required for coupling with G-proteins. J Cardiovasc Pharmacol 1998;31 Suppl 1:S230–S232
170. Katada S, Tanaka M, Touhara K. Structural determinants for membrane trafficking and G protein selectivity of a mouse olfactory receptor. J Neurochem 2004;90:1453–63
171. Tateyama M, Kubo Y. Coupling profile of the metabotropic glutamate receptor 1alpha is regulated by the C-terminal domain. Mol Cell Neurosci 2007;34:445–52
172. Mukhopadhyay S, Howlett AC. CB1 receptor-G protein association. Subtype selectivity is determined by distinct intracellular domains. Eur J Biochem 2001;268:499–505
173. Iida-Klein A, Guo J, Takemura M et al. Mutations in the second cytoplasmic loop of the rat parathyroid hormone (PTH)/PTH-related protein receptor result in selective loss of PTH-stimulated phospholipase C activity. J Biol Chem 1997;272:6882–9
174. Hiltscher R, Seuwen K, Boddeke HW, Sommer B, Laurie DJ. Functional coupling of human metabotropic glutamate receptor hmGlu1d: comparison to splice variants hmGlu1a and -1b. Neuropharmacology 1998;37:827–37
175. Joly C, Gomeza J, Brabet I, Curry K, Bockaert J, Pin JP. Molecular, functional, and pharmacological characterization of the metabotropic glutamate receptor type 5 splice variants: comparison with mGluR1. J Neurosci 1995;15:3970–81
176. Shyu JF, Inoue D, Baron R, Horne WC. The deletion of 14 amino acids in the seventh transmembrane domain of a naturally occurring calcitonin receptor isoform alters ligand binding and selectively abolishes coupling to phospholipase C. J Biol Chem 1996;271:31127–34
177. Mary S, Stephan D, Gomeza J, Bockaert J, Pruss RM, Pin JP. The rat mGlu1d receptor splice variant shares functional properties with the other short isoforms of mGlu1 receptor. Eur J Pharmacol 1997;335:65–72
178. Pindon A, van-Hecke G, van-Gompel P, Lesage AS, Leysen JE, Jurzak M. Differences in signal transduction of two 5-HT4 receptor splice variants: compound specificity and dual coupling with Galphas- and Galphai/o-proteins. Mol Pharmacol 2002;61:85–96
179. Namba T, Sugimoto Y, Negishi M et al. Alternative splicing of C-terminal tail of prostaglandin E receptor subtype EP3 determines G-protein specificity. Nature 1993;365:166–70

180. Negishi M, Namba T, Sugimoto Y et al. Opposite coupling of prostaglandin E receptor EP3C with Gs and G(o). Stimulation of Gs and inhibition of G(o). J Biol Chem 1993;268:26067–70
181. Hatae N, Aksentijevich N, Zemkova HW, Kretschmannova K, Tomic M, Stojilkovic SS. Cloning and functional identification of novel endothelin receptor type A isoforms in pituitary. Mol Endocrinol 2007;21:1192–204
182. Germano PM, Le SV, Oh DS et al. Differential coupling of the PAC1 SV1 splice variant on human colonic tumors to the activation of intracellular cAMP but not intracellular Ca2+ does not activate tumor proliferation. J Mol Neurosci 2004;22:83–92
183. Perez DM, Hwa J, Gaivin R, Mathur M, Brown F, Graham RM. Constitutive activation of a single effector pathway: evidence for multiple activation states of a G protein-coupled receptor. Mol Pharmacol 1996;49:112–22
184. Swaminath G, Xiang Y, Lee TW, Steenhuis J, Parnot C, Kobilka BK. Sequential binding of agonists to the beta2 adrenoceptor. Kinetic evidence for intermediate conformational states. J Biol Chem 2004;279:686–91
185. Abadji V, Lucas LJ, Chin C, Kendall DA. Involvement of the carboxyl terminus of the third intracellular loop of the cannabinoid CB1 receptor in constitutive activation of Gs. J Neurochem 1999;72:2032–8
186. Barroso S, Richard F, Nicolas-Etheve D, Kitabgi P, Labbe-Jullie C. Constitutive activation of the neurotensin receptor 1 by mutation of Phe(358) in Helix seven. Br J Pharmacol 2002;135:997–1002
187. Chaipatikul V, Loh HH, Law PY. Ligand-selective activation of mu-oid receptor: demonstrated with deletion and single amino acid mutations of third intracellular loop domain. J Pharmacol Exp Ther 2003;305:909–18
188. Surratt CK, Johnson PS, Moriwaki A et al. -mu opiate receptor. Charged transmembrane domain amino acids are critical for agonist recognition and intrinsic activity. J Biol Chem 1994;269:20548–53
189. Hosohata Y, Varga EV, Stropova D et al. Mutation W284L of the human delta opioid receptor reveals agonist specific receptor conformations for G protein activation. Life Sci 2001;68:2233–42
190. Kenakin T. New concepts in drug discovery: collateral efficacy and permissive antagonism. Nat Rev Drug Discov 2005;4:919–27
191. Kobilka BK, Deupi X. Conformational complexity of G-protein-coupled receptors. Trends Pharmacol Sci 2007;28:397–406
192. Hoffmann C, Zurn A, Bunemann M, Lohse MJ. Conformational changes in G-protein-coupled receptors-the quest for functionally selective conformations is open. Br J Pharmacol 2008;153 Suppl 1:S358–S366
193. Vauquelin G, Van L, I. G protein-coupled receptors: a count of 1001 conformations. Fundam Clin Pharmacol 2005;19:45–56
194. Perez DM, Karnik SS. Multiple signaling states of G-protein-coupled receptors. Pharmacol Rev 2005;57:147–61
195. Lopez-Gimenez JF, Villazon M, Brea J et al. Multiple conformations of native and recombinant human 5-hydroxytryptamine(2a) receptors are labeled by agonists and discriminated by antagonists. Mol Pharmacol 2001;60:690–9
196. Evans PD, Robb S, Cheek TR et al. Agonist-specific coupling of G-protein-coupled receptors to second-messenger systems. Prog Brain Res 1995;1062:59–68
197. Kenakin T. Agonist-receptor efficacy. II. Agonist trafficking of receptor signals. Trends Pharmacol Sci 1995;16:232–8
198. Urban JD, Clarke WP, von Zastrow M et al. Functional selectivity and classical concepts of quantitative pharmacology. J Pharmacol Exp Ther 2007;320:1–13
199. MacKinnon AC, Waters C, Jodrell D, Haslett C, Sethi T. Bombesin and substance P analogues differentially regulate G-protein coupling to the bombesin receptor. Direct evidence for biased agonism. J Biol Chem 2001;276:28083–91
200. Robb S, Cheek TR, Hannan FL, Hall LM, Midgley JM, Evans PD. Agonist-specific coupling of a cloned Drosophila octopamine/tyramine receptor to multiple second messenger systems. EMBO J 1994;13:1325–30

201. Spengler D, Waeber C, Pantaloni C et al. Differential signal transduction by five splice variants of the PACAP receptor. Nature 1993;365:170–5
202. Takasu H, Gardella TJ, Luck MD, Potts JT, Bringhurst FR. Amino-terminal modifications of human parathyroid hormone (PTH) selectively alter phospholipase C signaling via the type 1 PTH receptor: Implications for design of signal-specific PTH ligands. Biochemistry 1999;38:13453–60
203. Brink CB, Wade SM, Neubig RR. Agonist-directed trafficking of porcine alpha(2A)-adrenergic receptor signaling in Chinese hamster ovary cells: l-isoproterenol selectively activates G(s). J Pharmacol Exp Ther 2000;294:539–47
204. Eason MG, Jacinto MT, Liggett SB. Contribution of ligand structure to activation of alpha 2-adrenergic receptor subtype coupling to Gs. Mol Pharmacol 1994;45:696–702
205. Rudling JE, Richardson J, Evans PD. A comparison of agonist-specific coupling of cloned human alpha(2)-adrenoceptor subtypes. Br J Pharmacol 2000;131:933–41
206. Pauwels PJ, Rauly I, Wurch T. Dissimilar pharmacological responses by a new series of imidazoline derivatives at precoupled and ligand-activated alpha 2A-adrenoceptor states: evidence for effector pathway-dependent differential antagonism. J Pharmacol Exp Ther 2003;305:1015–23
207. Galandrin S, Oligny-Longpre G, Bonin H, Ogawa K, Gales C, Bouvier M. Conformational rearrangements and signaling cascades involved in ligand-biased MAPK signaling through the {beta}1-adrenergic receptor. Mol Pharmacol 2008;74:162–72.
208. Ghanouni P, Gryczynski Z, Steenhuis JJ et al. Functionally different agonists induce distinct conformations in the G protein coupling domain of the beta 2 adrenergic receptor. J Biol Chem 2001;276:24433–6
209. Rathz DA, Brown KM, Kramer LA, Liggett SB. Amino acid 49 polymorphisms of the human beta(1)-adrenergic receptor affect agonist-promoted trafficking. J Cardiovasc Pharmacol 2002;39:155–60
210. Xiao RP. Beta-adrenergic signaling in the heart: dual coupling of the beta2-adrenergic receptor to G(s) and G(i) proteins. Sci STKE 2001;2001:RE15
211. Watson C, Chen G, Irving P, Way J, Chen WJ, Kenakin T. The use of stimulus-biased assay systems to detect agonist-specific receptor active states: implications for the trafficking of receptor stimulus by agonists. Mol Pharmacol 2000;58:1230–8
212. Bosier B, Hermans E, Lambert DM. Differential modulation of AP-1- and CRE-driven transcription by cannabinoid agonists emphasizes functional selectivity at the CB(1) receptor. Br J Pharmacol 2008
213. Bonhaus DW, Chang LK, Kwan J, Martin GR. Dual activation and inhibition of adenylyl cyclase by cannabinoid receptor agonists: Evidence for agonist-specific trafficking of intracellular responses. J Pharmacol Exp Ther 1998;287:884–8
214. Shoemaker JL, Ruckle MB, Mayeux PR, Prather PL. Agonist-directed trafficking of response by endocannabinoids acting at CB2 receptors. J Pharmacol Exp Ther 2005;315:828–38
215. Fitzsimons CP, Gompels UA, Verzijl D et al. Chemokine-directed trafficking of receptor stimulus to different g proteins: selective inducible and constitutive signaling by human herpesvirus 6-encoded chemokine receptor U51. Mol Pharmacol 2006;69:888–98
216. Beyermann M, Heinrich N, Fechner K et al. Achieving signalling selectivity of ligands for the corticotropin-releasing factor type 1 receptor by modifying the agonist's signalling domain. Br J Pharmacol 2007;151:851–9
217. Reale V, Hannan F, Hall LM, Evans PD. Agonist-specific coupling of a cloned Drosophila melanogaster D1-like dopamine receptor to multiple second messenger pathways by synthetic agonists. J Neurosci 1997;17:6545–53
218. Panchalingam S, Undie AS. SKF83959 exhibits biochemical agonism by stimulating [S-35] GTP gamma S binding and phosphoinositide hydrolysis in rat and monkey brain. Neuropharmacology 2001;40:826–37
219. Ryman-Rasmussen JP, Nichols DE, Mailman RB. Differential activation of adenylate cyclase and receptor internalization by novel dopamine D1 receptor agonists. Mol Pharmacol 2005;68:1039–48

220. Urban JD, Vargas GA, von Zastrow M, Mailman RB. Aripiprazole has functionally selective actions at dopamine D2 receptor-mediated signaling pathways. Neuropsychopharmacology 2007;32:67–77
221. Kilts JD, Connery HS, Arrington EG et al. Functional selectivity of dopamine receptor agonists. II. Actions of dihydrexidine in D2L receptor-transfected MN9D cells and pituitary lactotrophs. J Pharmacol Exp Ther 2002;301:1179–89
222. Gay EA, Urban JD, Nichols DE, Oxford GS, Mailman RB. Functional selectivity of D2 receptor ligands in a Chinese hamster ovary hD2L cell line: evidence for induction of ligand-specific receptor states. Mol Pharmacol 2004;66:97–105
223. Moniri NH, Covington-Strachan D, Booth RG. Ligand-directed functional heterogeneity of histamine H1 receptors: novel dual-function ligands selectively activate and block H1-mediated phospholipase C and adenylyl cyclase signaling. J Pharmacol Exp Ther 2004;311:274–81
224. Gurwitz D, Haring R, Heldman E, Fraser CM, Manor D, Fisher A. Discrete activation of transduction pathways associated with acetylcholine m1 receptor by several muscarinic ligands. Eur J Pharmacol 1994;267:21–31
225. Akam EC, Challiss RAJ, Nahorski SR. G(q/11) and G(i/o) activation profiles in CHO cells expressing human muscarinic acetylcholine receptors: dependence on agonist as well as receptor-subtype. Br J Pharmacol 2001;132:950–8
226. Nickolls SA, Fleck B, Hoare SR, Maki RA. Functional selectivity of melanocortin 4 receptor peptide and nonpeptide agonists: evidence for ligand-specific conformational states. J Pharmacol Exp Ther 2005;313:1281–8
227. Allouche S, Polastron J, Hasbi A, Homburger V, Jauzac P. Differential G-protein activation by alkaloid and peptide opioid agonists in the human neuroblastoma cell line SK-N-BE. Biochem J 1999;342 (Pt 1):71–8
228. Pineyro G, Azzi M, De Lean A, Schiller P, Bouvier M. Short-term inverse-agonist treatment induces reciprocal changes in delta-opioid agonist and inverse-agonist binding capacity. Mol Pharmacol 2001;60:816–27
229. Reversi A, Rimoldi V, Marrocco T et al. The oxytocin receptor antagonist atosiban inhibits cell growth via a "biased agonist" mechanism. J Biol Chem 2005;280:16311–8
230. Negishi M, Irie A, Sugimoto Y, Namba T, Ichikawa A. Selective coupling of prostaglandin E receptor EP3D to Gi and Gs through interaction of alpha-carboxylic acid of agonist and arginine residue of seventh transmembrane domain. J Biol Chem 1995;270:16122–7
231. Heusler P, Pauwels PJ, Wurch T et al. Differential ion current activation by human 5-HT(1A) receptors in Xenopus oocytes: evidence for agonist-directed trafficking of receptor signalling. Neuropharmacology 2005;49:963–76
232. Palmer TM, Gettys TW, Jacobson KA, Stiles GL. Desensitization of the canine A2a adenosine receptor: delineation of multiple processes. Mol Pharmacol 1994;45:1082–94
233. Kurrasch-Orbaugh DM, Watts VJ, Barker EL, Nichols DE. Serotonin 5-hydroxytryptamine 2A receptor-coupled phospholipase C and phospholipase A2 signaling pathways have different receptor reserves. J Pharmacol Exp Ther 2003;304:229–37
234. Rosenzweig-Lipson S, Zhang J, Mazandarani H et al. Antiobesity-like effects of the 5-HT2C receptor agonist WAY-161503. Brain Res 2006;1073–1074:240–51
235. Berg KA, Maayani S, Goldfarb J, Scaramellini C, Leff P, Clarke WP. Effector pathway-dependent relative efficacy at serotonin type 2A and 2C receptors: evidence for agonist-directed trafficking of receptor stimulus. Mol Pharmacol 1998;54:94–104
236. Berg KA, Cropper JD, Niswender CM, Sanders-Bush E, Emeson RB, Clarke WP. RNA-editing of the 5-HT(2C) receptor alters agonist-receptor-effector coupling specificity. Br J Pharmacol 2001;134:386–92
237. Malmberg A, Strange PG. Site-directed mutations in the third intracellular loop of the serotonin 5-HT1A receptor alter G protein coupling from G(i) to G(s) in a ligand-dependent manner. J Neurochem 2000;75:1283–93
238. Siehler S, Nunn C, Zupanc GK, Hoyer D. Fish somatostatin sst3 receptor: comparison of radioligand and GTPgammaS binding, adenylate cyclase and phospholipase C activities

reveals different agonist-dependent pharmacological signatures. Auton Autacoid Pharmacol 2005;25:1–16
239. McLaughlin JN, Shen L, Holinstat M, Brooks JD, Dibenedetto E, Hamm HE. Functional selectivity of G protein signaling by agonist peptides and thrombin for the protease-activated receptor-1. J Biol Chem 2005;280:25048–59
240. Azzi M, Charest PG, Angers S et al. Beta-arrestin-mediated activation of MAPK by inverse agonists reveals distinct active conformations for G protein-coupled receptors. Proc Natl Acad Sci USA 2003;100:11406–11
241. Arey BJ, Stevis PE, Deecher DC et al. Induction of promiscuous G protein coupling of the follicle-stimulating hormone (FSH) receptor: a novel mechanism for transducing pleiotropic actions of FSH isoforms. Mol Endocrinol 1997;11:517–26
242. Nichols DE. Hallucinogens. Pharmacol Ther 2004;101:131–81
243. Galandrin S, Bouvier M. Distinct signaling profiles of beta1 and beta2 adrenergic receptor ligands toward adenylyl cyclase and mitogen-activated protein kinase reveals the pluridimensionality of efficacy. Mol Pharmacol 2006;70:1575–84
244. Lefkowitz RJ, Pierce KL, Luttrell LM. Dancing with different partners: protein kinase a phosphorylation of seven membrane-spanning receptors regulates their g protein-coupling specificity. Mol Pharmacol 2002;62:971–4
245. Engstrom M, Savola JM, Wurster S. Differential efficacies of somatostatin receptor agonists for G-protein activation and desensitization of somatostatin receptor subtype 4-mediated responses. J Pharmacol Exp Ther 2006;316:1262–8
246. Nemeth K, Chollet A. A single mutation of the neurokinin-2 (NK2) receptor prevents agonist-induced desensitization. Divergent conformational requirements for NK2 receptor signaling and agonist-induced desensitization in Xenopus oocytes. J Biol Chem 1995;270:27601–5
247. Stout BD, Clarke WP, Berg KA. Rapid desensitization of the serotonin(2C) receptor system: Effector pathway and agonist dependence. J Pharmacol Exp Ther 2002;302:957–62
248. Roettger BF, Ghanekar D, Rao R et al. Antagonist-stimulated internalization of the G protein-coupled cholecystokinin receptor. Mol Pharmacol 1997;51:357–62
249. Holloway AC, Qian H, Pipolo L et al. Side-chain substitutions within angiotensin II reveal different requirements for signaling, internalization, and phosphorylation of type 1A angiotensin receptors. Mol Pharmacol 2002;61:768–77
250. Este JA. Virus entry as a target for anti-HIV intervention. Curr Med Chem 2003;10:1617–32
251. Gray JA, Roth BL. Paradoxical trafficking and regulation of 5-HT(2A) receptors by agonists and antagonists. Brain Res Bull 2001;56:441–51
252. Maillet EL, Pellegrini N, Valant C et al. A novel, conformation-specific allosteric inhibitor of the tachykinin NK2 receptor (NK2R) with functionally selective properties. FASEB J 2007;21:2124–34
253. Hoare SR, Fleck BA, Gross RS, Crowe PD, Williams JP, Grigoriadis DE. Allosteric ligands for the corticotropin releasing factor type 1 receptor modulate conformational states involved in receptor activation. Mol Pharmacol 2008;73:1371–80

Chapter 4
Ligand-Selective Receptor Desensitization and Endocytosis

Jennifer L. Whistler

Abstract Following activation by its endogenous agonist ligand(s), nearly every G protein-coupled receptor (GPCR) studied to date undergoes regulation by a cascade of events that includes receptor desensitization, receptor endocytosis, receptor recycling, and/or receptor degradation. However, not all GPCR agonists promote receptor desensitization and endocytosis. Indeed, the ability of an agonist to promote endocytosis does not always vary in a linear fashion with agonist activity, and even GPCR antagonists, in some cases, can promote endocytosis, indicating that receptor desensitization and endocytosis are independent functional properties that can display ligand-selective effects, or functional selectivity. Therefore, for any GPCR-ligand pair there are at least two trafficking properties that must be assessed to fully understand signal transduction: (1) whether or not the ligand promotes endocytosis of the receptor and (2) where that receptor goes following its endocytosis. For the vast majority of drugs that target GPCRs neither of these things are known. In this chapter, we outline several cases in which ligands show functional selectivity for receptor trafficking with an emphasis on receptors in which altered ligand-induced trafficking has been shown to be relevant to human disease.

Keywords Trafficking, Endocytosis, Desensitization, Resensitization, Down-regulation, GPCR, Ligand, Biased agonism

4.1 Introduction

Following activation by its endogenous agonist ligand(s), nearly every G protein-coupled receptor (GPCR) studied to date undergoes regulation by a cascade of events that includes receptor desensitization, receptor endocytosis, receptor

---

J.L. Whistler
Department of Neurology, Ernest Gallo Clinic and Research Center, University of California
San Francisco, CA 94608
e-mail: jwhistler@gallo.ucsf.edu

K. Neve (ed.), *Functional Selectivity of G Protein-Coupled Receptor Ligands*,
DOI: 10.1007/978-1-60327-335-0_4,

recycling, and/or receptor degradation (for review see *(1)*). For the purposes of this chapter, we will define receptor desensitization as any process that alters the functional coupling of a receptor to its G protein/second messenger-signaling pathway. This can be mediated through receptor phosphorylation by any number of kinases as well as by several other mechanisms. Endocytosis/internalization will be defined as the translocation of receptors from the cell surface to an intracellular compartment. Receptor recycling or receptor resensitization will be defined as receptors returning to the cell surface following their endocytosis. Lastly, receptor degradation or downregulation will be defined as any process that decreases the number of ligand-binding sites. Desensitization, by definition then, turns off signal transduction. Endocytosis of desensitized receptors can, thereby, serve several different roles depending on the post-endocytic fate of the receptor. Specifically, following endocytosis, receptors can be recycled to the plasma membrane, retained in an intracellular compartment, or targeted for degradation. For receptors that are recycled, endocytosis serves as the first step toward resensitizing receptors to the presence of agonist. In contrast, for receptors that are degraded, endocytosis serves as the first step toward receptor downregulation. This dynamic cycle of receptor regulation may be designed to mediate the actions of native GPCR ligands, which are typically released in a phasic or pulsatile manner. However, not all GPCR agonists promote receptor desensitization and endocytosis. Indeed, the ability of a ligand to promote endocytosis does not always vary in a linear fashion with agonist activity (see for example *(2,3)*), and even GPCR antagonists, in some cases, can promote endocytosis *(4)*, indicating that receptor desensitization and endocytosis are independent functional properties that can display ligand-selective effects.

As is evidenced from this book, the magnitude of signal transduction from a single receptor-ligand complex is controlled by many factors. These include innate properties of the ligand itself such as affinity, potency, efficacy, bioavailability, and half-life. Together, these ligand properties constitute the "relative activity" of a particular ligand, which can differ depending on cellular context and the downstream effector being monitored. Importantly, the magnitude of signal transduction through the receptor-ligand pair is also controlled by the length of time the receptor remains coupled to downstream effectors. Although this can be dependent on the pharmacokinetic properties of the ligand, it is also regulated by the degree and rate of receptor desensitization and endocytosis that occurs. It follows that, because ligand activity and the degree of desensitization and endocytosis are not always linear, the degree of desensitization and endocytosis can be viewed as a specific "effector" of an individual receptor–ligand pair. Hence, the relative amount of signal transduced to a cell is a function of both the relative activity of the particular receptor–ligand pair, and the amount of desensitization and endocytosis that occurs, which regulates both the quantitative (how much) and qualitative (how long and in what pattern) effects of the ligand (for review see *(5)*).

One could hypothesize that agonists with identical intrinsic activities at the receptor, but varying abilities to facilitate desensitization and endocytosis, would thereby differ dramatically in the actual amount of signal transmitted to the cell during the lifetime of the ligand and receptor. Furthermore, the role of receptor endocytosis in modulating signal transduction would be dramatically different depending on whether the GPCR being studied was recycled or degraded following its internalization. In short, for any GPCR-ligand pair, there are at least two trafficking properties that must be assessed to fully understand signal transduction: (1) whether or not the ligand promotes endocytosis of the receptor and (2) where that receptor goes following its endocytosis.

There are numerous examples, some of which are detailed in this chapter, whereby one or the other or both of these trafficking properties shows functional ligand selectivity at a target GPCR. In the broader picture, more than 50% of the drugs used to treat human disease target one or more GPCR. For the vast majority of these drugs, the effects of the ligand on the endocytic and post-endocytic trafficking properties of the target receptor are unknown. This chapter is, thus, written to emphasize that obtaining this information could be critical to effective GPCR drug design for many GPCR families.

4.2 Specific Examples of Ligand-Biased Receptor Desensitization and Endocytosis

4.2.1 Opioid Receptors

Some of the most extensive studies examining the relationship between agonist activity and receptor endocytosis have used the opioid receptors as a model. This receptor class is reviewed at length in the second part of this book (see Chap. 12). Briefly, for the μ opioid receptor (MOR), the endogenous peptide ligands and several small molecule agonists induce rapid desensitization, endocytosis, and recycling of the receptor. In contrast, morphine induces only weak or partial desensitization and little to no receptor endocytosis both *in vitro (6–11)* and *in vivo (12–15)*. Some studies have concluded that the low degree of endocytosis promoted by morphine simply reflects its lower intrinsic efficacy at the MOR. Indeed, there is evidence from both *ex vivo* and heterologous systems that there is a linear relationship between efficacy and endocytosis for some opioid ligands (see for example *(16,17)*). However, mutants of the MOR with apparently unaltered morphine activity, nevertheless, do undergo rapid desensitization and endocytosis *(2,18,19)*. Furthermore, G protein coupling is not required for endocytosis of the opioid receptors *(20,21)*, as pretreatment of cells with pertussis toxin does not inhibit receptor endocytosis. Hence, coupling strength/signal strength *per se* is likely not the decisive factor determining whether a receptor is efficiently shuttled to the endocytic pathway.

As a consequence of differences in receptor desensitization and endocytosis, signal transduction promoted by morphine is both qualitatively and quantitatively different than that induced by the endogenous peptide ligands and the other opioid drugs that promote endocytosis of the MOR. The resulting imbalance of desensitization–endocytosis–resensitization has at least two consequences: (1) In cell types where morphine induces desensitization but not endocytosis and/or resensitization, desensitization is protracted; (2) In cell types where morphine induces neither desensitization nor endocytosis, prolonged signaling through the receptor leads to multiple cellular adaptations downstream of receptor-G protein coupling. Both protracted MOR desensitization and adaptive cellular changes likely contribute to the pronounced *in vivo* tolerance and dependence that occurs with chronic morphine treatment (for review see *(5)*). As a consequence, facilitating receptor endocytosis, using either genetic *(19)* or pharmacological *(13,14)* approaches, can restore the balance of signaling through the receptor and delay the development of tolerance and dependence.

Morphine at the MOR is not the only example of an agonist ligand that fails to promote receptor endocytosis. For example, etorphine is a potent agonist at all three opioid receptors: the MOR, the δ opioid receptor (DOR), and the κ opioid receptor (KOR) (see for example *(22)*). Although etorphine promotes robust endocytosis of both MOR *(6,12)* and DOR *(23)*, it does not promote endocytosis of the KOR *(24)*. Hence, etorphine shows functional selectivity for G protein signal transduction vs. endocytosis at the KOR.

Although the MOR is recycled after its endocytosis, the DOR can be transported deeper into the endocytic pathway and rapidly degraded by the lysosome following their endocytosis *(11,25)*. This regulated sorting of GPCRs between recycling and degradation obviously has dramatic effects on signal transduction (mediating resensitization vs. downregulation, respectively). Furthermore, there is evidence that some DOR ligands favor the degradative pathway while others favor the recycling pathway *(26,27)*.

4.2.2 Serotonin Receptors

Functional selectivity for signaling at the serotonin (5HT) receptors is likewise reviewed in the second half of this book (see Chap. 9). Most relevant with regard to this chapter, 5HT receptors provide a clear example that ligand efficacy *per se* is not what determines whether or not the receptor is rapidly endocytosed. This is evidenced by the observation that not only agonists, but also antagonists, at the 5HT2A receptor can promote receptor endocytosis *(4,28)*.

Drugs that activate 5HT receptors include lysergic acid diethylamide (LSD), 2,5-dimethoxy-4-iodoamphetamine (DOI), and 1-(2,5-dimethoxy-4-methylphenyl)-2-aminopropane (DOM). In addition, serotonin selective reuptake inhibitors (SSRIs) such as fluoxetine, and the serotonin transporter selective amphetamine 3,4 methylenedioxymethamphetamine (MDMA) also activate 5HT receptors by increasing serotonin tone. The 5HT2A receptor has been implicated in the hallucinogenic effects of all of these drugs (for review see *(29)*). Indeed, both hallucinogenic and non-hallucinogenic agonists at the 5HT2A receptor can activate phospholipase C (PLC), phospholipase A2, extracellular signal-related kinase (ERK)1/2 phosphorylation, and p38 phosphorylation to varying degrees exemplifying "ligand-selective" 5HT2A receptor activation (see Chap. 9 and for review see *(30)*). The ligand-selective profiles for these effectors, however, does not correlate with the hallucinogenic properties of each of these drugs *(31)* (and for review see *(32)*), suggesting that a property other than ligand activity is contributing to the hallucinogenic effects and is differentially mediated by these varying 5HT2A agonists.

Intriguingly, as reviewed in Chap. 5, some of these agonists at the 5HT2A receptor show a ligand-selective ability to promote arrestin recruitment and thereby receptor endocytosis. Specifically, while both serotonin and DOI activate the 5HT2A receptor, only serotonin promotes substantial arrestin-mediated desensitization and receptor endocytosis *(33)*. Hence, either endocytosis *per se*, or differential signal transduction as a consequence of arrestin recruitment, could alleviate some of the hallucinogenic side effects associated with serotinergic drugs.

Importantly, the post-endocytic sorting fate of the serotonin receptors has not been elucidated. Therefore, whether there are ligand-selective differences in recycling or degradation of the serotonin receptors with different ligands is not known.

4.2.3 Dopamine Receptors

The dopamine receptor family comprises five distinct subtypes of dopamine receptors the D1-like (D1 and D5) that are G_s-coupled ("s" for stimulatory) and the D2-like (D2, D3, and D4) that are G_i-coupled ("i" for inhibitory). All of these receptors are activated by the endogenous ligand dopamine, albeit to varying degrees (see Chap. 10). Intriguingly, dopamine also has differing abilities to promote endocytosis at the different dopamine receptors. For example, although dopamine promotes robust internalization of the D1 and D2 dopamine receptor *(34–37)*, it promotes little endocytosis of the D3 receptor *(38)*. However, activation of PKC promotes endocytosis of the D3 receptor both in the presence and absence of dopamine *(39)*. Hence, dopamine itself shows "ligand-selectivity" at the D3 receptor: in the absence of PKC activation it activates receptor-mediated signal transduction without promoting receptor endocytosis, and, in the presence of active PKC, dopamine activates both signal transduction and endocytosis.

There are several structurally distinct agonists at the D1 receptor. At least four agonist ligands from two structurally distinct classes have been shown to potently activate adenylyl cyclase without promoting receptor endocytosis *(40)*. These examples, emphasize, once again, that the ability of a ligand to promote receptor endocytosis cannot be predicted based on its activity at another effector, such as G protein. As a further complication, while both the D1 and the D2 receptor are endocytosed in response to activation by dopamine, the D1 receptor is recycled after endocytosis, while the D2 receptor is degraded *(36)*. In addition, the rate of recycling of the D1 receptor can differ depending on the ligand *(41)*.

Dysregulation of dopaminergic signaling is associated with several different neuropsychiatric diseases including depression, bipolar disorder, schizophrenia, addiction, and Parkinson's disease. The molecular mechanisms responsible for these alterations in dopamine signaling are largely unknown. However, dopamine receptors, in particular D2-like receptors, are significantly downregulated in animals with increased dopamine tone *(42)*, in untreated schizophrenic and bipolar patients *(43–45)*, in patients with chronic depression *(46–48)*, and in socially subordinate primates *(49,50)*. Drugs of abuse also cause a significant downregulation of D2-like receptors in humans *(51–54)*, monkeys *(55–58)*, and rodents *(59,60)*. The large percentage of drug abusers with comorbidities of depression, anxiety, and bipolar symptoms has led to the hypothesis that downregulation of D2-like receptors contributes to many modalities of neuropsychiatric disease (see *(61)* for example). As a consequence, numerous therapeutics have been developed that target, at least in part, the D2 dopamine receptor. In the vast majority of cases, the effects of these drugs on the endocytic and post-endocytic trafficking of the receptor is unknown.

Somewhat paradoxically, both agonist and antagonist drugs at the D2 receptors have shown efficacy in human subjects – even for the same disease (for review see *(62)*). However, if one assumes that loss of D2 receptors is contributing to the disease state, then both a D2 receptor antagonist that blocked endocytosis (and thereby downregulation) of receptor by endogenous dopamine, as well as a D2 receptor agonist that increased the amount of signal transduction through the D2 receptor, could have the same net effect: increased signal transduction through the D2 receptor. However, even if a D2 receptor antagonist drug does prevent downregulation of D2 receptors (and as noted above, not all antagonists do), their effectiveness would likely be weak, because the D2 receptors would not signal efficiently in the presence of the antagonist, but only once the antagonist had cleared to make room for dopamine.

Interestingly, agonists at the D2-like receptors have shown some promise as antidepressants, in addition to their efficacy in Parkinson's disease. However, we are compelled by the observation that their utility – for both these indications – is often transient (see for example *(63)*). This could reflect the fact that full agonists at the D2 receptor might be expected to facilitate endocytosis and consequently downregulation of D2 receptors. Hence, the target for these drugs (the D2 receptor) would be decreased over time. It is intriguing to speculate, then, that D2 receptor agonists that did not facilitate endocytosis of the D2R, or that facilitated endocytosis but did not promote receptor degradation, would be more therapeutically effective than either antagonists or full agonists. Such a drug would represent a functionally selective agonist that promoted signal transduction without promoting receptor endocytosis and thus downregulation. In no instance have the drugs developed against the D2-like receptors been specifically evaluated for their endocytic and post-endocytic properties on the receptor target. However, there is some hint that there could be trafficking differences among these drugs. For example, both bromocrytine and piribedil are agonists relatively specific for the D2 receptor *(64)*. However, bromocryptine shows better clinical efficacy over time *(65)* than does piribedil *(63)*. Perhaps, then, the trafficking pattern of the D2 receptor is different when the receptors are activated by these two drugs – downregulated by piribedil and to a lesser extent by bromocryptine.

4.2.4 Chemokine Receptors

There are several examples among the chemokine receptors of ligand-selective differences in receptor trafficking. For example, the CCR5 receptor, a chemokine receptor expressed on T cells and macrophages, and the principal coreceptor for HIV-1 infection, shows ligand-selective differences in signaling vs. endocytosis. Both RANTES (regulated on activation, normal T cell expressed and secreted), an agonist at the CCR5 receptor, and aminooxypentane-RANTES (AOP-RANTES), an antagonist at the CCR5 receptor, promote receptor endocytosis. Once again, as for the 5HT2A receptor (see above), this is an example of a GPCR "antagonist" that, nevertheless, functions as an agonist for receptor endocytosis. However, the post-endocytic trafficking of the CCR5 receptor is different depending on whether the receptor is occupied by RANTES or AOP-RANTES. Specifically, while RANTES-activated receptors are recycled, AOP-RANTES activated receptors are not *(66)*. Importantly, AOP-RANTES shows potent inhibition of macrophage infection by HIV-1 under conditions where RANTES is barely effective *(67)*, suggesting that ligand-selective differences in post-endocytic trafficking of the receptor can dramatically effect the therapeutic utility of CCR5 ligands.

4.2.5 Ligand-Selective Trafficking of Other GPCRs

The above sections outline a few examples of GPCR families where there appear to be ligand-selective effects on receptor desensitization and endocytosis. We highlighted these examples, in part because the existence of ligand-selective trafficking for these particular receptors has clear implications for drug development and human disease. This list is not meant to be exhaustive, but rather illustrative of the importance of receptor trafficking as an independent effector of a ligand-GPCR pair. Importantly, this phenomenon of ligand-selective endocytic and/or post-endocytic trafficking does not appear to be selective for a single family of GPCR. Among the Class A GPCRs, opioid and chemokine receptors are members of the peptide receptor family, whereas both dopamine and serotonin receptors are members of the amine receptor family.

In a separate Class A family, there is evidence that the CB1 cannabinoid receptor may also show ligand-selective endocytosis and post-endocytic sorting. Specifically, both the endogenous ligand anandamide and the exogenous agonist WIN55,212-2 promote robust internalization of the CB1 receptor, whereas delta-9-tetrahydrocannabinol, the psychoactive ingredient of cannabis, promotes little receptor internalization

(68–70). Furthermore, although brief exposure to low doses of either anandamide or WIN55,212-2 results in recycling of the endocytosed pool of receptor, high doses of these drugs, or low doses for a prolonged period, promote CB1 receptor degradation *(71)*. Importantly, correct axonal targeting of neuronal CB1 receptors depends on receptor recycling *(72)*. This would be the expected fate of receptors in the "drug-free" state, when endogenous ligand is released in a pulsatile manner and has a short half-life. It is possible, therefore, that CB1 receptors could be mistargeted and/or missorted in the presence of prolonged agonist exposure – for example, during exogenous drug use.

Among the Class B receptors the parathyroid hormone (PTH) 1 receptor, which is a major regulator of extracellular calcium homeostasis, shows ligand-selective endocytosis. Both the full length peptide PTH(1-34) and the C-terminally truncated version PTH (1-31) activate adenylyl cyclase and PLC. However, only the full length peptide PTH(1-34) promotes receptor endocytosis *(73)*. Perhaps more surprisingly, the N-terminally truncated PTH(7-34) promotes receptor internalization without activating either adenylyl cyclase or PLC *(73)*.

Among the Class C GPCRs, there has been no direct demonstration of an agonist-selective effect on receptor desensitization. Nevertheless, among this class of GPCR, several allosteric agonists and antagonists have been identified. It is intriguing to speculate that these allosteric ligands could preferentially affect signal transduction but not endocytosis (or vice versa) and would thereby show ligand-selective effects on receptor trafficking – especially in the presence of endogenous ligand at the orthosteric site. The mGluR7 receptor may be one example of this. An allosteric agonist at the mGluR7 receptor, *N,N'*-dibenzhydrylethane-1,2-diamine dihydrochloride (AMN082), promotes both receptor activation via G protein and receptor endocytosis. In the presence of the orthosteric agonist glutamate, the effects of AMN082 and glutamate are additive. However, in the presence of an antagonist at the orthosteric site AMN082 continues to promote endocytosis *(74)*. Although AMN082 also continues to activate G protein in the presence of the orthosteric antagonist, signal transduction appears to be more adversely affected by the antagonist than does endocytosis *(74)*.

4.3 Constitutive Desensitization and Endocytosis

Several GPCRs have been shown to undergo constitutive endocytosis in the absence of any ligand. In the majority of cases, these receptors have been found to be constitutively active. These include the CB1 cannabinoid receptor *(70)*, bradykinin B1 receptor *(75)*, melanocortin 4 receptor *(76)*, and several members of the metabotropic glutamate receptor family *(77–79)*, to name a few. However, mutational analysis of several constitutively active receptors has demonstrated that constitutive activity at the G protein and constitutive endocytosis are separable phenomena. For example, US28, one of the four chemokine receptors encoded by human cytomegalovirus, is constitutively active and is constitutively endocytosed. However, a mutant of US28, R129A, which is no longer constitutively active, is, nevertheless, constitutively endocytosed *(80)*. Conversely, several mutants of US28 that are no longer constitutively endocytosed remain constitutively active *(80)*.

In another example, two mutants of the complement factor 5a (C5a) chemotactic receptor, F251A and I124N/L127Q (NQ), show constitutive activity but only the NQ mutant is constitutive endocytosed *(81)*. Mutations that eliminate the constitutive activity of the NQ receptor, however, do not inhibit receptor endocytosis *(81)*. In short, these two examples further illustrate that desensitization and endocytosis represent independent functional properties of specific receptor conformations, often separable from other activities.

4.4 GPCR Heterodimerization as a Mechanism to Alter Ligand-Biased Desensitization and Endocytosis

A growing body of biochemical and functional evidence supports the existence of GPCR dimers/oligomers. Indeed, in some cases, receptor oligomerization is essential for receptor function, i.e., for the $GABA_B$ *(82)*, metabotropic glutamate *(83)*, taste receptors *(84)*, and rhodopsin *(85)*. In other cases, receptor dimerization has been shown to play a modulatory role, altering ligand affinity and/or potency and altering receptor trafficking (see for example *(86–88)*).

GPCR homodimerization has been shown to occur with several receptors, including the D2 dopamine receptor *(89)*, the MOR *(14)*, several of the somatostatin receptors *(90)*, and the CCR5 receptors *(91)*, for example. Heterodimerization has also been reported, not only between receptors within the same GPCR family such as the MOR and DOR opioid receptors *(92)*, the m2 and m3 muscarinic receptors *(93)* and the CCR2 and CCR5 chemokine receptors *(91)*, but also between diverse GPCR families such as opioid receptors and β2 adrenergic receptors *(94,95)*, adenosine A2a receptors and dopamine D1 receptors *(96)*, somatostatin SSTR5 and D2 dopamine receptors *(97)*, and CB1 receptors and D2 dopamine receptors *(98)*.

Although the existence of these diverse heterodimers is difficult to demonstrate *in vivo*, the potential for their existence has profound implications for the biology of GPCR signaling and also for drug design. This point is illustrated by the observation that the opioid ligand 6′-GNTI is a highly potent antagonist for G protein signaling at the DOR, a weak partial agonist at the KOR, and a full agonist at the DOR/KOR heterodimer *(99)*. 6′-GNTI is also a potent, tissue-specific analgesic in rodents, thereby confirming the existence of the heterodimer target *in vivo (99)*.

Few studies have examined the effects of receptor heterodimerization on altering ligand functional selectivity for signal transduction or trafficking. However, one can imagine that a GPCR heterodimer would be an ideal mechanism by which to create functional selectivity and/or an ideal target in which to exploit it. In one example, as mentioned earlier, etorphine shows functional selectivity at the KOR (where it stimulates signal transduction but not endocytosis), but not at the DOR (where it promotes both signal transduction and endocytosis). However, when the DOR is coexpressed with the KOR, and presumably forms heterodimers, etorphine no longer promotes DOR endocytosis *(100)*. Hence, etorphine shows functional selectivity for the DOR in the presence of the KOR, but not when expressed alone.

The lack of many other examples of GPCR heterodimerization affecting functional ligand selectivity is likely due to failure to look for it rather than failure of it to occur. In conclusion, one needs only to use a little imagination to envision the numerous ways in which receptor heterodimerization could affect ligand functional selectivity.

4.5 Concluding Remarks

In conclusion, ligand-induced desensitization, endocytosis, and post-endocytic sorting of GPCRs appear to be independent functional properties of specific receptor–ligand complexes, distinguishable from other activities at the receptor. This chapter clearly illustrates that one cannot predict the endocytic or post-endocytic fate of a receptor–ligand pair merely based on pharmacological studies of activity at effectors such as G proteins. In the vast majority of cases, the effects of GPCR-based drugs on the trafficking properties of the targeted receptor are unknown. However, for several receptors, it has now been shown that the trafficking fate of the receptor has substantial influence on drug responsiveness.

References

1. Ferguson SS, Zhang J, Barak LS, Caron MG. Molecular mechanisms of G protein-coupled receptor desensitization and resensitization. Life Sci 1998;62(17–18):1561–5.
2. Whistler JL, Chuang HH, Chu P, Jan LY, von Zastrow M. Functional dissociation of mu opioid receptor signaling and endocytosis: implications for the biology of opiate tolerance and addiction. Neuron 1999;23(4):737–46.
3. Alvarez VA, Arttamangkul S, Dang V, et al. mu-Opioid receptors: ligand-dependent activation of potassium conductance, desensitization, and internalization. J Neurosci 2002;22(13):5769–76.
4. Willins DL, Berry SA, Alsayegh L, et al. Clozapine and other 5-hydroxytryptamine-2A receptor antagonists alter the subcellular distribution of 5-hydroxytryptamine-2A receptors in vitro and in vivo. Neuroscience 1999;91(2):599–606.
5. Martini L, Whistler JL. The role of mu opioid receptor desensitization and endocytosis in morphine tolerance and dependence. Curr Opin Neurobiol 2007;17(5):556–64.
6. Keith DE, Murray SR, Zaki PA, et al. Morphine activates opioid receptors without causing their rapid internalization. J Biol Chem 1996;271(32):19021–4.
7. Arden JR, Segredo V, Wang Z, Lameh J, Sadee W. Phosphorylation and agonist-specific intracellular trafficking of an epitope-tagged mu-opioid receptor expressed in HEK 293 cells. J Neurochem 1995;65(4):1636–45.
8. Koch T, Schulz S, Pfeiffer M, et al. C-terminal splice variants of the mouse mu-opioid receptor differ in morphine-induced internalization and receptor resensitization. J Biol Chem 2001;276(33):31408–14.
9. Whistler JL, von Zastrow M. Morphine-activated opioid receptors elude desensitization by beta-arrestin. Proc Natl Acad Sci USA 1998;95(17):9914–9.

10. Yu Y, Zhang L, Yin X, Sun H, Uhl GR, Wang JB. Mu opioid receptor phosphorylation, desensitization, and ligand efficacy. J Biol Chem 1997;272(46):28869–74.
11. Whistler JL, Enquist J, Marley A, et al. Modulation of postendocytic sorting of G protein-coupled receptors. Science 2002;297(5581):615–20.
12. Keith DE, Anton B, Murray SR, et al. mu-Opioid receptor internalization: opiate drugs have differential effects on a conserved endocytic mechanism in vitro and in the mammalian brain. Mol Pharmacol 1998;53(3):377–84.
13. He L, Whistler JL. An opiate cocktail that reduces morphine tolerance and dependence. Curr Biol 2005;15(11):1028–33.
14. He L, Fong J, von Zastrow M, Whistler JL. Regulation of opioid receptor trafficking and morphine tolerance by receptor oligomerization. Cell 2002;108(2):271–82.
15. Trafton JA, Abbadie C, Marek K, Basbaum AI. Postsynaptic signaling via the [mu]-opioid receptor: responses of dorsal horn neurons to exogenous opioids and noxious stimulation. J Neurosci 2000;20(23):8578–84.
16. Kovoor A, Celver JP, Wu A, Chavkin C. Agonist induced homologous desensitization of mu-opioid receptors mediated by G protein-coupled receptor kinases is dependent on agonist efficacy. Mol Pharmacol 1998;54(4):704–11.
17. Selley DE, Liu Q, Childers SR. Signal transduction correlates of mu opioid agonist intrinsic efficacy: receptor-stimulated [35S]GTP gamma S binding in mMOR-CHO cells and rat thalamus. J Pharmacol Exp Ther 1998;285(2):496–505.
18. Finn AK, Whistler JL. Endocytosis of the mu opioid receptor reduces tolerance and a cellular hallmark of opiate withdrawal. Neuron 2001;32(5):829–39.
19. Kim JA, Bartlett S, He L, et al. Morphine-induced receptor endocytosis in a novel knockin mouse reduces tolerance and dependence. Curr Biol 2008;18(2):129–35.
20. Zaki PA, Keith DE Jr., Thomas JB, Carroll FI, Evans CJ. Agonist-, antagonist-, and inverse agonist-regulated trafficking of the delta-opioid receptor correlates with, but does not require, G protein activation. J Pharmacol Exp Ther 2001;298(3):1015–20.
21. Remmers AE, Clark MJ, Liu XY, Medzihradsky F. Delta opioid receptor down-regulation is independent of functional G protein yet is dependent on agonist efficacy. J Pharmacol Exp Ther 1998;287(2):625–32.
22. Raynor K, Kong H, Chen Y, et al. Pharmacological characterization of the cloned kappa-, delta-, and mu-opioid receptors. Mol Pharmacol 1994;45(2):330–4.
23. Von Zastrow M, Keith DE Jr., Evans CJ. Agonist-induced state of the delta-opioid receptor that discriminates between opioid peptides and opiate alkaloids. Mol Pharmacol 1993;44(1):166–72.
24. Li JG, Zhang F, Jin XL, Liu-Chen LY. Differential regulation of the human kappa opioid receptor by agonists: etorphine and levorphanol reduced dynorphin A- and U50,488H-induced internalization and phosphorylation. J Pharmacol Exp Ther 2003;305(2):531–40.
25. Tsao PI, von Zastrow M. Type-specific sorting of G protein-coupled receptors after endocytosis. J Biol Chem 2000;275(15):11130–40.
26. Marie N, Lecoq I, Jauzac P, Allouche S. Differential sorting of human delta-opioid receptors after internalization by peptide and alkaloid agonists. J Biol Chem 2003;278(25):22795–804.
27. Audet N, Paquin-Gobeil M, Landry-Paquet O, Schiller PW, Pineyro G. Internalization and Src activity regulate the time course of ERK activation by delta opioid receptor ligands. J Biol Chem 2005;280(9):7808–16.
28. Gray JA, Roth BL. Paradoxical trafficking and regulation of 5-HT(2A) receptors by agonists and antagonists. Brain Res Bull 2001;56(5):441–51.
29. Meltzer HY. Action of atypical antipsychotics. Am J Psychiatry 2002;159(1):153–4; author reply 154–5.
30. Nichols DE. Hallucinogens. Pharmacol Ther 2004;101(2):131–81.
31. Kurrasch-Orbaugh DM, Parrish JC, Watts VJ, Nichols DE. A complex signaling cascade links the serotonin2A receptor to phospholipase A2 activation: the involvement of MAP kinases. J Neurochem 2003;86(4):980–91.
32. Urban JD, Clarke WP, von Zastrow M, et al.. Functional selectivity and classical concepts of quantitative pharmacology. J Pharmacol Exp Ther 2007;320(1):1–13.

33. Schmid CL, Raehal KM, Bohn LM. Agonist-directed signaling of the serotonin 2A receptor depends on beta-arrestin2 interactions in vivo. Proc Natl Acad Sci USA 2008;105(3):1079–84.
34. Ariano MA, Sortwell CE, Ray M, Altemus KL, Sibley DR, Levine MS. Agonist-induced morphologic decrease in cellular D1A dopamine receptor staining. Synapse 1997;27(4):313–21.
35. Vickery RG, von Zastrow M. Distinct dynamin-dependent and -independent mechanisms target structurally homologous dopamine receptors to different endocytic membranes. J Cell Biol 1999;144(1):31–43.
36. Bartlett SE, Enquist J, Hopf FW, et al. Dopamine responsiveness is regulated by targeted sorting of D2 receptors. Proc Natl Acad Sci USA 2005;102(32):11521–6.
37. Paspalas CD, Rakic P, Goldman-Rakic PS. Internalization of D2 dopamine receptors is clathrin-dependent and select to dendro-axonic appositions in primate prefrontal cortex. Eur J Neurosci 2006;24(5):1395–403.
38. Kim KM, Valenzano KJ, Robinson SR, Yao WD, Barak LS, Caron MG. Differential regulation of the dopamine D2 and D3 receptors by G protein-coupled receptor kinases and beta-arrestins. J Biol Chem 2001;276(40):37409–14.
39. Cho EY, Cho DI, Park JH, Kurose H, Caron MG, Kim KM. Roles of protein kinase C and actin-binding protein 280 in the regulation of intracellular trafficking of dopamine D3 receptor. Mol Endocrinol 2007;21(9):2242–54.
40. Ryman-Rasmussen JP, Nichols DE, Mailman RB. Differential activation of adenylate cyclase and receptor internalization by novel dopamine D1 receptor agonists. Mol Pharmacol 2005;68(4):1039–48.
41. Ryman-Rasmussen JP, Griffith A, Oloff S, et al. Functional selectivity of dopamine D1 receptor agonists in regulating the fate of internalized receptors. Neuropharmacology 2007;52(2):562–75.
42. Jones SR, Gainetdinov RR, Hu XT, et al. Loss of autoreceptor functions in mice lacking the dopamine transporter. Nat Neurosci 1999;2(7):649–55.
43. Attarbaschi T, Sacher J, Geiss-Granadia T, et al. Striatal D(2) receptor occupancy in bipolar patients treated with olanzapine. Eur Neuropsychopharmacol 2007;17(2):102–7.
44. Kasper S, Tauscher J, Willeit M, et al. Receptor and transporter imaging studies in schizophrenia, depression, bulimia and Tourette's disorder--implications for psychopharmacology. World J Biol Psychiatry 2002;3(3):133–46.
45. Kasper S, Tauscher J, Kufferle B, et al. Sertindole and dopamine D2 receptor occupancy in comparison to risperidone, clozapine and haloperidol - a 123I-IBZM SPECT study. Psychopharmacology (Berl) 1998;136(4):367–73.
46. Parsey RV, Oquendo MA, Zea-Ponce Y, et al. Dopamine D(2) receptor availability and amphetamine-induced dopamine release in unipolar depression. Biol Psychiatry 2001;50(5):313–22.
47. Schneier FR, Liebowitz MR, Abi-Dargham A, Zea-Ponce Y, Lin SH, Laruelle M. Low dopamine D(2) receptor binding potential in social phobia. Am J Psychiatry 2000;157(3):457–9.
48. Schneier FR, Martinez D, Abi-Dargham A, et al. Striatal dopamine D(2) receptor availability in OCD with and without comorbid social anxiety disorder: preliminary findings. Depress Anxiety 2008;25(1):1–7.
49. Morgan D, Grant KA, Prioleau OA, Nader SH, Kaplan JR, Nader MA. Predictors of social status in cynomolgus monkeys (Macaca fascicularis) after group formation. Am J Primatol 2000;52(3):115–31.
50. Morgan D, Grant KA, Gage HD, et al. Social dominance in monkeys: dopamine D2 receptors and cocaine self-administration. Nat Neurosci 2002;5(2):169–74.
51. Fowler JS, Volkow ND, Kassed CA, Chang L. Imaging the addicted human brain. Sci Pract Perspect 2007;3(2):4–16.
52. Volkow ND, Chang L, Wang GJ, et al. Low level of brain dopamine D2 receptors in methamphetamine abusers: association with metabolism in the orbitofrontal cortex. Am J Psychiatry 2001;158(12):2015–21.
53. Volkow ND, Wang GJ, Maynard L, et al. Effects of alcohol detoxification on dopamine D2 receptors in alcoholics: a preliminary study. Psychiatry Res 2002;116(3):163–72.

54. Volkow ND, Fowler JS, Wang GJ. The addicted human brain: insights from imaging studies. J Clin Invest 2003;111(10):1444–51.
55. Czoty PW, Morgan D, Shannon EE, Gage HD, Nader MA. Characterization of dopamine D1 and D2 receptor function in socially housed cynomolgus monkeys self-administering cocaine. Psychopharmacology (Berl) 2004;174(3):381–8.
56. Czoty PW, Gage HD, Nader MA. PET imaging of striatal dopamine D2 receptors in nonhuman primates: increases in availability produced by chronic raclopride treatment. Synapse 2005;58(4):215–9.
57. Nader MA, Morgan D, Gage HD, et al. PET imaging of dopamine D2 receptors during chronic cocaine self-administration in monkeys. Nat Neurosci 2006;9(8):1050–6.
58. Moore RJ, Vinsant SL, Nader MA, Porrino LJ, Friedman DP. Effect of cocaine self-administration on dopamine D2 receptors in rhesus monkeys. Synapse 1998;30(1):88–96.
59. Chen JF, Aloyo VJ, Weiss B. Continuous treatment with the D2 dopamine receptor agonist quinpirole decreases D2 dopamine receptors, D2 dopamine receptor messenger RNA and proenkephalin messenger RNA, and increases mu opioid receptors in mouse striatum. Neuroscience 1993;54(3):669–80.
60. Subramaniam S, Lucki I, McGonigle P. Effects of chronic treatment with selective agonists on the subtypes of dopamine receptors. Brain Res 1992;571(2):313–22.
61. Palomo T, Archer T, Kostrzewa RM, Beninger RJ. Comorbidity of substance abuse with other psychiatric disorders. Neurotox Res 2007;12(1):17–27.
62. Sikich L. Efficacy of atypical antipsychotics in early-onset schizophrenia and other psychotic disorders. J Clin Psychiatry 2008;69 Suppl 4:21–5.
63. Shopsin B, Gershon S. Dopamine receptor stimulation in the treatment of depression: piribedil (ET-495). Neuropsychobiology 1978;4(1):1–14.
64. Millan MJ, Maiofiss L, Cussac D, Audinot V, Boutin JA, Newman-Tancredi A. Differential actions of antiparkinson agents at multiple classes of monoaminergic receptor. I. A multivariate analysis of the binding profiles of 14 drugs at 21 native and cloned human receptor subtypes. J Pharmacol Exp Ther 2002;303(2):791–804.
65. Bouras N, Bridges PK. Bromocriptine in depression. Curr Med Res Opin 1982;8(3):150–3.
66. Mack M, Luckow B, Nelson PJ, et al. Aminooxypentane-RANTES induces CCR5 internalization but inhibits recycling: a novel inhibitory mechanism of HIV infectivity. J Exp Med 1998;187(8):1215–24.
67. Simmons G, Clapham PR, Picard L, et al. Potent inhibition of HIV-1 infectivity in macrophages and lymphocytes by a novel CCR5 antagonist. Science 1997;276(5310):276–9.
68. Hsieh C, Brown S, Derleth C, Mackie K. Internalization and recycling of the CB1 cannabinoid receptor. J Neurochem 1999;73(2):493–501.
69. Coutts AA, Anavi-Goffer S, Ross RA, et al. Agonist-induced internalization and trafficking of cannabinoid CB1 receptors in hippocampal neurons. J Neurosci 2001;21(7):2425–33.
70. Leterrier C, Bonnard D, Carrel D, Rossier J, Lenkei Z. Constitutive endocytic cycle of the CB1 cannabinoid receptor. J Biol Chem 2004;279(34):36013–21.
71. Martini L, Waldhoer M, Pusch M, et al. Ligand-induced down-regulation of the cannabinoid 1 receptor is mediated by the G-protein-coupled receptor-associated sorting protein GASP1. Faseb J 2007;21(3):802–11.
72. Leterrier C, Laine J, Darmon M, Boudin H, Rossier J, Lenkei Z. Constitutive activation drives compartment-selective endocytosis and axonal targeting of type 1 cannabinoid receptors. J Neurosci 2006;26(12):3141–53.
73. Sneddon WB, Magyar CE, Willick GE, et al. Ligand-selective dissociation of activation and internalization of the parathyroid hormone (PTH) receptor: conditional efficacy of PTH peptide fragments. Endocrinology 2004;145(6):2815–23.
74. Pelkey KA, Yuan X, Lavezzari G, Roche KW, McBain CJ. mGluR7 undergoes rapid internalization in response to activation by the allosteric agonist AMN082. Neuropharmacology 2007;52(1):108–17.
75. Enquist J, Skroder C, Whistler JL, Leeb-Lundberg LM. Kinins promote B2 receptor endocytosis and delay constitutive B1 receptor endocytosis. Mol Pharmacol 2007;71(2):494–507.

76. Mohammad S, Baldini G, Granell S, Narducci P, Martelli AM, Baldini G. Constitutive traffic of melanocortin-4 receptor in Neuro2A cells and immortalized hypothalamic neurons. J Biol Chem 2007;282(7):4963–74.
77. Lavezzari G, Roche KW. Constitutive endocytosis of the metabotropic glutamate receptor mGluR7 is clathrin-independent. Neuropharmacology 2007;52(1):100–7.
78. Fourgeaud L, Bessis AS, Rossignol F, Pin JP, Olivo-Marin JC, Hemar A. The metabotropic glutamate receptor mGluR5 is endocytosed by a clathrin-independent pathway. J Biol Chem 2003;278(14):12222–30.
79. Dale LB, Bhattacharya M, Seachrist JL, Anborgh PH, Ferguson SS. Agonist-stimulated and tonic internalization of metabotropic glutamate receptor 1a in human embryonic kidney 293 cells: agonist-stimulated endocytosis is beta-arrestin1 isoform-specific. Mol Pharmacol 2001;60(6):1243–53.
80. Waldhoer M, Casarosa P, Rosenkilde MM, et al. The carboxyl terminus of human cytomegalovirus-encoded 7 transmembrane receptor US28 camouflages agonism by mediating constitutive endocytosis. J Biol Chem 2003;278(21):19473–82.
81. Whistler JL, Gerber BO, Meng EC, Baranski TJ, von Zastrow M, Bourne HR. Constitutive activation and endocytosis of the complement factor 5a receptor: evidence for multiple activated conformations of a G protein-coupled receptor. Traffic 2002;3(12):866–77.
82. Pin JP, Kniazeff J, Binet V, et al. Activation mechanism of the heterodimeric GABA(B) receptor. Biochem Pharmacol 2004;68(8):1565–72.
83. Kniazeff J, Bessis AS, Maurel D, Ansanay H, Prezeau L, Pin JP. Closed state of both binding domains of homodimeric mGlu receptors is required for full activity. Nat Struct Mol Biol 2004;11(8):706–13.
84. Nelson G, Chandrashekar J, Hoon MA, et al. An amino-acid taste receptor. Nature 2002;416(6877):199–202.
85. Filipek S, Krzysko KA, Fotiadis D, et al. A concept for G protein activation by G protein-coupled receptor dimers: the transducin/rhodopsin interface. Photochem Photobiol Sci 2004;3(6):628–38.
86. Jordan BA, Devi LA. G-protein-coupled receptor heterodimerization modulates receptor function. Nature 1999;399(6737):697–700.
87. George SR, Fan T, Xie Z, et al.. Oligomerization of mu- and delta-opioid receptors. Generation of novel functional properties. J Biol Chem 2000;275(34):26128–35.
88. Gomes I, Gupta A, Filipovska J, Szeto HH, Pintar JE, Devi LA. A role for heterodimerization of mu and delta opiate receptors in enhancing morphine analgesia. Proc Natl Acad Sci USA 2004;101(14):5135–9.
89. Lee SP, O'Dowd BF, Rajaram RD, Nguyen T, George SR. D2 dopamine receptor homodimerization is mediated by multiple sites of interaction, including an intermolecular interaction involving transmembrane domain 4. Biochemistry 2003;42(37):11023–31.
90. Pfeiffer M, Koch T, Schroder H, Laugsch M, Hollt V, Schulz S. Heterodimerization of somatostatin and opioid receptors cross-modulates phosphorylation, internalization, and desensitization. J Biol Chem 2002;277(22):19762–72.
91. Mellado M, Rodriguez-Frade JM, Vila-Coro AJ, et al. Chemokine receptor homo- or heterodimerization activates distinct signaling pathways. Embo J 2001;20(10):2497–507.
92. Gomes I, Jordan BA, Gupta A, Trapaidze N, Nagy V, Devi LA. Heterodimerization of mu and delta opioid receptors: A role in opiate synergy. J Neurosci 2000;20(22):RC110.
93. Maggio R, Vogel Z, Wess J. Coexpression studies with mutant muscarinic/adrenergic receptors provide evidence for intermolecular "cross-talk" between G-protein-linked receptors. Proc Natl Acad Sci USA 1993;90(7):3103–7.
94. Jordan BA, Trapaidze N, Gomes I, Nivarthi R, Devi LA. Oligomerization of opioid receptors with beta 2-adrenergic receptors: a role in trafficking and mitogen-activated protein kinase activation. Proc Natl Acad Sci USA 2001;98(1):343–8.
95. Rios C, Gomes I, Devi LA. Interactions between delta opioid receptors and alpha-adrenoceptors. Clin Exp Pharmacol Physiol 2004;31(11):833–6.

96. Gines S, Hillion J, Torvinen M, et al. Dopamine D1 and adenosine A1 receptors form functionally interacting heteromeric complexes. Proc Natl Acad Sci USA 2000;97(15):8606–11.
97. Rocheville M, Lange DC, Kumar U, Patel SC, Patel RC, Patel YC. Receptors for dopamine and somatostatin: formation of hetero-oligomers with enhanced functional activity. Science 2000;288(5463):154–7.
98. Kearn CS, Blake-Palmer K, Daniel E, Mackie K, Glass M. Concurrent stimulation of cannabinoid CB1 and dopamine D2 receptors enhances heterodimer formation: a mechanism for receptor cross-talk? Mol Pharmacol 2005;67(5):1697–704.
99. Waldhoer M, Fong J, Jones RM, et al. A heterodimer-selective agonist shows in vivo relevance of G protein-coupled receptor dimers. Proc Natl Acad Sci USA 2005;102(25):9050–5.
100. Chu P, Murray S, Lissin D, von Zastrow M. Delta and kappa opioid receptors are differentially regulated by dynamin-dependent endocytosis when activated by the same alkaloid agonist. J Biol Chem 1997;272(43):27124–30.

Chapter 5
Selectivity for G Protein or Arrestin-Mediated Signaling

Laura M. Bohn

Abstract G protein-coupled receptors (GPCR) are expressed throughout the body in various cell types and organs. The cellular complement of proteins and the immediate scaffolds assembled in proximity to the receptor can determine how a receptor will signal in response to a given agonist. Furthermore, a particular agonist may promote receptor activation such that it favors interactions with different signaling partners within a cell. Therefore, a given receptor may have diverse functions depending on where it is expressed or what ligand is binding to it. β-arrestins are ubiquitously expressed cellular regulatory proteins that can play multifaceted roles in GPCR signaling. This chapter focuses on how different ligands can reveal differential roles of β-arrestins in determining GPCR signaling and regulation. We will focus on the neurotransmitter receptors for serotonin and opioids for specific examples of such functional selectivity.

Keywords β-arrestin, G protein, Opioid receptors, Serotonin receptors, G protein-coupled receptor

5.1 Introduction

G protein-coupled receptors (GPCR) are expressed throughout the body on various cell types and organs. The cellular complement of proteins and the immediate scaffolds assembled in proximity to the receptor can determine how a receptor will signal in response to a given agonist. Furthermore, a particular agonist may promote receptor activation such that it favors interactions with different signaling partners within a cell. Therefore, a given receptor may have diverse functions depending on

---

L.M. Bohn
Departments of Pharmacology and Psychiatry, The Ohio State University
College of Medicine
Columbus, OH 43210-1239
e-mail: Laura.Bohn@osumc.edu

K. Neve (ed.), *Functional Selectivity of G Protein-Coupled Receptor Ligands*,
DOI: 10.1007/978-1-60327-335-0_5,

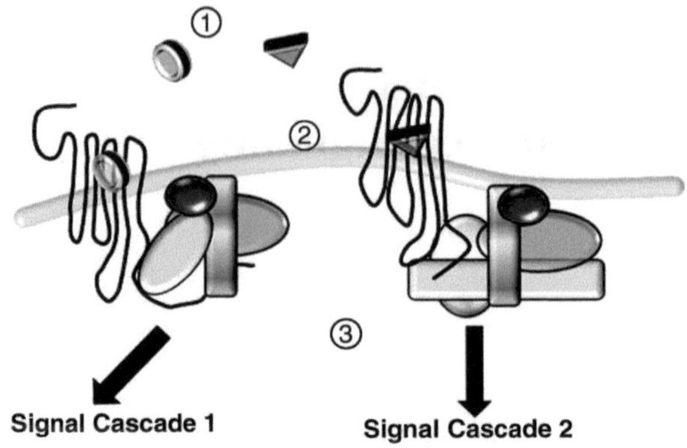

Fig. 5.1 Functional selectivity of GPCR signaling. *Ligands:* The chemical nature of the ligands will determine how they bind to receptors and directly affect the shape of the receptor. *Receptor:* The ligand-induced receptor conformation will influence interactions with other proteins. *Cell:* The protein complement differs between cellular environments; therefore, functional selectivity of agonist-induced receptor interactions may vary between cell types

where it is expressed or what ligand is binding to it (Fig. 5.1). β-arrestins are ubiquitously expressed cellular regulatory proteins that can play multifaceted roles in GPCR signaling. There has been a recent rapid proliferation of studies demonstrating that receptors can be desensitized, internalized, resensitized, ubiquitinated, down-regulation, assembled into signaling scaffolds, and trafficked depending on their ability to interact with β-arrestins. Most recently, it is becoming evident that the nature of the β-arrestin interaction with the receptor can be a function of the agonist. In this chapter, we will discuss how the ligand can determine how the cell responds using particular examples of serotonin and opioid receptors.

5.2 Agonist Directed-Signaling to G Proteins

5.2.1 Diversity of Heterotrimeric G Protein Coupling

For decades, GPCRs have been pharmaceutically targeted for the treatment of diverse diseases and disorders. In many cases, the goal has been to either block the effect of endogenous neurotransmitters that normally (or excessively) activate the receptor or to mimic the effects of neurotransmitters at the receptors to produce a desired biological response. These seven transmembrane spanning receptors (7TMR) transmit signal transduction cascades by initially coupling with a heterotrimeric G protein, which subsequently goes on to activate downstream effectors.

The heterotrimeric G proteins consist of alpha, beta, and gamma subunits. In early receptor pharmacology studies, each receptor was characterized for its ability to couple to a particular class of G proteins. For example; the mu opioid receptor couples to the $G\alpha_{i/o}$ proteins, and the β2 adrenergic receptor couples to $G\alpha_s$ proteins. However, it is relatively well accepted that these receptors can couple to other G protein subtypes. The concept of "functional selectivity" has evolved to explain the agonist-driven diversity in signaling observed downstream from a single receptor *(1–3)*. An attractive point of agonist-initiated signaling divergence is at the level of G protein coupling, and this aspect of functional selectivity is reviewed in Chap. 3 of this volume. Evidence continues to accumulate supporting G protein independent signaling downstream of GPCRs. Accordingly the potential for ligand-directed, functionally selective signaling events expands. Here we will focus on agonist-driven coupling to β-arrestins and how this can impact on receptor signaling and physiological function in vivo.

5.3 Receptor Regulation as a Function of Ligand

5.3.1 Receptor Regulation by GRKs and Arrestins

The potential for receptor signaling complexity extends beyond the potential for agonist-induced differences in G protein coupling as receptor regulatory events may also be determined by properties of the ligand *(4)*. Upon agonist stimulation of receptors, G protein signaling can be diminished by homologous desensitization events, which involve receptor phosphorylation by GPCR kinases (GRKs) and subsequent binding of β-arrestin proteins. There are 7 currently identified GRKs, 2 of which are found exclusively in the eye *(5,6)*. There are also two visual arrestins and two ubiquitously expressed β-arrestins *(7,8)*. Furthermore, although they will not be discussed here, there is a family of evolutionarily-related arrestin-like proteins, referred to as arrestin-domain containing proteins that include six human homologues *(9)*. Whether these proteins assume some of the regulatory and signaling roles of β-arrestins remains to be determined. Incapacitation of receptor signaling by β-arrestins can involve the prevention of interactions between receptors and signaling proteins, such as G proteins; β-arrestins can also initiate the internalization of receptors leading to either recycling events or receptor degradation *(7,8)*. Therefore, the ability of the ligand to induce the GRK-mediated phosphorylation and β-arrestin recruitment can ultimately impact on its G protein-mediated signaling events *(10)*. This principle has been especially evident in cellular studies wherein receptor responses are modified by GRK and β-arrestin overexpression and ablation by genetic or RNA interference manipulations. For example, overexpression of GRKs can increase the rate and extent of receptor phosphorylation, β-arrestin recruitment, internalization and desensitization in cellular systems *(11–17)* as well as in vivo *(18–21)*. Removal of GRKs and β-arrestins can have the opposite effect, delaying receptor trafficking and decreasing receptor desensitization rates *(22–26)*. Disruptions in desensitization and internalization are also revealed in mice where the GRK or β-arrestin has been genetically deleted *(6,27–29)*.

5.3.2 Agonist-Directed MOR Regulation

The physiological actions of morphine are mediated principally through the mu opioid receptor (MOR). Already supported by extensive pharmacological experiments, this has been further strengthened by the finding that mice lacking MOR do not experience morphine-mediated antinociception, respiratory suppression, inhibition of gastrointestinal transit, or increased locomotor activity *(30–32)*. The MOR has been widely studied for its somewhat unusual regulation in response to one of the most clinically relevant agonists, morphine. Morphine, as well as several of its chemical cousins, differs from other opioid agonists initially in its ability to induce GRK-mediated phosphorylation *(4,15,33–36)*. In cellular cultures (predominantly studied in the HEK-293 human cell line), the morphine-bound receptor is not robustly phosphorylated. Other agonists, including etorphine, fentanyl, methadone, and an enkephalin analog, robustly promote receptor phosphorylation *(15,37,38)*. Therefore, agonists such as etorphine could be considered functionally selective for inducing MOR phosphorylation, while agonists such as morphine introduce a bias against this particular posttranslational modification. Importantly, overexpression of GRKs can lead to morphine-induced receptor phosphorylation to similar levels observed with etorphine *(15)*. These observations demonstrate that the nature of the ligand directly impacts on ability of the receptor to interact with intracellular signaling components. In this case, the nature of the ligand can determine the affinity of the GRK for the agonist-bound receptor. An overexpression of the intracellular component, here the GRK, can shift the balance such that even in a nonfavorable orientation, phosphorylation of the receptor can occur. Moreover, the amount of GRK immediately available to regulate the receptor, whether due to gross overexpression or enriched at receptor domains, may depend on the particular cell type or neuronal population in which the receptor is expressed and will ultimately determine whether morphine actually differs from etorphine in respect to promoting receptor phosphorylation.

Phosphorylation of GPCRs facilitates the binding of β-arrestins and the assembly of trafficking proteins *(39–41)*. For the MOR, β-arrestin recruitment and receptor internalization profiles parallel the ligand-induced phosphorylation of the receptor in cellular model systems. Accordingly, morphine-bound receptors recruit little β-arrestin and are weakly internalized relative to receptors activated by methadone, etorphine, or fentanyl *(15,33,35,42–45)*. Overexpression of GRKs facilitates morphine-induced β-arrestin recruitment and internalization suggesting that the phosphorylation of the receptor is key to these subsequent trafficking events *(15,33,35,45)*.

5.3.2.1 Agonist-Directed MOR Internalization and Desensitization

Considering that signal transduction events can occur during the internalization process, making a distinction between receptor "internalization" and "desensitization" becomes critical. In some cases, the two events have been referred to as synonymous and the two terms have been used interchangeably, although they are distinct cellular signaling events. The confounding factor is that while a receptor that no longer couples to G proteins may indeed be desensitized and internalized; a receptor that no longer couples to G protein, but is not internalized is still desensitized (with respect its ability to couple to G proteins). The internalized receptor may transmit signal (as discussed below) and therefore, the internalized receptor cannot be collectively considered a desensitized receptor. The contribution of internalization and desensitization to receptor function in terms of clinical conditions such as opiate tolerance and physical dependence has been a considerable area of discussion in the opioid receptor signaling field *(46,47)*. It is clear that there are qualitatively distinct differences in the ability of certain opioid agonists to regulate receptor phosphorylation, β-arrestin-recruitment, and receptor internalization that cannot be explained by differences in drug efficacy in respect to G protein coupling or adenylyl cyclase inhibition *(35,45)*. The signaling state of the internalized receptor is also becoming an important consideration as more receptors are found to spend most of their time in intracellular vesicles. Therefore, making the distinction of internalization apart from desensitization is critical in considering receptor function and regulation.

5.3.2.2 Agonist-Directed MOR Regulation In Vivo

MOR regulation and desensitization may contribute to certain aspects of opioid tolerance, which is an important clinical problem. When considering the long-term consequences of opiate treatment, we must bear in mind that the fate of the receptor may not only be determined by agonist occupancy but also by the scaffolds surrounding the receptor. For example, MORs expressed in the striatum may not be subject to the same regulation as MORs in the thalamus, spinal cord, or gastrointestinal tract due to differences in the cellular environments. Haberstock-Debic et al. *(48)* nicely demonstrated the contribution of the immediate environment to MOR regulation within the same cell; wherein they report a lack of MOR internalization in nucleus accumbens neuron cell bodies upon morphine treatment and at the same time, in the same neurons, internalization of the receptor in dendritic projections. Further evidence of differential regulation specific to different brain regions has been seen using ^{35}S-GTPγS binding to rat brain slices following chronic morphine or heroin treatment. Sim et al. *(49)* demonstrated that chronic morphine treatment leads to uncoupling of the MOR from G protein predominantly in brainstem regions (such as dorsal raphe nucleus, locus coeruleus, and the parabrachial nuclei), while no effect was seen in other regions including striatum, thalamus, and hypothalamus. Upon chronic treatment with heroin, which is metabolized to morphine in brain, this same group reported that the MOR was less coupled to G protein in dorsal raphe nucleus, locus coeruleus, lateral parabrachial nucleus, rostral nucleus accumbens, periaqueductal grey (PAG), and medial and mediodorsal thalamus *(50)*. Noble and Cox *(51)* observed that chronic treatment with morphine in rats led to desensitization of the MOR as assessed by its ability to inhibit adenylyl cyclase wherein the receptor was desensitized in thalamus and PAG but not in caudate putamen or nucleus accumbens. These observations support that the location of the receptor may determine how a particular drug may induce different degrees of desensitization and internalization, which should be an important consideration for studying the contribution of the ligand to receptor regulation in vivo.

A recent study in locus coeruleus neurons in brain slices simultaneously compared MOR desensitization and trafficking in response to morphine, etorphine, met-enkephalin, and methadone treatments *(52)*. This group found that agonists, such as the enkephalin, methadone, and etorphine, that promote receptor internalization in cell culture also promoted MOR internalization in neurons while morphine did not. Further, they found that all of the compounds led to MOR desensitization in respect to the receptor's ability to suppress neuronal firing in electrophysiological recordings from adult mouse brain neurons. Together these studies suggest that MOR desensitization and internalization are distinct events and that specific ligands can direct different trafficking profiles within the same cellular population.

5.3.2.3 Agonist-Directed MOR Regulation in the Absence of β-Arrestins

The contribution of the cellular environment can also impact on the agonist's ability to trigger responses in vivo. For example, when compared with wild-type (WT) mice, {β}arrestin2 knockout ({β}arr2-KO) mice display enhanced morphine-induced antinociception in the hot plate test *(53–55)*; however, methadone, fentanyl, and etorphine effects do not differ between the two genotypes *(35)*. There is an inverse correlation between the agonist's ability to promote receptor phosphorylation, β-arrestin recruitment and receptor internalization in cell culture studies, and the impact of β-arrestin2 deletion in vivo; if the agonist induces robust phosphorylation/β-arrestin2 recruitment/internalization then the deletion of β-arrestin2 has little impact on the response to agonist in vivo. Likewise, if the agonist induces weak phosphorylation/β-arrestin2 recruitment/internalization, such as seen with morphine and heroin, then β-arrestin2 deletion has a significant impact on in vivo responsiveness. Although this particular study suggested that β-arrestin1 may substitute for β-arrestin2 when the receptor is highly phosphorylated, definitive studies determining such compensations or selectivity have yet to be reported *(35)*. Furthermore, a more comprehensive assessment of diverse agonists must be considered before such a conclusive correlation between the events can be made.

More recently, an agonist that has been characterized for its bias toward G protein signaling over β-arrestin recruitment regulation and trafficking *(45,56,57)*. This ligand, termed herkinorin, and several chemical derivatives, activates MOR-G protein coupling as well as downstream signaling yet fails to induce β-arrestin recruitment and receptor internalization – even in the presence of overexpressed GRKs *(45, 57)*. Since β-arrestins are involved in desensitizing and regulating opioid receptors and the removal β-arrestin2 in mice increases morphine's analgesic potency while attenuating morphine tolerance, it may be beneficial to develop opioid ligands that have bias against MOR-β-arrestin interactions.

5.4 Agonist-Directed β-Arrestin Signaling

In addition to coupling to G proteins (and to multiple G proteins), GPCRs (or perhaps in this case, more aptly termed 7TMRs) can couple to β-arrestins to transmit signals. In this scenario, β-arrestins appear to act as multifunctional scaffolding proteins to aid in the assemblage of signaling complexes in proximity to the receptor *(10,58)*. For some receptors, the scaffold moves with the receptor during the internalization event, while in other cases, the receptor-β-arrestin scaffold can transmit signals from the cell surface. It is somewhat perplexing that a particular GPCR can be both desensitized by β-arrestin interactions and still transmit signals through β-arrestin interactions – two events that would seem to be in opposition. Take, for example, the most-studied receptor in GPCR signaling is the β_2-adrenergic receptor. Early studies focused upon this receptor have led to the initial description of β-arrestins and GRKs (then β-adrenergic receptor kinases or βARK) as critical GPCR regulatory proteins. This work gave rise to the canonical model of β-arrestin as a damper of GPCR signaling, playing a major role in homologous desensitization whereby it binds to the receptor following GRK phosphorylation and prevents further G protein coupling *(7)*. The same laboratory, more than a decade later, discovered that β-arrestins mediate signaling from receptors to intracellular kinases *(58–60)*. The question arises: how can a receptor that is turned off by β-arrestins utilize β-arrestins to signal? Perhaps in a linear world, where events can only occur in linear progression such signaling would be impossible. But biology is complex, and compartmentalization, redundancy, and temporal regulation ultimately define the components of the immediate cellular environment that determine receptor interactions. Therefore, it is not surprising that many 7TMRs have functions that extend beyond coupling to G proteins.

5.4.1 β-Arrestin-Mediated MOR Signaling

β-arrestins have been shown to play a role in determining the temporal patterns of MOR-mediated ERK1/2 activation similar to regulation described for the parathyroid hormone receptor and the β2 adrenergic receptor *(61–63)*. Studies from Lakshmi Devi's laboratory suggest that the MOR, as a heterodimer with the delta opioid receptor, may utilize β-arrestins to activate ERK1/2 *(64)*. In vivo, mice lacking β-arrestin2 display enhanced antinociceptive responses to the MOR agonist, morphine, but some of the morphine-induced physiological responses are actually diminished. The morphine-induced inhibition of gastrointestinal transit (or constipation) as well as respiratory suppression is less severe in the βarr2-KO mice compared with WT mice *(65)*. Pharmacological studies using peripherally restricted agonists and intracerebroventricular injections suggest that the differential effects on constipation may be due to activation of MOR in the periphery and specifically in the colon *(65,66)*. Whether the signaling mechanisms mediated by MOR in the colon differ from that in the CNS is not known; however, similar to the CNS, MOR is internalized in response to agonist in the neurons of the enteric nervous system *(44)*. Chronic morphine leads to tolerance in the CNS and spinal cord when the drug is no longer effective at blocking pain perception, yet morphine still produces constipation in patients under long-term therapy suggesting that the mechanisms underlying morphine-induced MOR regulation may differ between the two neuronal populations. In the case of the βarr2-KO mice, it is attractive to speculate that β-arrestin2 may play a positive role in MOR signaling, perhaps involving receptor internalization, although extensive studies are still needed to test this hypothesis *(66)*.

Interestingly, loperamide, a peripherally restricted MOR agonists that is used as an over-the-counter treatment for diarrhea (immodium), differs from morphine in that it promotes robust MOR-β-arrestin interactions and induces MOR internalization in HEK cells (unpublished observations: L.M.B.). When administered to mice, loperamide delays gastrointestinal transit; however, loperamide has no inhibitory effect on transit times in mice lacking β-arrestin2 *(65)*. These observations underscore that the functional selectivity of agonists observed in cell culture may not always translate into functional selectivity in vivo as the two agonists promote different MOR regulatory profiles in regard to β-arrestin-mediated events in cells yet β-arrestin2 appears to be equally important for mediating the effects of both drugs in gut. The regulation of MOR in respect to opioid-induced constipation may prove to be a valuable therapeutic target *(67–68)*.

5.4.2 Functional Selectivity of 5-HT2AR Signaling and Trafficking

5.4.2.1 Agonist-Directed 5-HT2AR Signaling

Agonist-directed 5-HT2AR signaling has been demonstrated at the cellular level where it has been shown to dissociate at the level of phospholipase C (PLC) signaling and arachidonic acid release (presumably via activation of phospholipase A2, PLA_2) *(69)*. Activation of 5-HT2AR by a panel of hallucinogens and non-hallucinogens results in different profiles for inducing PLC vs. PLA_2 activation that is not correlated with the hallucinogenic properties of the drugs *(70–72)*. 5-HT2AR activation leads to

ERK1/2 phosphorylation as well as p38 Map Kinase phosphorylation in what appears to be via two separate pathways. Phosphorylation of these MAP kinases is upstream of PLA_2 activation and subsequent arachidonic acid release, and is independent of PLC activation in NIH3T3 cells *(70)*. Other studies examining the downstream effects of stimulating the 5-HT2AR suggest that hallucinogens and nonhallucinogens may differ based on their ability to induce early intermediate gene responses *(73)*; however, such downstream measures are difficult to trace back to the initial effector coupling to the receptor. Although it is not clear at what point receptor-mediated signaling diverges, it is apparent that the properties of the ligand can greatly influence the preference of one pathway over another *(3,70,71)*.

5.4.2.2 Agonist-Directed Trafficking of 5-HT2AR

Receptor internalization may also play a role in signaling of the 5-HT2AR; and certain ligands may play a significant part in determining that role. The 5-HT2AR is generally found in intracellular vesicles within cultured cells as well as in neurons *(74–76)*. Furthermore, antagonists as well as agonists promote 5-HT2AR internalization in vitro and in vivo *(77,78)*. Considering the opposing functions of serotonergic antagonists and agonists, particularly clozapine vs. serotonin, the mechanism of specialized internalization may ultimately play a role in determining receptor activity and may point to an alternative function of β-arrestins besides their classical role in receptor internalization. Several studies from Bryan Roth's laboratory have examined 5-HT2AR trafficking in response to diverse agonists and antagonists *(78–80)*. Such studies have demonstrated that βarr2 colocalizes with the 5-HT2AR in prefrontal cortical neurons of rat where some colocalization was noted within intracellular vesicles *(75)*. Further studies suggest that 5-HT2ARs can be internalized following agonist stimulation in a β-arrestin-independent but dynamin-dependent manner in HEK-293 cells *(74,81)*. However, these studies utilized the expression of dominant negative β-arrestins, which may not disrupt all aspects of β-arrestin function. Moreover, studies in C6 glioma cells indicated that β-arrestins were indeed important in regulating the trafficking of the 5-HT2AR implicating the importance of the cellular environment's contribution to receptor regulation *(74)*. Taken together, these studies support that β-arrestins may be important in regulating the activity of the 5-HT2AR in vivo and that the βarr2 molecule may not be limited to its traditional role of internalizing receptors.

β-Arrestin-Mediated 5-HT2AR Signaling and Trafficking

Recently, by utilizing mouse embryonic fibroblasts genetically devoid of β-arrestin1 and β-arrestin2, β-arrestins have been shown to be essential for mediating serotonin-induced 5-HT2AR internalization but not for mediating internalization induced by DOI (2,5-dimethoxy-4-iodoamphetamine), an amphetamine-derived hallucinogen and 5-HT2R selective agonist *(29)*. Signaling to ERK1/2 by 5-HT2AR was higher for serotonin in WT MEFs than DOI; removal of β-arrestins decreased ERK1/2 signals to that of DOI. Inhibition of PLC prevented a third of serotonin – and all of DOI-mediated ERK activation in WT MEFs, while it blocked all agonist stimulation of ERK1/2 in the β-arrestin null cells. These findings suggest that the two agonists promote receptor signaling to ERK cascades where DOI can promote a β-arrestin-independent cascade (presumably fully dependent on G protein and subsequent PLC activation) while serotonin can lead to activation of ERK via a β-arrestin-dependent mechanism.

β-arrestin2 also proves to be important in determining the trafficking profile of the 5-HT2AR in neuronal cultures. The 5-HT2AR is found in intracellular vesicles in cortical neurons prepared from WT mice, while most of the receptor is found on the soma surface in neuronal preparations from βarr2-KO mice. Whether β-arrestins play a role in 5-HT2AR trafficking that directly influences ERK activation remains to be determined. However, studies in vivo indicate that while DOI activates ERK1/2 in the frontal cortex of both WT and βarr2-KO mice; 5-HTP (5-hydroxytryptophan, a serotonin precursor) only activates ERK1/2 in the WT mice *(29)*.

5.4.2.3 Functional Selectivity of 5-HT2AR Agonists In Vivo

Studying the biological effects of serotonin in vivo can be difficult as there is not a high degree of receptor selectivity among agonists and antagonists. Furthermore, it is difficult in general to study the effects of ligand-directed signaling in vivo as multiple receptors are present and drug treatment can induce the release of other agonists (i.e., endogenous neurotransmitters) such that the physiological response can be difficult to designate to the tested compound. Therefore, it is essential to have robust physiological responses that have been both pharmacologically and genetically linked to the particular receptor of interest. In rodents, 5-HT2AR agonists and high doses of serotonin produce a "head twitch response," which has been used as a mouse model of hallucinogenic behaviors *(82)*. Extensive pharmacological studies in mice indicate that the 5-HT2AR is the target for serotonin and drug-induced head twitch responses (reviewed in: *(71,83)*). The most selective antagonist to the 5-HT2AR currently available is M100907 (R(+)-alpha-(2,3-dimethoxyphenyl)-1-[2-(4-fluorophenylethyl)]-4-piperidinemethanol); which has 200-fold selectivity over serotonin 2C receptors and is routinely used to assess agonist effects at 5-HT2ARs in vivo as it prevents agonist-induced head twitch responses *(71)*. Furthermore, mice lacking the 5-HT2AR do not display the head twitch response when treated with a wide range of hallucinogens *(73)*.

When treated with the serotonin precursor, 5-HTP, βarr2-KO mice do not display the head twitch response seen in normal WT mice; however, DOI produces an equivalent response in both genotypes *(29)*. This observation suggests that DOI's mechanism of action underlying the head twitch response does not depend on β-arrestin2 while serotonin (or some metabolic derivative) signaling at this receptor, to produce this behavior, requires β-arrestin2. A direct connection between β-arrestin-mediated signaling, trafficking, and behavior in response to the diverse agonists has yet to be made. The cellular regulation of the 5-HT2AR may contribute to determining behavioral sensitivity to both serotonergic agonists and endogenous serotonin. The role that β-arrestins play in this regulation, especially in determining the ability of diverse agonists to direct 5-HT2AR function may prove to be a fruitful avenue in new drug development. Moreover, the impact of such regulation on physiological processes may have clinical implications for the treatment of neuropsychiatric disorders.

5.5 Summary

Functional selectivity of GPCR signaling is becoming increasingly evident and the role of β-arrestins in determining the initial direction of ligand-prompted signaling may lead to more diverse potential paths than originally anticipated for these seven transmembrane spanning receptors. Just as the ligand may confer preferred interactions (whether with G proteins, GRKs, β-arrestins, or other signaling partners), it becomes clear that the composition of the cellular environment can also impact on how the receptor responds to the "functionally selective" ligands. Therefore, in assigning whether a ligand confers functional selectivity, the actions of the agonist must be considered as effecting not only the isolated receptor, but the receptor as a sum of its environment or as a "receptosome." The receptor can therefore be defined by its constituents including: the lipid composition of the membrane, the neighboring receptors (such as other GPCRs, ion channels, and kinase receptors), the scaffold of intracellular signaling proteins in the immediate vicinity of the receptor, the complement and levels of proteins expressed in the cell, the potential interactions with extracellular proteins, possible posttranslational modifications of the receptor, and the temporal relay of cellular events occurring within the cell at the time of ligand presentation. Therefore, although cellular studies continue to be important for identifying possible signaling pathways and potential divergences between ligands, the ultimate goal must be to determine whether functional selectivity is preserved in the cellular environment wherein the receptor mediates its physiological response.

The degree of complexity underlying receptor function becomes daunting from a drug development perspective. Demonstrating how receptors couple to different effectors in response to diverse ligands when they are expressed in their endogenous tissue environments remains the biggest challenge. Some of the greatest difficulties in evaluating agonist-directed G protein coupling in vivo is that signal to noise ratios are difficult to enhance and that, no matter how selective the pharmacological agent is, it remains difficult to assign multiple responses to single receptors as a function of the ligand in living organisms. However, recognizing that functional selectivity may also be impacted by the cellular environment may allow us to *fine-tune* receptor signaling to conserve agonism of some biological effects while preventing activation of others and could expand the pharmacological landscape greatly.

References

1. Gilchrist A. Modulating G-protein-coupled receptors: from traditional pharmacology to allosterics. Trends Pharmacol Sci 2007;28(8):431–7.
2. Mailman RB. GPCR functional selectivity has therapeutic impact. Trends Pharmacol Sci 2007;28(8):390–6.
3. Urban JD, Clarke WP, von Zastrow M, et al. Functional selectivity and classical concepts of quantitative pharmacology. J Pharmacol Exp Ther 2007;320(1):1–13.
4. Kelly E, Bailey CP, Henderson G. Agonist-selective mechanisms of GPCR desensitization. Br J Pharmacol 2008; 153 Supply 1: S379–88.
5. Premont RT, Inglese J, Lefkowitz RJ. Protein kinases that phosphorylate activated G protein-coupled receptors. Faseb J 1995;9(2):175–82.
6. Premont RT, Gainetdinov RR. Physiological roles of G protein-coupled receptor kinases and arrestins. Annu Rev Physiol 2007;69:511–34.
7. Pierce KL, Lefkowitz RJ. Classical and new roles of beta-arrestins in the regulation of G-protein-coupled receptors. Nat Rev Neurosci 2001;2(10):727–33.
8. Luttrell LM, Lefkowitz RJ. The role of beta-arrestins in the termination and transduction of G-protein-coupled receptor signals. J Cell Sci 2002;115(Pt 3):455–65.
9. ***Alvarez C. On the origins of arrestin and rhodopsin. BMC Evolutinary Biology 2008;8:222.
10. Violin JD, Lefkowitz RJ. Beta-arrestin-biased ligands at seven-transmembrane receptors. Trends Pharmacol Sci 2007;28(8):416–22.
11. Freedman NJ, Liggett SB, Drachman DE, Pei G, Caron MG, Lefkowitz RJ. Phosphorylation and desensitization of the human beta 1-adrenergic receptor. Involvement of G protein-coupled receptor kinases and cAMP-dependent protein kinase. J Biol Chem 1995;270(30):17953–61.
12. Freedman NJ, Ament AS, Oppermann M, Stoffel RH, Exum ST, Lefkowitz RJ. Phosphorylation and desensitization of human endothelin A and B receptors. Evidence for G protein-coupled receptor kinase specificity. J Biol Chem 1997;272(28):17734–43.
13. Ferguson SS, Downey WE, 3rd, Colapietro AM, Barak LS, Menard L, Caron MG. Role of beta-arrestin in mediating agonist-promoted G protein-coupled receptor internalization. Science 1996;271(5247):363–6.
14. Menard L, Ferguson SS, Barak LS, et al. Members of the G protein-coupled receptor kinase family that phosphorylate the beta2-adrenergic receptor facilitate sequestration. Biochemistry 1996;35(13):4155–60.
15. Zhang J, Ferguson SS, Barak LS, et al. Role for G protein-coupled receptor kinase in agonist-specific regulation of mu-opioid receptor responsiveness. Proc Natl Acad Sci USA 1998;95(12):7157–62.
16. Mundell SJ, Luty JS, Willets J, Benovic JL, Kelly E. Enhanced expression of G protein-coupled receptor kinase 2 selectively increases the sensitivity of A2A adenosine receptors to agonist-induced desensitization. Br J Pharmacol 1998;125(2):347–56.
17. Appleyard SM, Celver J, Pineda V, Kovoor A, Wayman GA, Chavkin C. Agonist-dependent desensitization of the kappa opioid receptor by G protein receptor kinase and beta-arrestin. J Biol Chem 1999;274(34):23802–7.
18. Koch WJ, Rockman HA, Samama P, et al. Cardiac function in mice overexpressing the beta-adrenergic receptor kinase or a beta ARK inhibitor. Science 1995;268(5215):1350–3.
19. Rockman HA, Choi DJ, Rahman NU, Akhter SA, Lefkowitz RJ, Koch WJ. Receptor-specific in vivo desensitization by the G protein-coupled receptor kinase-5 in transgenic mice. Proc Natl Acad Sci USA 1996;93(18):9954–9.
20. Iaccarino G, Tomhave ED, Lefkowitz RJ, Koch WJ. Reciprocal in vivo regulation of myocardial G protein-coupled receptor kinase expression by beta-adrenergic receptor stimulation and blockade. Circulation 1998;98(17):1783–9.
21. Rockman HA, Hamilton RA, Jones LR, Milano CA, Mao L, Lefkowitz RJ. Enhanced myocardial relaxation in vivo in transgenic mice overexpressing the beta2-adrenergic receptor is associated with reduced phospholamban protein. J Clin Invest 1996;97(7):1618–23.

22. Kohout TA, Lin FS, Perry SJ, Conner DA, Lefkowitz RJ. beta-Arrestin1 and 2 differentially regulate heptahelical receptor signaling and trafficking. Proc Natl Acad Sci USA 2001;98(4):1601–6.
23. Zhang X, Wang F, Chen X, et al. Beta-arrestin1 and beta-arrestin2 are differentially required for phosphorylation-dependent and -independent internalization of delta-opioid receptors. J Neurochem 2005;95(1):169–78.
24. Ren XR, Reiter E, Ahn S, Kim J, Chen W, Lefkowitz RJ. Different G protein-coupled receptor kinases govern G protein and beta-arrestin-mediated signaling of V2 vasopressin receptor. Proc Natl Acad Sci USA 2005;102(5):1448–53.
25. Paing MM, Stutts AB, Kohout TA, Lefkowitz RJ, Trejo J. beta -Arrestins regulate protease-activated receptor-1 desensitization but not internalization or Down-regulation. J Biol Chem 2002;277(2):1292–300.
26. Qiu Y, Loh HH, Law PY. Phosphorylation of the delta-opioid receptor regulates its beta-arrestins selectivity and subsequent receptor internalization and adenylyl cyclase desensitization. J Biol Chem 2007;282(31):22315–23.
27. Gainetdinov RR, Bohn LM, Walker JK, et al. Muscarinic supersensitivity and impaired receptor desensitization in G protein-coupled receptor kinase 5-deficient mice. Neuron 1999;24(4):1029–36.
28. Gainetdinov RR, Premont RT, Bohn LM, Lefkowitz RJ, Caron MG. Desensitization of G protein-coupled receptors and neuronal functions. Annu Rev Neurosci 2004;27:107–44.
29. Schmid CL, Raehal KM, Bohn LM. Agonist-directed signaling of the serotonin 2A receptor depends on beta-arrestin2 interactions in vivo. Proc Natl Acad Sci USA 2008;105(3):1079–84.
30. Matthes HW, Maldonado R, Simonin F, et al. Loss of morphine-induced analgesia, reward effect and withdrawal symptoms in mice lacking the mu-opioid-receptor gene. Nature 1996;383(6603):819–23.
31. Sora I, Takahashi N, Funada M, et al. Opiate receptor knockout mice define mu receptor roles in endogenous nociceptive responses and morphine-induced analgesia. Proc Natl Acad Sci USA 1997;94(4):1544–9.
32. Kieffer BL. Opioids: first lessons from knockout mice. Trends Pharmacol Sci 1999;20(1):19–26.
33. Whistler JL, von Zastrow M. Morphine-activated opioid receptors elude desensitization by beta-arrestin. Proc Natl Acad Sci USA 1998;95(17):9914–9.
34. Kovoor A, Celver JP, Wu A, Chavkin C. Agonist induced homologous desensitization of mu-opioid receptors mediated by G protein-coupled receptor kinases is dependent on agonist efficacy. Mol Pharmacol 1998;54(4):704–11.
35. Bohn LM, Dykstra LA, Lefkowitz RJ, Caron MG, Barak LS. Relative opioid efficacy is determined by the complements of the G protein-coupled receptor desensitization machinery. Mol Pharmacol 2004;66(1):106–12.
36. Johnson EA, Oldfield S, Braksator E, et al. Agonist-selective mechanisms of mu-opioid receptor desensitization in human embryonic kidney 293 cells. Mol Pharmacol 2006;70(2):676–85.
37. Wolf R, Koch T, Schulz S, et al. Replacement of threonine 394 by alanine facilitates internalization and resensitization of the rat mu opioid receptor. Mol Pharmacol 1999;55(2):263–8.
38. Schulz S, Mayer D, Pfeiffer M, Stumm R, Koch T, Hollt V. Morphine induces terminal micro-opioid receptor desensitization by sustained phosphorylation of serine-375. Embo J 2004;23(16):3282–9.
39. Laporte SA, Oakley RH, Zhang J, et al. The beta2-adrenergic receptor/betaarrestin complex recruits the clathrin adaptor AP-2 during endocytosis. Proc Natl Acad Sci USA 1999;96(7):3712–7.
40. Zhang J, Ferguson SS, Barak LS, et al. Molecular mechanisms of G protein-coupled receptor signaling: role of G protein-coupled receptor kinases and arrestins in receptor desensitization and resensitization. Recept Channels 1997;5(3–4):193–9.
41. Claing A, Laporte SA, Caron MG, Lefkowitz RJ. Endocytosis of G protein-coupled receptors: roles of G protein-coupled receptor kinases and beta-arrestin proteins. Prog Neurobiol 2002;66(2):61–79.

42. Arden JR, Segredo V, Wang Z, Lameh J, Sadee W. Phosphorylation and agonist-specific intracellular trafficking of an epitope-tagged mu-opioid receptor expressed in HEK 293 cells. J Neurochem 1995;65(4):1636–45.
43. Keith DE, Murray SR, Zaki PA, et al. Morphine activates opioid receptors without causing their rapid internalization. J Biol Chem 1996;271(32):19021–4.
44. Sternini C, Spann M, Anton B, et al. Agonist-selective endocytosis of mu opioid receptor by neurons in vivo. Proc Natl Acad Sci USA 1996;93(17):9241–6.
45. Groer CE, Tidgewell K, Moyer RA, et al. An opioid agonist that does not induce micro-opioid receptor--arrestin interactions or receptor internalization. Mol Pharmacol 2007;71(2):549–57.
46. Koch T, Hollt V. Role of receptor internalization in opioid tolerance and dependence. Pharmacol Ther 2008;117(2):199–206.
47. Christie MJ. Cellular neuroadaptations to chronic opioids: tolerance, withdrawal and addiction. Br J Pharmacol 2008;154(2):384–96.
48. Haberstock-Debic H, Wein M, Barrot M, et al. Morphine acutely regulates opioid receptor trafficking selectively in dendrites of nucleus accumbens neurons. J Neurosci 2003; 23(10):4324–32.
49. Sim LJ, Selley DE, Dworkin SI, Childers SR. Effects of chronic morphine administration on mu opioid receptor-stimulated [35S]GTPgammaS autoradiography in rat brain. J Neurosci 1996;16(8):2684–92.
50. Sim-Selley LJ, Selley DE, Vogt LJ, Childers SR, Martin TJ. Chronic heroin self-administration desensitizes mu opioid receptor-activated G-proteins in specific regions of rat brain. J Neurosci 2000;20(12):4555–62.
51. Noble F, Cox BM. Differential desensitization of mu- and delta- opioid receptors in selected neural pathways following chronic morphine treatment. Br J Pharmacol 1996;117(1):161–9.
52. Arttamangkul S, Quillinan N, Low M, Vonzastrow M, Pintar J, Williams JT. Differential activation and trafficking of mu-opioid receptors in brain slices. Mol Pharmacol 2008;74:972–9.
53. Bohn LM, Lefkowitz RJ, Gainetdinov RR, Peppel K, Caron MG, Lin FT. Enhanced morphine analgesia in mice lacking beta-arrestin2. Science 1999;286(5449):2495–8.
54. Bohn LM, Gainetdinov RR, Lin FT, Lefkowitz RJ, Caron MG. Mu-opioid receptor desensitization by beta-arrestin2 determines morphine tolerance but not dependence. Nature 2000; 408(6813):720–3.
55. Bohn LM, Lefkowitz RJ, Caron MG. Differential mechanisms of morphine antinociceptive tolerance revealed in (beta)arrestin2 knock-out mice. J Neurosci 2002;22(23):10494–500.
56. Harding WW, Tidgewell K, Byrd N, et al. Neoclerodane diterpenes as a novel scaffold for mu opioid receptor ligands. J Med Chem 2005;48(15):4765–71.
57. Tidgewell K, Groer CE, Harding WW, et al. Herkinorin analogues with differential beta-arrestin2 interactions. J Med Chem 2008;51(8):2421–31.
58. Shenoy SK, Lefkowitz RJ. Seven-transmembrane receptor signaling through beta-arrestin. Sci STKE 2005;2005(308):cm10.
59. Luttrell LM, Ferguson SS, Daaka Y, et al. Beta-arrestin-dependent formation of beta2 adrenergic receptor-Src protein kinase complexes. Science 1999;283(5402):655–61.
60. Daaka Y, Luttrell LM, Ahn S, et al. Essential role for G protein-coupled receptor endocytosis in the activation of mitogen-activated protein kinase. J Biol Chem 1998;273(2):685–8.
61. Zheng H, Loh HH, Law PY. Beta-arrestin-dependent mu-opioid receptor-activated extracellular signal-regulated kinases (ERKs) Translocate to Nucleus in Contrast to G protein-dependent ERK activation. Mol Pharmacol 2008;73(1):178–90.
62. Gesty-Palmer D, Chen M, Reiter E, et al. Distinct beta-arrestin- and G protein-dependent pathways for parathyroid hormone receptor-stimulated ERK1/2 activation. J Biol Chem 2006;281(16):10856–64.
63. Shenoy SK, Drake MT, Nelson CD, et al. beta-arrestin-dependent, G protein-independent ERK1/2 activation by the beta2 adrenergic receptor. J Biol Chem 2006;281(2):1261–73.
64. Rozenfeld R, Devi LA. Receptor heterodimerization leads to a switch in signaling: beta-arrestin2-mediated ERK activation by mu-delta opioid receptor heterodimers. Faseb J 2007;21(10): 2455–65.

65. Raehal KM, Walker JK, Bohn LM. Morphine side effects in beta-arrestin2 knockout mice. J Pharmacol Exp Ther 2005;314(3):1195–201.
66. Bohn LM, Raehal KM. Opioid receptor signaling: relevance for gastrointestinal therapy. Curr Opin Pharmacol 2006;6(6):559–63.
67. Bruns IR, Chhum S, Dinh AT, et al. A potential novel strategy to separate therapeutic- and side-effects that are mediated via the same receptor: beta-arrestin2/G-protein coupling antagonists. J Clin Pharm Ther 2006;31(2):119–28.
68. Ross GR, Gabra BH, Dewey WL, Akbarali HI. Morphine tolerance in the mouse ileum and colon. J Pharmacol Exp Ther 2008;372:561–72.
69. Berg KA, Maayani S, Goldfarb J, Scaramellini C, Leff P, Clarke WP. Effector pathway-dependent relative efficacy at serotonin type 2A and 2C receptors: evidence for agonist-directed trafficking of receptor stimulus. Mol Pharmacol 1998;54(1):94–104.
70. Kurrasch-Orbaugh DM, Parrish JC, Watts VJ, Nichols DE. A complex signaling cascade links the serotonin2A receptor to phospholipase A2 activation: the involvement of MAP kinases. J Neurochem 2003;86(4):980–91.
71. Nichols DE. Hallucinogens. Pharmacol Ther 2004;101(2):131–81.
72. McLean TH, Parrish JC, Braden MR, Marona-Lewicka D, Gallardo-Godoy A, Nichols DE. 1-Aminomethylbenzocycloalkanes: conformationally restricted hallucinogenic phenethylamine analogues as functionally selective 5-HT2A receptor agonists. J Med Chem 2006;49(19): 5794–803.
73. Gonzalez-Maeso J, Weisstaub NV, Zhou M, et al. Hallucinogens recruit specific cortical 5-HT(2A) receptor-mediated signaling pathways to affect behavior. Neuron 2007;53(3):439–52.
74. Gray JA, Sheffler DJ, Bhatnagar A, et al. Cell-type specific effects of endocytosis inhibitors on 5-hydroxytryptamine(2A) receptor desensitization and resensitization reveal an arrestin-, GRK2-, and GRK5-independent mode of regulation in human embryonic kidney 293 cells. Mol Pharmacol 2001;60(5):1020–30.
75. Gelber EI, Kroeze WK, Willins DL, et al. Structure and function of the third intracellular loop of the 5-hydroxytryptamine2A receptor: the third intracellular loop is alpha-helical and binds purified arrestins. J Neurochem 1999;72(5):2206–14.
76. Xia Z, Hufeisen SJ, Gray JA, Roth BL. The PDZ-binding domain is essential for the dendritic targeting of 5-HT2A serotonin receptors in cortical pyramidal neurons in vitro. Neuroscience 2003;122(4):907–20.
77. Gray JA, Roth BL. Paradoxical trafficking and regulation of 5-HT(2A) receptors by agonists and antagonists. Brain Res Bull 2001;56(5):441–51.
78. Roth BL, Hanizavareh SM, Blum AE. Serotonin receptors represent highly favorable molecular targets for cognitive enhancement in schizophrenia and other disorders. Psychopharmacology (Berl) 2004;174(1):17–24.
79. Willins DL, Alsayegh L, Berry SA, et al. Serotonergic antagonist effects on trafficking of serotonin 5-HT2A receptors in vitro and in vivo. Ann N Y Acad Sci 1998;861:121–7.
80. Willins DL, Berry SA, Alsayegh L, et al. Clozapine and other 5-hydroxytryptamine-2A receptor antagonists alter the subcellular distribution of 5-hydroxytryptamine-2A receptors in vitro and in vivo. Neuroscience 1999;91(2):599–606.
81. Bhatnagar A, Willins DL, Gray JA, Woods J, Benovic JL, Roth BL. The dynamin-dependent, arrestin-independent internalization of 5-hydroxytryptamine 2A (5-HT2A) serotonin receptors reveals differential sorting of arrestins and 5-HT2A receptors during endocytosis. J Biol Chem 2001;276(11):8269–77.
82. Corne SJ, Pickering RW. A possible correlation between drug-induced hallucinations in man and a behavioural response in mice. Psychopharmacologia 1967;11(1):65–78.
83. Meltzer HY. Mechanism of Action of Atypical Antipsychotic Drugs. In: Davis KL, Charney D, Coyle JT, Nemeroff C, eds. Neuropsychopharmacology: The Fifth Generation of Progress. New York: Raven Press; 2002:819–32.

Chapter 6
In Vivo Evidence for and Consequences of Functional Selectivity

Kim A. Neve, Marc G. Caron, and Jean-Martin Beaulieu

Abstract Functional selectivity refers to the ability of some ligands to stimulate a subset of the possible consequences of activation of a receptor. This chapter addresses two related issues that are critical for consideration of the therapeutic utility of functional selectivity: the evidence that functional selectivity is a pharmacologically relevant phenomenon that can be observed in vivo, and characterization of the unique in vivo properties of functionally selective ligands. Topics reviewed include G protein-biased agonists for μ-opioid and serotonin 5-HT_{2A} receptors, arrestin-biased agonists for angiotensin II AT_1 receptors and β-adrenergic receptors, antagonists that cause internalization of 5-HT_{2A} receptors, and in vivo evidence for functionally selective dopamine receptor ligands. Barriers to using functional selectivity as a criterion for rational drug design are discussed.

Keywords μ-opioid receptor, Serotonin receptor, Dopamine receptor, β-adrenergic receptor, Adenylate cyclase, G protein, Arrestin, Phospholipase C, Signal transduction

6.1 Introduction

Functional selectivity refers to the ligand-induced activation of a subset of the possible consequences of receptor activation. For G protein-coupled receptors (GPCRs), these consequences may include signaling via multiple G protein-regulated pathways, including pathways regulated by either Gα or Gβγ subunits, as well as engaging mechanisms involved in receptor desensitization (phosphorylation of the receptor, binding of arrestin, internalization) and arrestin-mediated signaling. When a ligand activates one or more of those processes but does not have the ability

J.-M. Beaulieu (✉)
CRULRG, Research Institute, Department of Anatomy and Physiology, Faculty of Medicine Université Laval, Quebec City, Canada
e-mail: Martin.Beaulieu@CRULRG.ULAVAL.CA

K. Neve (ed.), *Functional Selectivity of G Protein-Coupled Receptor Ligands*,
DOI: 10.1007/978-1-60327-335-0_6,

to induce the full repertoire of responses, then that ligand is said to be functionally selective. Functional selectivity is thought to result from an interaction between the conformational flexibility of GPCRs and the unique structure of each GPCR ligand, giving rise to an unique set of ligand-bound conformations for each ligand-receptor pair. Each of the multitude of receptor-interacting proteins that regulate and mediate signaling (G proteins, kinases, arrestins, scaffolds, other receptors) has distinct binding determinants, many of them dependent on receptor activation, and different ligand-receptor pairs may stabilize receptor conformations that differ in how those binding determinants are displayed. As reflected throughout this volume, most unequivocal demonstrations of functional selectivity have used in vitro systems, typically involving heterologous expression of one receptor subtype to eliminate the possibility that the presence of two targets for the ligands being tested gives the false appearance of functional selectivity at one target. To achieve the goal of exploiting functional selectivity in drug development, it is important to consider two questions that are the topic of this chapter. Can functional selectivity be demonstrated in vivo, or is it only observed in artificial systems? If functional selectivity can be observed in vivo, what are the behavioral and physiological consequences; how do the effects of a functionally selective drug differ from those of an agonist or antagonist that is not functionally selective?

6.2 G Protein or Arrestin-Biased Agonists

Many ligands exhibit functional selectivity for G protein-mediated or arrestin-mediated processes, and are referred to as G protein or arrestin-biased ligands *(1)*. Because of the multifaceted role of arrestin in signal transduction *(2, 3)*, functional selectivity for or against arrestin-mediated processes could have effects on receptor desensitization and arrestin-mediated signaling. Desensitization/internalization and signaling are often investigated separately, so this review will often consider them separately, but it should be noted that in some cases these may be two aspects of the same phenomenon in that receptor-bound and receptor-activated arrestin may engage the full panoply of arrestin-mediated processes that are present in a particular cell. In other cases, the participation of distinct GPCR kinases (GRKs) in desensitization and signaling, for example GRK2/3 in desensitization and GRK5/6 in arrestin-mediated signaling *(4, 5)*, raises the possibility of functionally selective ligands that distinguish among different arrestin-mediated processes.

When considering the consequences for receptor desensitization of ligand selectivity for or against arrestin-mediated processes, it is important to consider the complex role of arrestins in the regulation of receptor responsiveness. Although there may be exceptions to each step of this process, the canonical pathway for homologous desensitization of GPCRs includes phosphorylation of the activated receptor by GRKs, binding of arrestin to the activated and phosphorylated receptor, arrestin-dependent internalization of the receptor, and either resensitization and recycling back to the membrane or trafficking to lysosomes for degradation *(2)*. For some

GPCRs, arrestin plays a dual role in the process of desensitization/resensitization by mediating desensitization (i.e., preventing receptor coupling to G proteins) *(6)*, while also mediating the internalization that is required for resensitization *(7, 8)*. On the one hand, arrestin-biased agonists would engage this pathway, in addition to mechanisms for arrestin-dependent signaling, without the necessity for a concomitant association with G proteins. On the other hand, in addition to lacking arrestin-dependent signaling, G protein-biased ligands would be less likely to induce receptor internalization, which could either reduce arrestin-mediated desensitization or enhance desensitization by preventing arrestin-mediated resensitization of G protein-mediated signaling. Because desensitization is perceived as a phenomenon that limits the therapeutic efficacy of many drugs, the identification of nondesensitizing agonists is an important goal.

6.2.1 G Protein-Biased Agonists

6.2.1.1 μ-Opioid Receptors

As reviewed in Chaps. 4 and 12, there is considerable evidence from both in vivo and in vitro studies that morphine is less likely than other μ-opioid receptor ligands such as etorphine to induce internalization of the $G\alpha_{i/o}$-coupled μ-opioid receptor *(9, 10)*. There are disagreements in the literature concerning the extent to which morphine causes receptor desensitization, with evidence that morphine is very effective at eliciting arrestin-dependent internalization of μ-opioid receptors in some neurons *(11, 12)*, and concerning the relationship between receptor desensitization and behavioral tolerance to opiate-induced antinociception (Chap. 12); resolving those issues is beyond the scope of this chapter. In vivo studies with arrestin3 (β-arrestin2) null mutant (KO) mice suggest that a consequence of the functional selectivity of morphine for G protein-mediated pathways and its stabilization of a receptor conformation with low affinity for arrestins is a long-term behavioral tolerance that is selectively reduced in the absence of arrestin3 *(13–15)*. In contrast, nociception induced by agonists that induce robust binding of arrestin2 and arrestin3 displays tolerance that is unaffected by loss of arrestin3 *(15)*. Although these data suggest that arrestin3 (and presumably arrestin-dependent internalization) is required for tolerance to morphine, and that a drug with even less ability to recruit arrestin to the μ-opioid receptor would induce less behavioral tolerance, a different approach using "knock-in" mice in which the gene for the μ-opioid receptor is replaced with a mutant receptor that undergoes robust morphine-induced internalization demonstrates that tolerance to morphine is selectively reduced in the mice *(16)*. The authors' interpretation of these results is that a drug similar to morphine but with *enhanced* ability to engage mechanisms for arrestin-mediated internalization would have improved antinociceptive effects, with less tolerance. The concept of reducing the negative bias of morphine with respect to arrestin-dependent desensitization/resensitization was previously explored in vitro;

overexpression of GRK2 along with the μ-opioid receptor in HEK293 cells enhances morphine-induced phosphorylation of the μ-opioid receptor and promotes arrestin interaction and receptor internalization *(17)*.

The novel compound herkinorin may help to resolve the differing roles for arrestin in tolerance that are suggested by the studies described earlier. Although descriptions of the effects of herkinorin in vivo have not yet been published, in vitro studies demonstrate that herkinorin is significantly more biased than morphine toward G protein-mediated signaling; unlike morphine, herkinorin does not induce internalization of heterologously expressed μ-opioid receptors even in the presence of overexpressed GRK2 *(18)*. Two side effects to morphine, constipation and respiratory suppression, are greatly reduced in arrestin3 KO mice *(19)*. One interpretation of these results is that the low ability of morphine to activate arrestin-mediated signaling is still sufficient to induce these apparently arrestin-mediated physiological responses, and that a functionally selective drug such as herkinorin that is even more biased toward G protein-mediated signaling would be associated with fewer side effects.

6.2.1.2 Serotonin 5-HT$_{2A}$ Receptors

5-HT$_{2A}$ receptors signal through Gα_q to activate phospholipase C (PLC) and mobilize calcium from internal stores. Many, but not all, 5-HT$_{2A}$ receptor agonists are hallucinogenic, and there has been considerable interest in identifying the basis for the selective effects of hallucinogenic 5-HT$_{2A}$ agonists *(20–22)*. Several lines of evidence in vivo and using neuronal cultures indicate that internalization of the 5-HT$_{2A}$ receptor is mediated by arrestin3, and the at stimulation of that 5-HT$_{2A}$ receptor by serotonin activates extracellular signal-regulated kinase (ERK) via both G protein and arrestin-mediated pathways *(23,24)*. Recently, Schmid et al. used arrestin3 KO mice to demonstrate that most of the serotonin-induced ERK1/2 activation in murine frontal cortex is mediated by arrestin3, as ERK activation is abolished in the mice that lack arrestin3 *(24)*. The synthetic hallucinogen 2,5-demethoxy-4-iodoamphetamine (DOI) also stimulates ERK1/2 activation in the frontal cortex, but in contrast to serotonin, the ERK response to DOI is not greatly affected by genetic deletion of arrestin3, suggesting that it is mediated by G proteins and that DOI is a G protein-biased ligand *(24)*. Serotonin signaling through arrestin3 is behaviorally relevant, since the head twitch response to the serotonin precursor 5-hydroxytryptophan, but not the response to DOI, is virtually eliminated in arrestin3 KO mice. The mouse head twitch response is thought to be specifically associated with the action of hallucinogenic compounds at the 5-HT$_{2A}$ receptor *(25)*.

Although in mouse embryonic fibroblasts lacking arrestin3 the ERK response to administration of DOI is mediated by Gα_q *(24)*, an alternative hypothesis for the selective effects of hallucinogenic drugs is suggested by the work of González-Maeso et al. *(25)*. Hallucinogenic drugs, such as DOI and lysergic acid diethylamide (LSD), induce a pattern of expressed genes in mouse somatosensory cortex that differs from the pattern induced by nonhallucinogenic 5-HT$_{2A}$ agonists such as lisuride; in particular, *c-fos* induction is observed for all 5-HT$_{2A}$ agonists, but only hallucinogenic agonists induce *egr-1* and *egr-2*. In cortical neuronal cultures, it was found that gene expression responses to LSD, but not lisuride, are partially inhibited by treatment with pertussis toxin, indicating that signaling by LSD requires both $G\alpha_q$ and pertussis toxin-sensitive G proteins $G\alpha_{i/o}$ *(25)*. Thus, functional selectivity between G proteins may be the basis for hallucinogenic or nonhallucinogenic properties of 5-HT_{2A} agonists.

6.2.2 Arrestin-Biased Agonists

6.2.2.1 Angiotensin II AT_1 Receptors

The angiotensin II type 1 (AT_1) receptor signals via both G protein-mediated ($G\alpha_q$, PLC) and arrestin-mediated (arrestin3, ERK1/2) pathways *(26)*. Evidence for a physiologically relevant role of AT_1 receptor signaling via arrestin comes from studies of mice with cardiac-specific expression of either wild type AT_1 receptor or a mutant receptor deficient in coupling to G proteins. Overexpression of the mutant receptor, which presumably signaled only through arrestin-mediated pathways, caused more severe cardiac hypertropy and bradycardia, whereas overexpression of the wild type receptor caused more severe fibrosis and apoptosis *(27)*. An analog of angiotensin II, Sar1,Ile4,Ile8-angiotensin II (SII), has been developed that is an antagonist for $G\alpha_q$-mediated signaling but an agonist for arrestin3-mediated activation of ERK1/2 *(28–30)*. A study using isolated cardiac myocytes demonstrated that angiotensin II and SII have similar positive ionotropic and lusitropic effects, but that the response to SII, but not angiotensin II, was dependent on arrestin3 and GRK6 *(31)*. In this in vivo-like system, then, angiotensin II and SII have similar effects via distinct pathways. Because of its antagonism of AT_1 receptor signaling via $G\alpha_q$, it is thought that SII, as well as potential SII-like ligands, would share the blood pressure-reducing effect of conventional AT_1 receptor blockers while perhaps having additional beneficial effects through activation of arrestin-mediated signaling pathways, which are frequently cytoprotective *(1)*. In vivo, administration of either angiotensin II or the G protein-biased ligand SII into the third cerebral ventricle activates ERK1/2 and increases intake of NaCl, but whereas angiotensin II administration also induces c-fos expression in hypothalamic nuclei adjacent to the ventricle and increases water consumption, SII lacks these latter effects and also antagonizes angiotensin II-induced water consumption *(32)*.

6.2.2.2 β-Adrenergic Receptors

β-adrenergic receptor stimulation causes $G\alpha_s$-mediated stimulation of adenylate cyclase and cyclic AMP accumulation and arrestin-dependent activation of ERK1/2. As reviewed in Chap. 7, there is considerable in vitro evidence for drug functional selectivity between these $G\alpha_s$-mediated and arrestin-mediated pathways. In particular, the β-blocker carvedilol has been described as unique among β-blockers because although it is an inverse agonist for stimulation of cyclic AMP accumulation, it stimulates presumably GRK-catalyzed phosphorylation of the β_2-adrenergic receptor, recruitment of arrestin, receptor internalization, and activation of ERK1/2 in human embryonic kidney (HEK) 293 cells *(33)*. The authors speculate that this functional selectivity contributes to the particular efficacy of carvedilol in the treatment of heart failure. Other studies, however, suggest that several β-blockers that are antagonists or inverse agonists for stimulation of adenylate cyclase are able to activate ERK1/2 through β_1-adrenergic and β_2-adrenergic receptors heterologously expressed in HEK 293 cells, and that at least in the case of the β_1-adrenergic receptor, the response is not mediated by arrestin *(34, 35)*.

6.2.2.3 Dopamine D_2 Receptors

Dopamine D_2 receptors are coupled to $G\alpha_{i/o}$ to inhibit the production of cyclic AMP and thus diminish PKA activity, also modulating other signaling pathways (potassium channels, calcium channels, PLC, ERK1/2) as a result of the Gβγ subunits that are liberated upon activation of $G\alpha_{i/o}$ *(36–38)*. Recently, a novel arrestin3-dependent signaling pathway has been identified in the mouse striatum that is independent of cyclic AMP and ERK1/2, but that involves inhibition of the protein kinase Akt and consequently activation of glycogen synthase kinase 3 (GSK3) *(39, 40)*. Arrestin3 participation in this pathway is demonstrated both by coimmunoprecipitation and GST pull-down studies showing that arrestin3 interacts with Akt, GSK3β, and the heterotrimeric protein phosphatase PP2A, and by the lack of D_2 receptor regulation of this pathway in arrestin3 KO mice *(41)*. Stimulation of D2-like receptors causes the formation of a protein complex composed of at least Akt, arrestin3, and PP2A, which facilitates the dephosphorylation/deactivation of Akt by PP2A *(42)* in response to dopamine (Fig. 6.1).

There is abundant evidence that this arrestin-based signaling complex has an important role in dopamine-dependent behaviors. Arrestin3 KO mice display reduced climbing and locomotor activity, respectively, in response to the direct dopamine agonist apomorphine and the indirect agonist amphetamine. Furthermore, mice lacking both arrestin3 and the dopamine transporter display a reduction of the novelty-induced hyperactivity that is characteristic of hyperdopaminergic transporter-KO (DAT-KO) mice *(41)*. GSK3 inhibitors also reduce hyperactivity in both amphetamine-treated wild type mice and in DAT KO mice *(39,43)*, whereas transgenic mice expressing a constitutively active GSK3β mutant develop a locomotor hyperactivity phenotype that is reminiscent of DAT-KO mice *(44)*.

Amphetamine-induced disruption of prepulse inhibition is a rodent model of psychosis used to screen for antipsychotic activity *(45)*. Mice lacking the Akt isoform Akt1 also show enhanced amphetamine-induced suppression of prepulse inhibition *(46)*. Data such as these, combined with the ability of lithium to disrupt the formation of the arrestin3/Akt/PP2A signaling complex and impair D2-like receptor signaling *(39, 47)*,

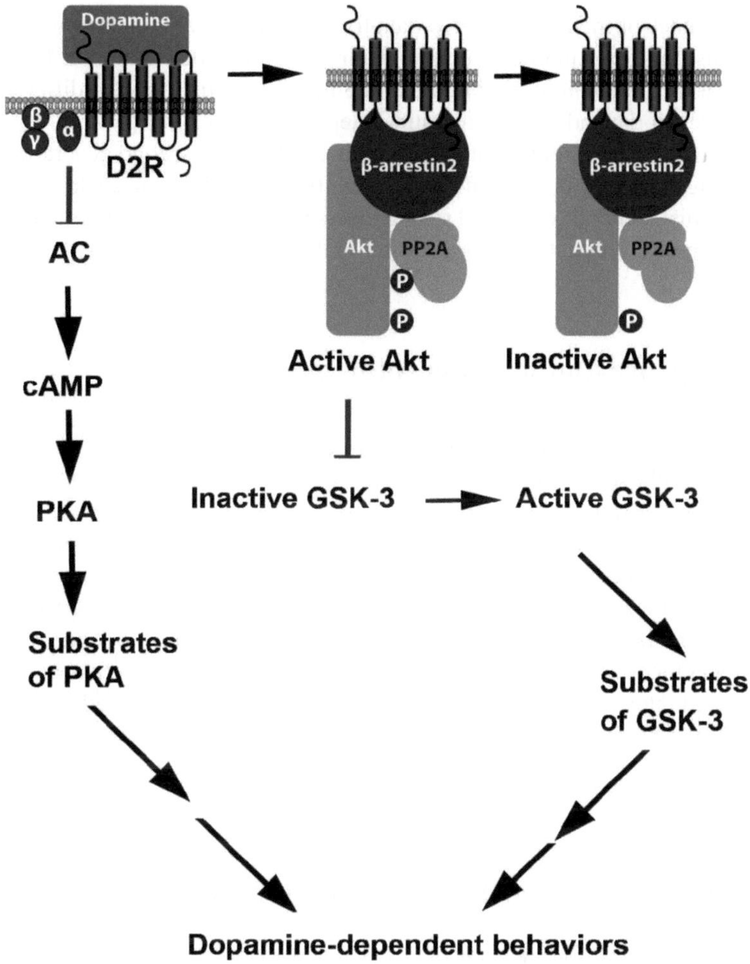

Fig. 6.1 G protein- and arrestin-mediated signaling by the dopamine D_2 recepor. The binding of dopamine to the D_2 receptor stimulates rapid signaling to adenylate cyclase and other G protein Gα- and Gβγ-regulated signaling pathways, and a more prolonged accumulation of a signaling complex, built on an arrestin3 (βarrestin2) scaffold, that leads to inactivation of the protein kinase Akt and activation of glycogen synthase kinase-3 (GSK-3)

suggest that this signaling pathway may contribute to psychosis *(48, 49)*, and that if a D_2 receptor agonist biased against this pathway would alleviate motor dysfunction in Parkinson's disease, the agonist would also be associated with a lower incidence of psychosis *(50)*. Furthermore, the ability of lithium to interfere with the formation of the arrestin3/Akt/PP2A complex *(47)* without affecting cAMP-mediated responses *(39)* suggests that functional selectivity may also be achieved by targeting signaling intermediates immediately downstream of receptors.

In the absence of in vitro or in vivo data for agonists that are functionally selective for or against this arrestin3-dependent signaling pathway, the creation of D_2 receptor mutants that are deficient in binding arrestin3 may be useful tools for differentiating between arrestin-dependent and arrestin-independent signaling pathways *(51, 52)*. Additionally, it may be significant that a number of first and second generation antipsychotic drugs that vary in their efficacy for inhibition of adenylate cyclase, including a partial agonist and both weak and strong inverse agonists, are all neutral antagonists for agonist-induced recruitment of arrestin3 to the D_2 receptor *(53)*.

6.3 Antagonist-Induced Receptor Internalization

GPCR internalization is typically an agonist-dependent process initiated by receptor activation and phosphorylation, as described in Sect. 2.1. For the serotonin 5-HT_{2A} receptor, however, most antagonists also cause receptor internalization, albeit to a lesser extent than agonist-induced internalization *(54–56)*, and antagonist-induced internalization is independent of arrestin, indicating that alternative mechanisms for internalization must be involved *(57)*. In vitro studies have also described antagonist-induced internalization of adenosine A_1 *(58)*, cholecystokinin *(59)*, and vasopressin V_2 *(60)* receptors. Determining the functional and behavioral significance of antagonist-induced internalization of these receptors will be aided by the identification of additional, noninternalizing antagonists with which the internalizing antagonists can be compared.

6.4 Functionally Selective Ligands for Dopamine Receptors

In contrast to some ligands discussed earlier where functional selectively was firmly established using in vitro systems in which single receptor subtypes are heterologously expressed, and there is often some understanding of the mechanism of functional selectivity (e.g., arrestin vs. G proteins), there are several dopamine receptor ligands that appear to be functionally selective, but where the mechanism of selectivity is uncertain.

6.4.1 Functional Selectivity Related to D1-Like Receptor Stimulation of Phospholipase C

Dopamine D_1 receptor coupling to the heterotrimeric G proteins $G\alpha_s$ or $G\alpha_{olf}$ induces the activation of adenyate cyclase and cyclic AMP accumulation *(38)*. In addition to this canonical D_1 receptor pathway, a D1-like receptor that is coupled to $G\alpha_q$ and that stimulates PLC to mobilize calcium has been described *(61–65)*. The existence of D1-like receptors that signal independently of $G\alpha_{s/olf}$ and adenylate cyclase has also been suggested by studies of mice genetically lacking either $G\alpha_{olf}$ or adenylate cyclase type 5, in which D_1 agonist-induced locomotor activity is preserved, even though D_1-like receptor stimulation of adenylate cyclase is absent *(66, 67)*, and by the observation that SKF83959, an antagonist for D_1 receptor stimulation of adenylate cyclase, has behavioral effects that are shared with most nonselective D_1 receptor agonists *(68–70)*. SKF83959 is one of two potential functionally selective ligands related to this PLC pathway; SKF83959 is selective for stimulation of PLC *(71–73)*, whereas SKF83822 stimulates adenylate cyclase but not PLC *(74)*.

SKF83959 and SKF83822 have very different behavioral effects. Like the prototypical D_1 agonist SKF38393, SKF83959 elicits rotational behavior in the unilateral 6-hydroxydopamine-lesioned rat model and oral dyskinesia in monkeys that had been chronically treated with dopamine receptor antagonists, two models of dopamine receptor supersensitivity *(70, 73)*. In naïve rodents, SKF83959 elicits the intense grooming and vertical jaw movements with tongue protrusions that are characteristic of D_1 agonists *(69, 75)*, as well as vacuous chewing that appears not to be mediated by D1-like receptors *(69)*. SKF83822, in contrast, elicits arousal and locomotor activation rather than oral dyskinesia in the monkey model, and sniffing, locomotion, and, at higher doses, intense seizures in mice *(70, 76)*.

As discussed in Chap. 10, there is controversy over whether SKF83959 and SKF83822 are functionally selective ligands or agonists at different receptor subtypes. Behavioral data addressing whether SKF83959 acts at the D_1 receptor are mixed, with D_1 receptor null mutant (D_1-KO) mice showing often substantial but generally not statistically significant reductions in SKF83959-induced grooming, locomotion, rearing, and sifting *(77)*, whereas SKF83959-induced vertical jaw movements and tongue protrusions are absent in D_1-KO mice *(78)*. SKF83959-mediated production of IP3 has been reported to be preserved in D_1-KO mice, indicating that the PLC-linked D1-like receptor is not a product of the *drd1* gene *(79)*. This result is, however, contradicted by a more recent finding that activation of $G\alpha_q$ by dopamine receptor agonists was abolished in D_1 and in D_2-KO mice *(80)*.

Recent evidence in heterologous expression systems and under some conditions in vivo has suggested that D_1 and D_2-receptors may be part of a heteromeric complex, and that activation of the heteromers by combined binding of D_1 and D_2 receptor ligands stimulates PLC and induces the release of calcium from intracellular stores through $G\alpha_q$ coupling *(80–82)*. Interestingly, unlike most adenylate cyclase-stimulating D_1 agonists, the adenylate cyclase-specific agonist SKF83822 does not increase Ca^{2+} mobilization in cells expressing D_1 and D_2 receptors even in the presence of quinpirole, and the PLC-specific D_1 agonist SKF83959 activates the heteromer and Ca^{2+} mobilization in the absence of a D_2 agonist *(80)*. Although some of the data are contradictory and cannot be reconciled into a consistent story, the parallel between the selective effects of these two drugs at the D_1/D_2 heteromer and at the "PLC-linked" and "adenylate cyclase-linked" D_1 receptors is intriguing and suggestive of functional selectivity based on receptor heteromerization

(Fig. 6.2). However, this intriguing possibility has to be further considered in the context of whether D_1 and D_2 dopamine receptors are expressed in the same dopaminoceptive neurons in the brain *(83)*.

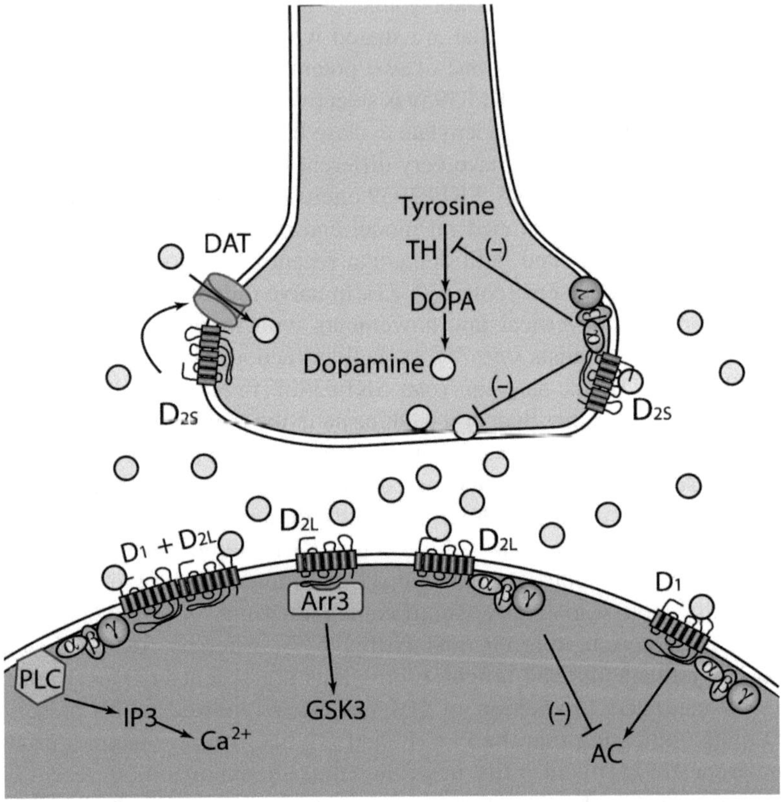

Fig. 6.2 Dopamine autoreceptors and postsynaptic receptors. D_1 and D_2 postsynaptic receptors activate G proteins to stimulate and inhibit, respectively, adenylate cyclase (AC). Activation of postsynaptic D_2 receptors also leads to arrestin-mediated signaling that activates glycogen synthase kinase 3 (GSK3). Binding of dopamine to both receptors in the D_1/D_2 heteromer, in contrast, activates phospholipase C (PLC), producing inositol trisphosphate (IP3) and leading to mobilization of Ca^{2+}. D_{2L} is the major postsynaptic D_2 receptor in the basal forebrain. In dopaminergic terminals, dopamine autoreceptors, believed to be primarily the D_{2S} receptor, inhibit the tyrosine hydroxylase- (TH-) catalyzed synthesis of dopamine, inhibit dopamine release, and enhance the reuptake of dopamine via the dopamine transporter (DAT). Dopamine autoreceptors are more sensitive than postsynaptic receptors, responding to lower concentrations of dopamine. There is abundant evidence for involvement of G protein-mediated pathways such as inhibition of adenylate cyclase and activation of K^+ channels in the regulation of dopamine synthesis and release, whereas the involvement of arrestin-mediated pathways has not been tested. SKF83959 activates the D_1/D_2 heteromer but not the D_1 receptor, and SKF83822 activates the D_1 receptor but not the heteromer. In many model systems, dihydrexidine is an agonist at D_2 autoreceptors but an antagonist at D_2 postsynaptic receptors, whereas the reverse is characteristic of aripiprazole (*See Color Plates*)

6.4.2 Functionally Selective D_2 Receptor Ligands

Dihydrexidine is a ligand originally developed by Nichols, Mailman, and colleagues as a full-efficacy D_1 receptor agonist, and subsequently determined to have roughly equivalent affinity at D_1 and D_2 receptors, as well as full efficacy for both D_1 receptor-stimulated and D_2 receptor-inhibited adenylate cyclase activity in cell expression systems *(84–87)*. As reviewed more thoroughly in Chap. 10, both dihydrexidine and its more D_2-selective analog *N*-n-propyldihydrexidine act as antagonists in models of D_2 autoreceptor function such as inhibition of dopamine release and dopamine neuron firing *(87, 88)*. Several of the assays of autoreceptor function were in vivo assays *(87)*. D_2 receptor agonists generally have biphasic effects on locomotor activity, causing inhibition at low doses that is thought to reflect activation of the more sensitive autoreceptors, and locomotor activation that reflects activation of postsynaptic D_2 receptors. One might expect, then, that a drug functionally selective for postsynaptic receptors would cause locomotor activation without the inhibition observed at low doses of nonfunctionally selective agonists. Dihydrexidine fits this profile, causing locomotor activation that is sensitive to D_2 receptor blockade *(89)*. In contrast, the more D_2-selective analog *N*-n-propyldihydrexidine caused inhibition of locomotor activity throughout a range of doses as well as vacuous chewing and yawning, with all responses sensitive to inhibition by a D_2 receptor antagonist *(90)*. Thus, not only is the mechanism of the functional selectivity of dihydrexidine and *N*-n-propyldihydrexidine unknown, but the behavioral consequences are also unclear because of the discrepancy between the effects of the two compounds, a discrepancy that might be due to a greater contribution of D_1 receptors to the effects of dihydrexidine. Nevertheless, the results with *N*-n-propyldihydrexidine indicate that the functional selectivity is more than just selectivity for postsynaptic receptors, and that the drugs probably also distinguish among postsynaptic signaling pathways. D_2 receptors do more than stimulate locomotor activity, and it will be instructive to learn more about which behavioral effects mediated by D_2 receptors are activated by these functionally selective ligands.

Aripiprazole (OPC-14597) was developed as a candidate antipsychotic drug, based on the hypothesis that a D_2 receptor weak partial agonist that inhibits dopamine release by stimulating more sensitive autoreceptors while serving as an antagonist for overstimulated of postsynaptic receptors and perhaps providing some low level of stimulation of the receptors would be an effective antipsychotic with reduced side effects (Fig. 6.2) *(91, 92)*, a hypothesis that appears to have been supported by the excellent therapeutic profile of the drug *(93)*. In vivo studies have demonstrated that aripiprazole is indeed an autoreceptor-selective agonist that inhibits tyrosine hydroxylase activity and that functions as an antagonist in behavioral models of postsynaptic D_2 receptor function *(91, 94)*. What is surprising about these results is that some of these models of postsynaptic receptor function (hyperlocomotion in reserpinized mice and contralateral rotation in unilaterally 6-hydroxydopamine-lesioned rats) are models of supersensitive receptors, in which partial agonists are typically more efficacious, just as they are at highly sensitive D_2 autoreceptors. To explain these results on the basis of simple partial agonism, one would have to conclude that aripiprazole is an extremely weak partial agonist.

Some subsequent in vitro studies have generally supported the hypothesis that aripiprazole is a conventional partial agonist *(95–97)*, whereas other studies have suggested that the variation in the efficacy of aripiprazole from one model system or signaling pathway to another is too great to be explained solely by partial agonism, and more likely reflects functionally selective signaling *(98, 99)*. Clearly, to determine whether the apparent functional selectivity of aripiprazole contributes to its therapeutic profile, it will be necessary to define more precisely its mechanism of functional selectivity.

A recent study using bioluminescence resonance energy transfer (BRET) indicates that aripiprazole can act as a partial D_2 receptor agonist for the regulation of adenylate cyclase while being a neutral antagonist for agonist-induced recruitment of arrestin3 to the D_{2L} receptor *(53)*. Interestingly, this contrasts with another report that aripiprazole is a partial agonist for recruitment of arrestin3 to the D_{2S} receptor *(100)*. These data suggest the intriguing hypothesis that aripiprazole is functionally selective for G protein-mediated signaling vs. arrestin-mediated signaling by the long alternatively spliced form of the D_2 receptor, D_{2L}, but not functionally selective between these signaling mechanisms at the short form of the D_2 receptor, D_{2S}. Although additional testing will be required, this hypothesis is consistent with the evidence that D_{2S} is the primary autoreceptor regulating dopamine synthesis and release *(101–103)*.

6.5 Conclusions

If functional selectivity is to be used in the rational design of therapeutic drugs, there are several challenges that must be met. First, in cases where the mechanism of functional selectivity is thought to be known as a result of work with in vitro systems (e.g., an arrestin-biased or G protein-biased ligand), to determine the behavioral or therapeutic consequences of that functional selectivity it will be necessary to test multiple drugs that are similarly functionally selective, ideally drugs based on multiple chemical backbones. Although many studies of the behavioral effects of functionally selective drugs in animal models have included excellent controls with selective antagonists and null mutant mice to determine that apparent functional selectivity is not mediated by known alternative sites of action, these controls generally cannot be used in assessing therapeutic effects in humans. Also, it is a truism of pharmacology that every drug has at least one site of action that is still unknown, and the best control for this is to assess the effects of multiple drugs with distinct structures. Second, when the mechanism of functional selectivity is unknown, it is critical that it be determined. It should be clear from this chapter and others in this volume that there are many ways that a ligand can display functional selectivity, so to attribute the therapeutic efficacy of any drug to functional selectivity without defining the nature and mechanisms of that property gives little guidance for the rational design of additional drugs and for screening drug candidates, or for testing the hypothesis of the relationship between functional selectivity and therapeutic efficacy. We believe that both of these challenges will be met over the next few years, and anticipate that drugs with improved therapeutic profiles will become available as a result of the deliberate screening for functional selective ligands.

References

1. Violin JD, Lefkowitz RJ. β-Arrestin-biased ligands at seven-transmembrane receptors. TIPS 2007;28:416–22.
2. Pierce KL, Lefkowitz RJ. Classical and new roles of β-arrestins in the regulation of G-protein-coupled receptors. Nat Rev Neurosci 2001;2:727–33.
3. Luttrell LM, Lefkowitz RJ. The role of β-arrestins in the termination and transduction of G-protein-coupled receptor signals. J Cell Sci 2002;115:455–65.
4. Ren XR, Reiter E, Ahn S, Kim J, Chen W, Lefkowitz RJ. Different G protein-coupled receptor kinases govern G protein and β-arrestin-mediated signaling of V2 vasopressin receptor. Proc Natl Acad Sci USA 2005;102:1448–53.
5. Reiter E, Lefkowitz RJ. GRKs and β-arrestins: roles in receptor silencing, trafficking and signaling. Trends Endocrinol Metab 2006;17:159–65.
6. Lohse MJ, Benovic JL, Codina J, Caron MG, Lefkowitz RJ. β-Arrestin: a protein that regulates β-adrenergic receptor function. Science 1990;248:1547–50.
7. Yu SS, Lefkowitz RJ, Hausdorff WP. β-Adrenergic receptor sequestration - a potential mechanism of receptor resensitization. J Biol Chem 1993;268:337–41.
8. Pippig S, Andexinger S, Lohse MJ. Sequestration and recycling of β_2-adrenergic receptors permit receptor resensitization. Mol Pharmacol 1995;47:666–76.
9. Arden JR, Segredo V, Wang Z, Lameh J, Sadée W. Phosphorylation and agonist-specific intracellular trafficking of an epitope-tagged μ-opioid receptor expressed in HEK 293 cells. J Neurochem 1995;65:1636–45.
10. Keith DE, Anton B, Murray SR et al. μ-opioid receptor internalization: Opiate drugs have differential effects on a conserved endocytic mechanism *in vitro* and in the mammalian brain. Mol Pharmacol 1998;53:377–84.
11. Dang VC, Williams JT. Morphine-Induced μ-opioid receptor desensitization. Mol Pharmacol 2005;68:1127–32.
12. Haberstock-Debic H, Kim KA, Yu YJ, von Zastrow M. Morphine promotes rapid, arrestin-dependent endocytosis of μ-opioid receptors in striatal neurons. J Neurosci 2005; 25:7847–57.
13. Bohn LM, Gainetdinov RR, Lin FT, Lefkowitz RJ, Caron MG. μ-Opioid receptor desensitization by β-arrestin2 determines morphine tolerance but not dependence. Nature 2000;408:720–3.
14. Bohn LM, Lefkowitz RJ, Caron MG. Differential mechanisms of morphine antinociceptive tolerance revealed in βarrestin2 knock-out mice. J Neurosci 2002;22:10494–500.
15. Bohn LM, Dykstra LA, Lefkowitz RJ, Caron MG, Barak LS. Relative opioid efficacy is determined by the complements of the G protein-coupled receptor desensitization machinery. Mol Pharmacol 2004;66:106–12.
16. Kim JA, Bartlett S, He L et al. Morphine-induced receptor endocytosis in a novel knockin mouse reduces tolerance and dependence. Curr Biol 2008;18:129–35.
17. Zhang J, Ferguson SS, Barak LS et al. Role for G protein-coupled receptor kinase in agonist-specific regulation of μ-opioid receptor responsiveness. Proc Natl Acad Sci USA 1998;95:7157–62.
18. Groer CE, Tidgewell K, Moyer RA et al. An opioid agonist that does not induce μ-opioid receptor--arrestin interactions or receptor internalization. Mol Pharmacol 2007; 71:549–57.
19. Raehal KM, Walker JK, Bohn LM. Morphine side effects in β-arrestin2 knockout mice. J Pharmacol Exp Ther 2005;314:1195–201.

20. Roth BL, Willins DL, Kristiansen K, Kroeze WK. 5-hydroxytryptamine2-family receptors (5-hydroxytryptamine2A, 5-hydroxytryptamine2B, 5-hydroxytryptamine2C): where structure meets function. Pharmacol Ther 1998;79:231–57.
21. Egan CT, Herrick-Davis K, Miller K, Glennon RA, Teitler M. Agonist activity of LSD and lisuride at cloned 5HT2A and 5HT2C receptors. Psychopharmacology (Berl) 1998; 136:409–14.
22. Nichols DE. Hallucinogens. Pharmacol Ther 2004;101:131–81.
23. Gelber EI, Kroeze WK, Willins DL et al. Structure and function of the third intracellular loop of the 5-hydroxytryptamine2A receptor: the third intracellular loop is α-helical and binds purified arrestins. J Neurochem 1999;72:2206–14.
24. Schmid CL, Raehal KM, Bohn LM. Agonist-directed signaling of the serotonin 2A receptor depends on β-arrestin2 interactions *in vivo*. Proc Natl Acad Sci USA 2008;105:1079–84.
25. González-Maeso J, Weisstaub NV, Zhou M et al. Hallucinogens recruit specific cortical 5-HT2A receptor-mediated signaling pathways to affect behavior. Neuron 2007;53:439–52.
26. Ahn S, Nelson CD, Garrison TR, Miller WE, Lefkowitz RJ. Desensitization, internalization, and signaling functions of β-arrestins demonstrated by RNA interference. Proc Natl Acad Sci USA 2003;100:1740–4.
27. Zhai P, Yamamoto M, Galeotti J et al. Cardiac-specific overexpression of AT1 receptor mutant lacking $G_{\alpha}q/G_{\alpha}i$ coupling causes hypertrophy and bradycardia in transgenic mice. J Clin Invest 2005;115:3045–56.
28. Holloway AC, Qian H, Pipolo L et al. Side-chain substitutions within angiotensin II reveal different requirements for signaling, internalization, and phosphorylation of type 1A angiotensin receptors. Mol Pharmacol 2002;61:768–77.
29. Wei H, Ahn S, Shenoy SK et al. Independent β -arrestin2 and G protein-mediated pathways for angiotensin II activation of extracellular signal-regulated kinases 1 and 2. Proc Natl Acad Sci USA 2003;100:10782–7.
30. Ahn S, Shenoy SK, Wei H, Lefkowitz RJ. Differential kinetic and spatial patterns of β-arrestin and G protein-mediated ERK activation by the angiotensin II receptor. J Biol Chem 2004;279:35518–25.
31. Rajagopal K, Whalen EJ, Violin JD et al. β-arrestin2-mediated inotropic effects of the angiotensin II type 1A receptor in isolated cardiac myocytes. Proc Natl Acad Sci USA 2006;103:16284–9.
32. Daniels D, Yee DK, aulconbridge LF, luharty SJ. Divergent behavioral roles of angiotensin receptor intracellular signaling cascades. Endocrinology 2005;146:5552–60.
33. Wisler JW, DeWire SM, Whalen EJ et al. A unique mechanism of β-blocker action: carvedilol stimulates β-arrestin signaling. Proc Natl Acad Sci USA 2007;104:16657–62.
34. Galandrin S, Bouvier M. Distinct signaling profiles of β_1 and β_2 adrenergic receptor ligands toward adenylyl cyclase and mitogen-activated protein kinase reveals the pluridimensionality of efficacy. Mol Pharmacol 2006;70:1575–84.
35. Galandrin S, Oligny-Longpré G, Bonin H, Ogawa K, Galés C, Bouvier M. Conformational rearrangements and signaling cascades involved in ligand-biased mitogen-activated protein kinase signaling through the β1-adrenergic receptor. Mol Pharmacol 2008;74:162–72.
36. Enjalbert A, Bockaert J. Pharmacological characterization of the D_2 dopamine receptor negatively coupled with adenylate cyclase in rat anterior pituitary. Mol Pharmacol 1983;23:576–84.
37. Missale C, Nash SR, Robinson SW, Jaber M, Caron MG. Dopamine receptors: from structure to function. Physiol Rev 1998;78:189–225.
38. Neve KA, Seamans JK, Trantham-Davidson H. Dopamine receptor signaling. J Recept Signal Transduct Res 2004;24:165–205.
39. Beaulieu JM, Sotnikova TD, Yao W-D et al. Lithium antagonizes dopamine-dependent behaviors mediated by an AKT/glycogen synthase kinase 3 signaling cascade. Proc Natl Acad Sci USA 2004;101:5099–104.
40. Beaulieu JM, Tirotta E, Sotnikova TD et al. Regulation of Akt signaling by D_2 and D_3 dopamine receptors *in vivo*. J Neurosci 2007;27:881–5.

41. Beaulieu JM, Sotnikova TD, Marion S, Lefkowitz RJ, Gainetdinov RR, Caron MG. An Akt/β-arrestin2/PP2A signaling complex mediates dopaminergic neurotransmission and behavior. Cell 2005;122:261–73.
42. Andjelkovic M, Jakubowicz T, Cron P, Ming XF, Han JW, Hemmings BA. Activation and phosphorylation of a pleckstrin homology domain containing protein kinase (RAC-PK/PKB) promoted by serum and protein phosphatase inhibitors. Proc Natl Acad Sci USA 1996; 93:5699–704.
43. Gould TD, Einat H, Bhat R, Manji HK. AR-A014418, a selective GSK-3 inhibitor, produces antidepressant-like effects in the forced swim test. Int J Neuropsychopharmacol 2004; 7:387–90.
44. Prickaerts J, Moechars D, Cryns K et al. Transgenic mice overexpressing glycogen synthase kinase 3β: a putative model of hyperactivity and mania. J Neurosci 2006;26:9022–9.
45. Powell SB, Geyer MA. Overview of animal models of schizophrenia. Curr Protoc Neurosci 2007;Chapter 9:Unit 9.24.
46. Emamian ES, Hall D, Birnbaum MJ, Karayiorgou M, Gogos JA. Convergent evidence for impaired AKT1-GSK3β signaling in schizophrenia. Nat Genet 2004;36:131–7.
47. Beaulieu JM, Marion S, Rodriguiz RM et al. A β-arrestin2 signaling complex mediates lithium action on behavior. Cell 2008;132:125–36.
48. Beaulieu JM, Gainetdinov RR, Caron MG. The Akt-GSK-3 signaling cascade in the actions of dopamine. TIPS 2007;28:166–72.
49. Tan HY, Nicodemus KK, Chen Q et al. Genetic variation in *AKT1* is linked to dopamine-associated prefrontal cortical structure and function in humans. J Clin Invest 2008;118:2200–8.
50. Fénelon G. Psychosis in Parkinson's disease: phenomenology, frequency, risk factors, and current understanding of pathophysiologic mechanisms. CNS Spectr 2008;13:18–25.
51. Lan H, Liu Y, Bell MI, Gurevich VV, Neve KA. A dopamine D2 receptor mutant capable of G protein-mediated signaling but deficient in arrestin binding. Mol Pharmacol 2009; 75:113–23.
52. Lan H, Teeter MM, Gurevich VV, Neve KA. An intracellular loop 2 amino acid residue determines differential binding of arrestin to the dopamine D2 and D3 receptors. Mol Pharmacol 2009;75:19–26.
53. Masri B, Salahpour A, Didriksen M et al. Antagonism of dopamine D2 receptor/β-arrestin2 interaction is a common property of clinically effective antipsychotics. Proc Natl Acad Sci USA 2008;105:13656–61.
54. Berry SA, Shah MC, Khan N, Roth BL. Rapid agonist-induced internalization of the 5-hydroxytryptamine2A receptor occurs via the endosome pathway *in vitro*. Mol Pharmacol 1996;50:306–13.
55. Willins DL, Berry SA, Alsayegh L et al. Clozapine and other 5-hydroxytryptamine-2A receptor antagonists alter the subcellular distribution of 5-hydroxytryptamine-2A receptors *in vitro* and *in vivo*. Neuroscience 1999;91:599–606.
56. Gray JA, Roth BL. Paradoxical trafficking and regulation of 5-HT2A receptors by agonists and antagonists. Brain Res Bull 2001;56:441–51.
57. Bhatnagar A, Willins DL, Gray JA, Woods J, Benovic JL, Roth BL. The dynamin-dependent, arrestin-independent internalization of 5-hydroxytryptamine 2A (5-HT_{2A}) serotonin receptors reveals differential sorting of arrestins and 5-HT_{2A} receptors during endocytosis. J Biol Chem 2001;276:8269–77.
58. Navarro A, Zapata R, Canela EI, Mallol J, Lluis C, Franco R. Epidermal growth factor (EGF)-induced up-regulation and agonist- and antagonist-induced desensitization and internalization of A1 adenosine receptors in a pituitary-derived cell line. Brain Res 1999; 816:47–57.
59. Roettger BF, Ghanekar D, Rao R et al. Antagonist-stimulated internalization of the G protein-coupled cholecystokinin receptor. Mol Pharmacol 1997;51:357–62.
60. Pfeiffer R, Kirsch J, Fahrenholz F. Agonist and antagonist-dependent internalization of the human vasopressin V2 receptor. Exp Cell Res 1998;244:327–39.
61. Undie AS, Friedman E. Stimulation of a dopamine D1 receptor enhances inositol phosphates formation in rat brain. J Pharmacol Exp Ther 1990;253:987–92.

62. Wang HY, Undie AS, Friedman E. Evidence for the coupling of G_q protein to D_1-like dopamine sites in rat striatum: Possible role in dopamine-mediated inositol phosphate formation. Mol Pharmacol 1995;48:988–94.
63. Pacheco MA, Jope RS. Comparison of [^{3}H]phosphatidylinositol and [^{3}H]phosphatidylinositol 4,5-bisphosphate hydrolysis in postmortem human brain membranes and characterization of stimulation by dopamine D1 receptors. J Neurochem 1997;69:639–44.
64. Jin L-Q, Wang H-Y, Friedman E. Stimulated D1 dopamine receptors couple to multiple Gα proteins in different brain regions. J Neurochem 2001;78:981–90.
65. Tang TS, Bezprozvanny I. Dopamine receptor-mediated Ca^{2+} signaling in striatal medium spiny neurons. J Biol Chem 2004;279:42082–94.
66. Lee KW, Hong JH, Choi IY et al. Impaired D2 dopamine receptor function in mice lacking type 5 adenylyl cyclase. J Neurosci 2002;22:7931–40.
67. Iwamoto T, Okumura S, Iwatsubo K et al. Motor dysfunction in type 5 adenylyl cyclase-null mice. J Biol Chem 2003;278:16936–40.
68. Arnt J, Hyttel J, Sánchez C. Partial and full dopamine D1 receptor agonists in mice and rats: relation between behavioural effects and stimulation of adenylate cyclase activity in vitro. Eur J Pharmacol 1992;213:259–67.
69. Deveney AM, Waddington JL. Pharmacological characterization of behavioural responses to SK&F 83959 in relation to 'D_1-like' dopamine receptors not linked to adenylyl cyclase. Br J Pharmacol 1995;116:2120–6.
70. Peacock L, Gerlach J. Aberrant behavioral effects of a dopamine D1 receptor antagonist and agonist in monkeys: evidence of uncharted dopamine D1 receptor actions. Biol Psychiatry 2001;50:501–9.
71. Panchalingam S, Undie AS. SKF83959 exhibits biochemical agonism by stimulating [(35)S] GTP gamma S binding and phosphoinositide hydrolysis in rat and monkey brain. Neuropharmacology 2001;40:826–37.
72. Jin LQ, Goswami S, Cai GP, Zhen XC, Friedman E. SKF83959 selectively regulates phosphatidylinositol-linked D_1 dopamine receptors in rat brain. J Neurochem 2003;85:378–86.
73. Zhen X, Goswami S, Friedman E. The role of the phosphatidyinositol-linked D_1 dopamine receptor in the pharmacology of SKF83959. Pharmacol Biochem Behav 2005;80:597–601.
74. Undie AS, Weinstock J, Sarau HM, Friedman E. Evidence for a distinct D_1-like dopamine receptor that couples to activation of phosphoinositide metabolism in brain. J Neurochem 1994;62:2045–8.
75. Tomiyama K, McNamara FN, Clifford JJ, Kinsella A, Koshikawa N, Waddington JL. Topographical assessment and pharmacological characterization of orofacial movements in mice: dopamine D_1-like vs. D_2-like receptor regulation. Eur J Pharmacol 2001;418:47–54.
76. O'Sullivan GJ, Roth BL, Kinsella A, Waddington JL. SK&F 83822 distinguishes adenylyl cyclase from phospholipase C-coupled dopamine D1-like receptors: behavioural topography. Eur J Pharmacol 2004;486:273–80.
77. Clifford JJ, Tighe O, Croke DT et al. Conservation of behavioural topography to dopamine D1-like receptor agonists in mutant mice lacking the D_{1A} receptor implicates a D_1-like receptor not coupled to adenylyl cyclase. Neuroscience 1999;93:1483–9.
78. Tomiyama K, McNamara FN, Clifford JJ et al. Phenotypic resolution of spontaneous and D_1-like agonist-induced orofacial movement topographies in congenic dopamine D_{1A} receptor 'knockout' mice. Neuropharmacology 2002;42:644–52.
79. Friedman E, Jin LQ, Cai GP et al. D_1-like dopaminergic activation of phosphoinositide hydrolysis is independent of D1A dopamine receptors: evidence from D_{1A} knockout mice. Mol Pharmacol 1997;51:6–11.
80. Rashid AJ, So CH, Kong MM et al. D1-D2 dopamine receptor heterooligomers with unique pharmacology are coupled to rapid activation of Gq/11 in the striatum. Proc Natl Acad Sci USA 2007;104:654–9.
81. Lee SP, So CH, Rashid AJ et al. Dopamine D1 and D2 receptor co-activation generates a novel phospholipase C-mediated calcium signal. J Biol Chem 2004;279:35671–8.

82. Dziedzicka-Wasylewska M, Faron-Górecka A, Andrecka J, Polit A, Kusmider M, Wasylewski Z. Fluorescence studies reveal heterodimerization of dopamine D_1 and D_2 receptors in the plasma membrane. Biochemistry 2006;45:8751–9.
83. Shuen JA, Chen M, Gloss B, Calakos N. *Drd1a*-tdTomato BAC transgenic mice for simultaneous visualization of medium spiny neurons in the direct and indirect pathways of the basal ganglia. J Neurosci 2008;28:2681–5.
84. Lovenberg TW, Brewster WK, Mottola DM et al. Dihydrexidine, a novel selective high potency full dopamine D-1 receptor agonist. Eur J Pharmacol 1989;166:111–3.
85. Brewster WK, Nichols DE, Riggs RM et al. trans-10,11-dihydroxyl-5,6,6a,7,8,12b-hexahydrobenxo[a]phenanthridine: a highly potent selective dopamine D1 full agonist. J Med Chem 1990;33:1756–64.
86. Mottola DM, Brewster WK, Cook LL, Nichols DE, Mailman RB. Dihydrexidine, a novel full efficacy D_1 dopamine receptor agonist. J Pharmacol Exp Ther 1992;262:383–93.
87. Mottola DM, Kilts JD, Lewis MM et al. Functional selectivity of dopamine receptor agonists. I. Selective activation of postsynaptic dopamine D_2 receptors linked to adenylate cyclase. J Pharmacol Exp Ther 2002;301:1166–78.
88. Kilts JD, Connery HS, Arrington EG et al. Functional selectivity of dopamine receptor agonists. II. Actions of dihydrexidine in D_{2L} receptor-transfected MN9D cells and pituitary lactotrophs. J Pharmacol Exp Ther 2002;301:1179–89.
89. Darney KJ, Jr., Lewis MH, Brewster WK, Nichols DE, Mailman RB. Behavioral effects in the rat of dihydrexidine, a high-potency, full-efficacy D1 dopamine receptor agonist. Neuropsychopharmacology 1991;5:187–95.
90. Smith HP, Nichols DE, Mailman RB, Lawler CP. Locomotor inhibition, yawning and vacuous chewing induced by a novel dopamine D2 post-synaptic receptor agonist. Eur J Pharmacol 1997;323:27–36.
91. Kikuchi T, Tottori K, Uwahodo Y et al. 7-{4-[4-(2,3-dichlorophenyl)-1-piperazinyl] butyloxy}-3,4-dihydro- 2(1*H*)-quinolinone (OPC-14597), a new putative antipsychotic drug with both presynaptic dopamine autoreceptor agonistic activity and postsynaptic D_2 receptor antagonistic activity. J Pharmacol Exp Ther 1995;274:329–36.
92. Tamminga CA. Partial dopamine agonists in the treatment of psychosis. J Neur Transm 2002;109:411–20.
93. Kane JM, Carson WH, Saha AR et al. Efficacy and safety of aripiprazole and haloperidol versus placebo in patients with schizophrenia and schizoaffective disorder. J Clin Psychiatry 2002;63:763–71.
94. Leite JV, Guimaraes FS, Moreira FA. Aripiprazole, an atypical antipsychotic, prevents the motor hyperactivity induced by psychotomimetics and psychostimulants in mice. Eur J Pharmacol 2008;578:222–7.
95. Lawler CP, Prioleau C, Lewis MM et al. Interactions of the novel antipsychotic aripiprazole (OPC-14597) with dopamine and serotonin receptor subtypes. Neuropsychopharmacology 1999;20:612–27.
96. Burris KD, Molski TF, Xu C et al. Aripiprazole, a novel antipsychotic, is a high-affinity partial agonist at human dopamine D2 receptors. J Pharmacol Exp Ther 2002;302:381–9.
97. Cosi C, Carilla-Durand E, Assié MB et al. Partial agonist properties of the antipsychotics SSR181507, aripiprazole and bifeprunox at dopamine D2 receptors: G protein activation and prolactin release. Eur J Pharmacol 2006;535:135–44.
98. Shapiro DA, Renock S, Arrington E et al. Aripiprazole, a novel atypical antipsychotic drug with a unique and robust pharmacology. Neuropsychopharmacology 2003;28:1400–11.
99. Urban JD, Vargas GA, von Zastrow M, Mailman RB. Aripiprazole has functionally selective actions at dopamine D2 receptor-mediated signaling pathways. Neuropsychopharmacology 2007;32:67–77.
100. Klewe IV, Nielsen SM, Tarpø L et al. Recruitment of β-arrestin2 to the dopamine D2 receptor: insights into anti-psychotic and anti-parkinsonian drug receptor signaling. Neuropharmacology 2008;54:1215–22.

101. Wang Y, Xu R, Sasaoka T, Tonegawa S, Kung MP, Sankoorikal EB. Dopamine D2 long receptor-deficient mice display alterations in striatum-dependent functions. J Neurosci 2000;20:8305–14.
102. Usiello A, Baik JH, Rouge-Pont F et al. Distinct functions of the two isoforms of dopamine D2 receptors. Nature 2000;408:199–203.
103. Khan ZU, Mrzljak L, Gutierrez A, De la Calle A, Goldman-Rakic PS. Prominence of the dopamine D2 short isoform in dopaminergic pathways. Proc Natl Acad Sci USA 1998;95:7731–6.

Part II
Subfamilies of Therapeutically Relevant Receptors

Chapter 7
Functional Selectivity at Adrenergic Receptors

Richard R. Neubig

Abstract The adrenergic receptors have been a model for understanding the structure, function, regulation, and biology of GPCRs. They play a similar role with respect to ligand functional selectivity (or ligand-dependent differential signaling). The β-adrenergic receptor has provided key structural information about ligand binding and also rich biophysical data demonstrating that different ligands can produce different conformational states of the receptor. There is also substantial data in cell systems showing that ligands can produce distinct spectrums of response including differential roles of the Gi and Gs-type G proteins as well as contrasting agonist and inverse agonist actions at G protein-dependent and G protein-independent (largely β-arrestin-dependent) signals. There is also in vivo evidence for physiologically important roles for these different signals. The clinical importance of β-adrenergic blockers in cardiovascular disease and emerging evidence for ligand-specific vs. class-specific effects in clinical studies makes a full understanding of ligand functional selectivity very important. Challenges in the future will be to determine the unique receptor conformations produced by the binding of different ligands, show how those conformations interact with the complex cellular context and regulatory processes to result in unique signal output patterns, and finally to translate that basic knowledge into novel therapeutics that take advantage of the additional specificity provided by ligand functional selectivity.

Keywords Adrenergic receptor, G protein, Gi, Gs, Adenylate cyclase, ERK, PI3K, Akt, GRK, Beta-arrestin, Potassium channel, Calcium channel, Cardiac myocyte, Heart failure

Abbreviations AC: Adenylyl cyclase; βArr: Beta arrestin; cAMP: 3′5′ cyclic adenosine monophopshpate; CHO: Chinese hamster ovary cell line; ERK:

R.R. Neubig
Department of Pharmacology, The University of Michigan Medical School, Ann Arbor MI 48109
e-mail: RNeubig@umich.edu

K. Neve (ed.), *Functional Selectivity of G Protein-Coupled Receptor Ligands*,
DOI: 10.1007/978-1-60327-335-0_7,

Extracellular signal regulated kinase; Gi: Inhibitory guanine nucleotide binding protein; GPCR: G protein coupled receptor; Gs: Stimulatory guanine nucleotide binding protein; HEK: Human embryonic kidney cell line; MAP kinase: Mitogen activated protein kinases; mBB: Monobromobimane; PDZ: Pleckstrin, discs large, Zo-1 homology domain; PI3K: Phosphatidyl inositol 3-kinase; PKA: cAMP dependent protein kinase (protein kinase A); RET: Resonance energy transfer

7.1 Introduction

Adrenergic receptors are key therapeutic targets and important prototypes for the study of signal transduction mechanisms. β-adrenergic blockers are among the most heavily used cardiovascular drugs because of their actions in the treatment of hypertension, congestive heart failure, and prevention of the recurrence of myocardial infarctions. Similarly, selective ligands for both β1 (metoprolol, atenolol, etc.) and β2 (terbutaline, salbutamol, etc.) adrenoceptors are heavily used clinically in cardiovascular disease, asthma, and other conditions. α1 blockers and α2 agonists are both beneficial in the treatment of hypertension, and each has important uses in urology, withdrawal syndromes, etc. Thus tremendous efforts have been made in both the academic and pharmaceutical sectors to learn about and understand the pharmacology, structure, function, and regulation of these receptors.

Consequently, the adrenergic receptors have been prototypes in our understanding of GPCRs from the initial classification of α and β-adrenergic receptors by Ahlquist and the development of β-blockers by Sir James Black to their important role in defining the GPCR superfamily when elucidation of their sequence made the unexpected connection with rhodopsin and subsequently the nearly 1,000 other GPCR genes *(1,2)*. Furthermore, biochemical analysis of the activation of adenylyl cyclase (AC) by β-adrenergic receptors and other GPCRs led to the discovery of the four families of heterotrimeric G proteins *(3,4)*, which gave the GPCRs their name. Studies of the regulation of β-adrenergic receptors by the G protein-coupled receptor kinases and β-arrestin, along with similar studies of rhodopsin, revealed a novel and important regulatory mechanism for GPCRs in general that we now know can also lead to distinct signaling outputs *(5)*. These latter two discoveries play a central role our emerging understanding of ligand-dependent functional selectivity through GPCRs.

The β-adrenergic receptors, in particular, provide a very important model for understanding functional selectivity by GPCR ligands. The β-adrenergic receptors are the only ligand-activated receptors for which there are crystal structures and ground-breaking biophysical studies from the Kobilka group, and others, provide compelling evidence for unique conformational changes resulting from the binding of different ligands. This latter work helps put all work on GPCR functional selectivity on a firmer theoretical footing.

Furthermore, the adrenergic receptors couple to multiple distinct signal outputs. There is well-established coupling to two different G protein stystems (Gs and Gi) for both the β-receptors and the α2 adrenergic receptors. Although the former provide a strong input to Gs they also, in certain circumstances, can activate Gi-dependent signaling outputs. Conversely, the α2 aderenergic receptors strongly activate Gi pathways but are also able to stimulate AC through Gs. Beyond the dual coupling to Gs and Gi, there is substantial evidence for G protein independent signaling through recruitment of β-arrestin to phosphorylated receptors *(5)* and probably through assembly of other signal transduction scaffolds *(6,7)*.

This confluence of critical structural and biophysical information with an array of different signaling outputs has made the adrenergic receptors, particularly the β-adrenergic receptors, highly informative models in the understanding of functional selectivity by GPCR ligands. The present chapter will examine these ideas in more detail.

7.2 Structural/Biophysical Basis of Functional Selectivity

7.2.1 Structure of the β-Adrenergic Receptors

The recently published crystal structures of β-adrenergic receptors *(8–10)* have provided exciting new insights into ligand binding and potential structural bases for conformational changes. Most importantly, they suggest that direct observations of the structural determinants of ligand functional selectivity may be just around the corner. An early analysis of differential docking of functional selective ligands into the β2-receptor structure has suggested specific hypotheses that can be tested *(11)*.

7.2.2 Biophysical Studies of Ligand-Induced Conformational Changes

In a more than decade-long, oeuvre using challenging techniques of membrane protein purification, clever site-selective fluorescence labeling, and careful kinetic studies, the Kobilka lab has provided unambiguous evidence for distinct conformational changes in the β2-adrenergic receptor induced by an array of ligands. Much of this work was reviewed recently *(12)*.

A key observation is that the full agonist isoproterenol causes a two-phase change in fluorescence, with both fast and slow components, from a tetramethylrhodamine probe attached to cys256$^{6.27}$ in the critical i3 loop near TM6. Ligands with a catechol moiety, including catechol itself, are able to induce a fast fluorescence change. Ligands that have the amine moiety but no catechol, such as salbutamol, induce only the slow fluorescence increase. Interestingly, the combined addition of catechol and salbutamol mimics the full effect of epinephrine – the combined slow and fast increase. These data are interpreted in the context of a model in which there are two key intramolecular switches that are modified upon binding of ligands termed the ionic lock ($R^{3.50}$ in the TM3 DRY motif and an acidic amino acid $E^{6.30}$ near TM6) and the rotamer toggle switch ($W^{6.48}$, $P^{6.50}$, and $F^{6.52}$). The fast change is interpreted to represent the breaking of the ionic lock, while the slow change represents the flip of the rotamer toggle switch.

A method to specifically probe the conformation of the ionic lock involved placing a fluorophore (monobromobimane, mBB) at amino acid position $271^{6.33}$ at the cytoplasmic end of TM6, and a quenching amino acid (tryptophan) at position $135^{3.45}$ at the cytoplasmic end of TM3. Under basal conditions, the amino acids of the closed ionic lock keep the tryptophan away from the mBB and fluorescence is high. When the receptor is activated, the lock is opened and the tryptophan and mBB come closer together and the mBB fluorescence is quenched. Agonists, including salbutamol and the weak partial agonist halostanchine, fully induce the fluorescence change with this probe, although they only partially replicate the effects of isoproterenol on the cys$256^{6.27}$ TMR-labeled receptor. In contrast, catechol, which produces a substantial but incomplete effect on cys$256^{6.27}$ TMR, has no effect on the mBB quenching. The recent crystal structures of the β-adrenergic receptors *(8–10)* suggest that the original concept of the ionic lock as an explicit hydrogen bond between $R^{3.50}$ and $E^{6.30}$ is probably not correct but there are clearly substantial conformational changes taking place near the TM3 DRY motif and the adjacent TM6 residues.

Thus, compounds with a catechol group can open the ionic lock but this is not sufficient to lead to efficacy as measured by Gs protein activation ([^{35}S]GTPγS binding or AC activation). In contrast, agonists that can induce the slow change attributed to the rotamer toggle switch are good agonists in Gs activation.

7.3 Cellular Functional Selectivity at β-Adrenergic Receptors

By definition, functional selectivity of ligands at a receptor requires that there be at least two different signaling outputs produced by the receptor. In the case of the β-adrenergic receptors, there are multiple G protein-dependent signals but also some G protein-independent outputs. β-adrenergic receptors are quite effective in activating the Gs protein and stimulating AC. Surprisingly the β2-adrenergic receptor, like the α2a-adrenergic receptor discussed below, is able to activate Gi signaling as well *(13,14)*. The dual Gs/Gi coupling has been best documented for β2 and β3-adrenergic receptors, although β1 responses mediated through Gi have also been demonstrated *(15)*. The usual approach to demonstrate a role for Gi is the use of pertussis toxin to uncouple it from the receptor.

It is not clear why a receptor would be designed to both activate AC via Gs and to inhibit it via Gi. Two potential explanations could be considered. First, the high efficiency of the β-receptor to activate Gs often leads to a receptor reserve in which only a small fraction of receptors needs to be activated to maximally turn on AC. Consequently, ligand concentrations well below the K_d for receptor binding are able to produce considerable stimulation of AC. In contrast, the poor coupling of the β2-receptor to Gi necessitates greater fractional receptor occupancy and higher ligand concentrations to achieve its effect. Thus the AC inhibition effect occurs only at high agonist concentrations and only for high efficacy agonists, providing a concentration-dependent differential on-then-off signal to AC. A second, perhaps more likely, explanation is that the important effect of Gi is not to inhibit AC but to regulate some other signaling output such as ERK or Akt activation *(4)*.

In addition to the Gs and Gi effects, the β-adrenergic receptors, like many other GPCRs, are able to activate signals that appear to be independent of G proteins. Consequently, it has been proposed that this family of receptors should be termed 7TM receptors instead of GPCRs. The most frequently seen G protein-independent signaling output is activation of the MAP kinase pathway – specifically ERK. The mechanisms by which 7TMRs activate ERK are quite complex and cell-type-dependent (see *(5,16,17)* for review). At least two different mechanisms have been invoked for β-adrenergic receptor-mediated activation of ERK. The first involves receptor phosphorylation, often by GRKs, and recruitment of β-arrestin. β-arrestin serves as a scaffold to recruit ERK and other ERK pathway components to the membrane leading to a cytosol-specific ERK phosphorylation *(18,19)*. A second mechanism involves activation of c-src. This may involve a Gi-dependent pathway that is still poorly defined or it could involve direct receptor src recruitment to the cell membrane in a G protein and β-arrestin independent manner.

In each of the following sections discussing ligand functional selectivity, I will first outline some of the evidence for distinct signal outputs by the β-adrenergic receptors then present the data on ligands that can selectively control those different outputs. The majority of this work has focused on the β2-adrenergic receptor but similar observations have been made for the β1 and β3-receptors as well.

7.4 β2-Adrenergic Receptors

7.4.1 Dual Gs/Gi Signaling

7.4.1.1 Regulation of AC

System: The β2-adrenergic receptor in S49 lymphoma cells provided one of the first systems for characterizing the mechanism of G protein signaling to AC and led to the identification of the first heterotrimeric G protein $G\alpha_s$r. The observation that β-agonist stimulation of AC was enhanced by pertussis toxin *(20,21)* suggested that the β-adrenergic receptor could also couple to the inhibitory Gi protein. This was firmly established by Molinoff and colleagues *(22)* who showed that, in the $G\alpha_s$ deficient cyc^- variant of S49 cells, high affinity binding of the β-receptor agonist hydroxybenzyl isoproterenol could be virtually eliminated by treatment with pertussis toxin.

Ligand selectivity: Surprisingly, despite the considerable evidence for ligand functional selectivity at β2-receptor in receptor conformational changes and in signaling to myocardial contractility and to kinase cascades (e.g. ERK), there is no clear demonstration of a functionally selective β2-receptor ligand at the level of adenylyl cyclase regulation. As described later for α2-adrenergic receptors, one might expect biphasic regulation of adenylate cyclase by those ligands that efficiently couple to Gi. A simple test of this question – and also direct tests of Gs vs. Gi activation using RET probes or nucleotide binding – would be useful.

7.4.1.2 Myocardial Contractility/Beating Rate

System: One of the best-characterized, and potentially most clinically important, systems for dual β2-adrenergic receptor signaling via Gs and Gi is cardiac myocytes. Human myocytes express about 20–30% β2 receptors, as a proportion of their total β-receptor complement. Rodent myocytes express an even greater proportion of β2-receptors (30–40%). The dual signaling by Gs and Gi in this system was first described by Ui *(21)* then studied extensively by Xiao, Lakatta, and colleagues *(13,23)*. Xiao et al. found clear differences in the coupling of β1 and β2-receptors to cAMP regulation, cardiac cell contractility, and beating rate. Most importantly, β2-adrenergic stimulation of contractility by the selective agonist zinterol or by isoproterenol in the presence of the β1-blocker CGP 20712A was markedly enhanced by pretreatment with pertussis toxin, indicating dual coupling to Gs and Gi. In contrast, β1 stimulation with norepinephrine or with isoproterenol plus the β2-blocker ICI 118,551 was not affected by pertussis toxin treatment. This suggested an unique effect of β2-receptors to engage negatively inotropic actions through Gi. The enhanced contractility in the presence of pertussis toxin was also accompanied by increased activation of inward Ca^{++} currents. This result has been replicated using a wide variety of systems and functional readouts including β2-receptor expressing transgenic mice *(24)*, β1 and β2-receptor deficient knockouts *(25)*, $G\alpha_{i2}$ loss *(26)* or gain *(27)* of function systems, and heart failure models *(28)*.

Ligand selectivity: Two ligands have been reported to show a selective action in this system. Norepinephrine, in measures of myocyte contractility and beating rates, was recently shown to lack the Gi coupling exhibited by epinephrine *(29,30)*. These results were due to effects on the β2-receptor since the effect was observed in β2-receptor transgenic overexpressing mice and in β1-receptor knockout myocytes. The beating rate mechanism appears to involve an apparent agonist switching phenomenon as the effect of PTX was only observed after about 3–5 min and was lost with receptors that were mutated to remove the GRK phosphorylation sites (Ser355, Ser356) *(30)*. This contrasts with the PKA-site dependence previously reported by Daaka et al. *(14)*. There also appears to be a contribution of differential receptor internalization in mediating this selectivity as eliminating the C-terminal PDZ ligand from the β2-adrenergic receptor abolished the difference between epinephrine and norepinephrine. In separate studies, fenoterol, a β2-selective agonist, differed from other β2-agonists in showing no increase in contractility with pertussis toxin *(31)* suggesting selective Gs activation. This was confirmed in a model using a cAMP-dependent inhibition of phenylephrine-induced protein synthesis in neonatal myocytes as the functional readout *(32)*. Both terbutaline and salbutamol showed significant Gi coupling in that system but fenoterol showed no pertussis-dependent enhancement of response. These results were somewhat complicated by a possible contribution of β1-receptors (despite the use of "selective" β2-agonists) but fenoterol clearly showed unique properties.

Interpreting results from such system-based readouts is complex. It will be of interest to know if these ligand-specific differences are reflected simply in distinct conformational changes of the receptor. Given the potential role of receptor trafficking, it is possible that this may not be observable in well-defined receptor/G protein reconstitution systems. Given the physiological and clinical significance of β-receptor signaling in the heart, such effects, if present in human adult heart tissue, could have important therapeutic implications even with the complex mechanism.

7.4.1.3 PI3K and Myocardial Cell Survival

System: In addition to the effects of Gi signaling in control of contractility and heart rate, activation of the PI3K-Akt pathway through Gi has also been shown to be important for myocardial survival and prevention of apoptosis *(33)* (see *(23)* for review). β1-receptors induce cardiac cell apoptosis, while β2-receptors do not. Indeed they can inhibit apoptosis induced by β1 activation. The ability of β2-receptors to produce a survival advantage has also been shown in transgenic overexpressing mouse model systems *(34,35)* and in knockout mouse myocytes *(33)*.

Ligand selectivity: There are no data yet showing selectivity of agonists for either PI3K/Akt activation or myocardial cell survival over cAMP signaling but it seems plausible that any ligands found with a preference for activating Gi over Gs through the β2-adrenergic receptor might show such a response. Clearly such a selective signal would be of potential interest therapeutically.

7.4.2 Extracellular Signal-Regulated Kinase (ERK)

System: Besides the classical roles of Gs and Gi in regulation of ACs, they also signal to numerous other intracellular effectors including the protein kinases ERK and Akt. The mechanisms leading to ERK activation by GPCRs, including the β-adrenergic receptors, are very complex and often cell-type dependent. Numerous review articles have discussed this topic *(16–18,36–38)*, so I will not discuss it in detail here. Two main categories of mechanisms are those that utilize G proteins ($G\alpha_s$, $G\alpha_q$, or $G\alpha_i$) and those that do not. In the latter case, a number of systems have been shown to depend on recruitment of β-arrestin (βArr) *(39–41)* to the receptor, which most likely recruits other components leading to ERK activation.

Ligand specificity: This is probably the best studied system for functional selectivity at β-adrenergic receptors with significant data on β2-receptors and also (see below) on β1 and β3-receptors. In a striking result demonstrating ligand-based functional selectivity, two groups reported seemingly contradictory agonist and inverse agonist actions in two different assays with the classical β-antagonist propranolol *(39,42)*. They showed that, in measures of cAMP stimulation by the β2-adrenergic receptor, propranolol behaved as an inverse agonist. In contrast, when measuring receptor-mediated activation of ERK phosphorylation, it produced a clear stimulatory signal. The signal was substantial (~14–30% of that of isoproterenol). Similarly other inverse agonists, alprenolol and ICI118551, showed unexpected ERK stimulation. Neither group found a role for Gi in this response, as it was not blocked by pertussis toxin. Azzi et al. *(39)* showed that siRNA-mediated suppression of βArr2 blocked the ERK activation signal.

This result on propranolol has been extended by several groups. Baker et al. *(42)* in their initial report showed that several ligands, including alprenolol and carvedilol, with minimal efficacy for cAMP stimulation produced marked activation of CRE luciferase activity with at least some dependence on ERK signaling. These results are potentially complicated by the dual cAMP and ERK mechanisms of the CRE readout. In 2006, Bouvier and colleagues *(43)* proposed that functionally selective action of β2-receptor ligands on cAMP vs. ERK signaling can be described as "pluridimensional efficacy" and prepared a Cartesian representation of the dual efficacy of several ligands. Plotting activation of cAMP production on the *x*-axis and ERK phosphorylation on the *y*-axis (Fig. 7.1), they illustrate that several ligands show greater relative efficacy for ERK activation than for Gs-mediated AC stimulation. Although propranolol is the most striking example (inverse agonist for cAMP production and agonist for ERK activation), several other ligands including bucindolol and carvedilol have near neutral activity at Gs/cAMP responses but clear ERK activation. In contrast to the results from the Hill lab *(42)*, Bouvier and colleagues found that carvedilol was a neutral antagonist and alprenolol was an inverse agonist. Lefkowitz and coworkers *(40)* found that alprenolol was a weak agonist and carvedilol an inverse agonist for cAMP. To date, no ligand has been found for β2-receptors that is an inverse agonist for ERK and an agonist for Gs/cAMP.

In light of the potential role of βArr in these non-G protein-dependent signaling pathways, more direct measures of βArr recruitment to receptor have been established. Drake et al. *(44)* used a CFP/YFP FRET pair to detect binding of βArr to the β2-adrenergic receptor in cells. They focused on compounds that showed clear agonist activity for cAMP production (also measured with a FRET method) and found that most agonists showed comparable activation of the two signals. Three catecholamine analogs that had an ethyl moiety on the α-carbon showed modest increases in βArr recruitment compared with the cAMP response detected.

7.4.3 β2-Adrenergic Receptor Summary

There are examples at the β2-adrenergic receptor where ligands have clearly different efficacies for distinct signaling outputs. The mechanisms involve both differential coupling to Gs and Gi pathways (cAMP and PI3K) and non-G protein-dependent ERK activation. The most striking example is probably propranolol where three

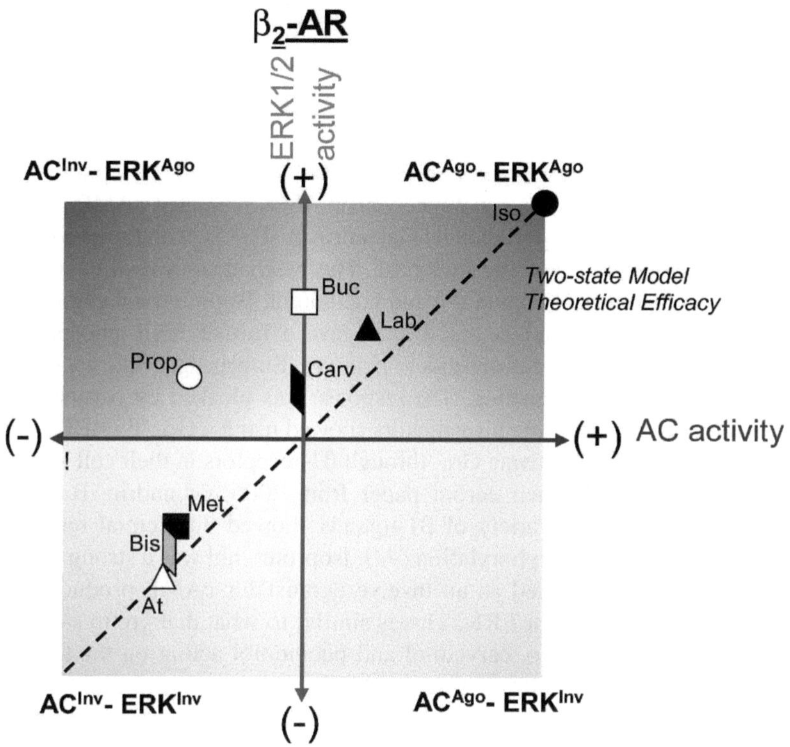

Fig. 7.1 Pluridimensional efficacy – β2-adrenergic receptor. An illustration adapted from *(43)* shows differential signal outputs to cAMP production and activation of ERK illustrating the range of behaviors of different β2-adrenergic ligands. Compounds listed in the *upper right quadrant* are agonists for both outputs while compounds in the *lower left* are inverse agonists for both. Propranolol in the *upper left quadrant* is an inverse agonist for cAMP production but is an agonist for ERK activation. This provides one of the clearest cases of ligand functional selectivity. At a more quantitative level, any compound that is off of the diagonal line has potential for functional selectivity (*See Color Plates*)

different groups report that it has inverse agonist activity for cAMP production but is a reasonable agonist for activating ERK signaling. This is potentially quite exciting for the development of novel therapeutics with improved specificity and unique therapeutic effects. It is important, however, to recognize that these systems are quite complex and that results will likely vary substantially from one cell type to another. Indeed, with the same receptor (β2) in a common cell line (HEK) there are differences from lab to lab (e.g., alprenolol as either a weak agonist or an inverse agonist for cAMP production) and even for distinct clones of HEK cells within a single lab there are marked differences in the relative contribution of Gi signaling *(38)*. Some of these effects are likely also to depend on the localization of receptors in both plasma membrane and intracellular compartments. Because of this strong context dependence, the analysis of functionally selective signaling properties will require careful analysis of underlying mechanisms and will also need to be moved rapidly to native tissues to evaluate their significance.

7.5 Cellular Functional Selectivity at *β*1-Adrenergic Receptors

7.5.1 ERK

System: The β1-adrenergic receptor in HEK cells, like the β2-receptor, is able to activate ERK in addition to its usual mechanism of Gs-mediated cAMP production. The mechanism was recently studied by Galandrin et al. *(15)*, who found a G protein-independent mechanism involving src kinase. This mechanism was stimulated by two classical β1 "antagonists," propranolol and bucindolol. Isoproterenol engaged this G protein-independent mechanism but it also gave a further ERK phosphorylation signal by means of a G protein-dependent pathway. Surprisingly, this appears to use Gi rather than Gs for its signaling. The response was blocked by pertussis but not cholera toxin. Furthermore, the investigators showed using a $G\alpha_{i2}/\beta\gamma$ BRET pair that isoproterenol was able to activate $G\alpha_{i2}$ through β1-receptors in their cell system.

Ligand specificity: In their earlier paper from 2006, Galandrin, Bouvier, and colleagues showed that a variety of β1-ligands showed differential regulation of cAMP levels and ERK phosphorylation *(43)*. Isoproterenol was a strong agonist for both, while propranolol acted as an inverse agonist for cAMP production and a reasonably strong agonist for ERK. This is similar to what that group saw with the β2-adrenergic receptor. Also, carvedilol and bucindolol acting on the β1-receptor were partial agonists in both pathways with carvedilol having a slight preference for ERK and bucindolol giving a relatively bigger cAMP response. Interestingly, labetolol was a good agonist for cAMP production but gave a very small ERK response (only 5% of that of Iso). Thus labetolol provides an example where ERK signaling is weak to rule out the possibility that there is more efficient coupling to ERK than cAMP as an explanation for the signaling differences. In contrast to the β2-receptor, there were no ligands that acted as inverse ERK agonists at β1-adrenergic receptor, despite the strong inverse efficacy of bisoprol, atenolol, and metoprolol for cAMP responses. This may be due in part to the very low basal ERK activity making it difficult to identify inverse agonists. Thus, the β1-adrenergic receptor and the β2-adrenergic receptor show a clear pattern of functional selectivity – or pluri-dimensional efficacy for cAMP and ERK responses.

7.5.2 Secondary Allosteric Site

The β1-adrenergic receptor also shows differential ligand effects through a second (allosteric) binding site. It was noted in the late 1990s by several groups that the β1-receptor blocker CGP 12177 also showed agonist properties. Hill and colleagues *(45,46)* showed that it is a potent inhibitor of classic β-agonist actions ($Ki \sim 10^{-9}$) but that it could also, with about a 2-log lower potency, act as an agonist for cAMP production. Besides CGP 12177, carvedilol was also a mixed ligand – antagonist at the catecholamine site and agonist at the CGP 12177 site. Other ligands had different combinations of properties. CGP 20712A and atenolol are pure antagonists at the catecholamine site while alprenolol and pindolol are agonists at both sites. There were also discrepancies in relative potencies of several compounds for effects on [^{3}H]cAMP production and regulation of a CRE gene reporter system. As noted for the β2-receptor above, the latter response may involve both PKA and ERK-dependent CREB phosphorylation, which complicates the interpretation of ligand functional selectivity. To better understand these data and the mechanism of signaling from the second site, it will be necessary to fully separate the cAMP and other signal outputs from the β1 AR in response to these novel ligands.

7.6 Cellular Functional Selectivity at β3-Adrenergic Receptors

System: The evidence for two distinct signal outputs from the β3-adrenergic receptor is less well-defined than for the β1 and β2-receptors. As with the β1 and β2-systems, there seem to be differences between the various reports *(47–49)*. All used CHO cells expressing β3-receptors but there were differences in the readout (cAMP, ERK, p38, CRE reporter, and extracellular acidification). The most direct demonstration of dual signaling was in an early report by Issad and coworkers *(48)*, who showed that β3-receptors mediate increases in ERK and p38 phosphorylation as well as cAMP levels. ERK activation was blocked by pertussis toxin implicating a Gi family G protein and also by PI3K inhibitors wortmannin and LY 294002. Sato et al. *(50)* measured cAMP and extracellular acidification (ECAR), while Baker *(47)* used cAMP and a CRE expression reporter.

Ligand selectivity: Because the output signals for ERK and cAMP were clearly separated (i.e., ERK activation was not induced by forskolin or blocked by H89), the evidence for ligand functional selectivity is easier to interpret in the Gerhardt et al. study *(48)*. In one of the stronger types of evidence for functional selectivity, they saw a swap in potency order and efficacy for norepinephrine and epinephrine at the two responses. The EC50 differences were substantial (8- to 20-fold). Furthermore, BRL 37344 was an antagonist for ERK activation and a potent full agonist for cAMP stimulation. Baker *(47)* found a number of ligands, including carvedilol and propranolol, which showed substantially greater relative intrinsic activity for the CRE reporter than for cAMP production when compared with isoproterenol. This does not appear to be due to spare receptors for the former response as the EC50 values were quite similar for the two outputs. In contrast to the Gerhardt et al. study *(48)*, she did not see any real change in potency or efficacy for norepinephrine and epinephrine. Similar to results with the β1-receptor, she also found agonist-dependent antagonist Ki values and identified a ligand ZD 7114, which showed a clear biphasic response in both the cAMP and CRE reporter outputs, reminiscent of the two sites for CGP 12177 at the β1-receptor. Finally, Summers and coworkers *(49)* found differential intrinsic activities at cAMP and ECAR for SR59230A, a ligand for which Baker also found anomalous behavior. Thus, the β3-receptor also is likely to have some functional selectivity but additional work will be required to fully sort out the details and mechanisms.

7.7 α1-Adrenergic Receptors

System: Much of the focus on α1-adrenergic receptors has been on defining the three subtypes and their properties. There has been, however, some information on differential signaling outputs (see *(51)* for review). α1-adrenergic receptors primarily couple to Gq-PLC-Ca^{++} signaling pathways but α1b-adrenergic receptors have also been reported to activate pertussis-toxin sensitive signals *(52–56)*. In particular, phospholipase A2 *(52,54)* and PI3K *(53)* signals from α1b receptors have been shown to be pertussis-toxin sensitive as have inhibitory effects on cardiac voltage-gated Ca^{++} channels *(56)* and a negative inotropic effect *(55)*.

Ligand selectivity: This differential signaling output by α1-adrenergic receptors (particularly α1b receptors) suggests the possibility of functional selectivity but there is no comprehensive comparison of different α1-adrenergic receptor ligands to test this concept. Perez and collaborators, however, did report a constitutively active mutant α1b receptor (C128F) that had differential effects on signal outputs (greater effect on PLC signaling vs. PLA2) and also differential effects on catecholamines (epinephrine) compared with imidazoline ligands (cirazoline). This is highly suggestive that there may be ligands that can selectively activate PLC vs PLA2 and vice versa, but to date there is no clear evidence of functional selectivity by α1-adrenergic receptor ligands

7.8 α2-Adrenergic Receptors

The α2-adrenergic receptors (α2a, α2b, and α2c) signal primarily through Gi/o family G proteins. They mediate a number of signals through this mechanism including inhibition of AC, activation of potassium (Kir3.x) currents, inhibition of voltage-gated Ca^{++} currents (Cav2.x), and activation of MAP kinase pathways (ERK) *(57–59)*. Most of the work on signaling mechanisms and ligand functional selectivity has been done on the α2a-adrenergic receptor.

7.8.1 *cAMP: Biphasic Gs/Gi Signaling*

System: Like the β2-adrenergic receptor, the α2a-adrenergic receptor consistently shows dual coupling to Gi and Gs proteins *(60–62)*. With increasing concentrations of full agonists, there is first inhibition then stimulation of AC. After pertussis toxin treatment, the inhibitory component is eliminated leaving only the stimulatory component. The coupling to Gi and inhibition of AC is much more efficient (ca. 100-fold) than that for Gs with EC_{50} values of the full α2 agonist UK14304 for AC inhibition of about 0.2 nM compared with 40 nM for Gs and AC activation *(62)*. Similar differences were seen with epinephrine and other ligands *(61,63)*.

Interestingly, Gi activation is regulated by distinct intracellular sites in the α2a receptor than is Gs activation. Eason and Liggett *(61)* showed that mutations in the N-terminal end of the i3 loop of the α2a-adrenergic receptor resulted in loss of AC stimulation but a minimal change in AC inhibition. The key residues were amino acids 218–228. Wade et al. *(62)* identified two regions in the C-terminal end of the i3 loop that each selectively modulate Gi or Gs signaling. Near TM6 of the α2a-adrenergic receptor are two basic regions that had been proposed to contribute to G protein activation. The REKR site immediately next to TM6 and the RWRGR site two residues further out in the i3 loop were both mutated to AEAA (B3) or REAA (B2) and AWAGA (R3) to eliminate positively charged amino acids. Surprisingly, the B2 mutation markedly reduced Gi signaling, while the R3 mutation essentially eliminated Gs signaling. In both cases, the response through the other G protein was essentially intact. These two studies clearly show that the structural determinants for Gi and Gs activation are different and indicate that ligands selectively exposing these putative "effector" regions of the i3 loop would selectively activate the two downstream signals.

Ligand functional selectivity: An early analysis *(64)* of ligand functional selectivity reported that catecholamines were able to activate Gs effectively but that imidazoline ligands such as BHT-920 and BHT-933 showed selective activation of Gi vs Gs. This analysis, however, is complicated by the much greater coupling efficiency of the receptor for Gi than for Gs and might represent spare receptors for Gi signaling. A subsequent analysis *(63)* carefully corrected for spare receptors by comparing a cell line expressing a low amount of receptor for measures of Gi signaling and a higher amount for Gs signaling. Under these conditions, the maximum responses for both Gi and Gs were similarly reduced by an irreversible α receptor blocker thus ruling out any complications of spare receptors. Under these conditions, 12 ligands were tested and only two showed any discrepancy between potency and efficacy at Gi vs. Gs responses. Oxymetazoline showed enhanced Gi signaling but this was shown to be due to endogenous 5HT1 receptors. Isoproterenol, the classic β-agonist, showed a markedly greater Gs signal than it did for the Gi response. Clearly the risk of an artifact from endogenous β-receptors was a problem but the signal was blocked by yohimbine but not propranolol. Also, no response was seen in untransfected CHO cells. Thus the classic β-adrenergic agonist isoproterenol appeared to induce a Gs-specific conformation of the normally Gi-coupled α2-adrenergic receptor.

7.8.2 Ion Channel Regulation

System: In addition to regulation of AC, α2-adrenergic receptors produce profound inhibitory signals in the brain and other excitable tissues through regulation of ion channels. Activation of the G protein-coupled inwardly rectifying potassium channels (GIRK or Kir3.x) is generally mediated by $G\alpha_i$ proteins. The voltage-gated Ca^{++} channels (N-type family, Cav2.1 and 2.2) are inhibited by α2 receptor activation – most likely through $G\alpha_o$. Both mechanisms generally lead to inhibition of neural signals. Surprenant and colleagues *(65)* showed that a D79N mutation in the α2a-adrenergic receptor led to a striking loss of K^+ channel activation, while inhibition of Ca^{++} channels was largely intact; however, this effect may be due to the efficiency of coupling to the two signal outputs (see later). Also, the relative roles of different Gi/o subunit subtypes in physiological functions *(27)* provide another potential locus for functional selectivity from α2 receptors.

Ligand functional selectivity: To date, there is not clear evidence for agonists that can differentiate between these two signaling pathways except insofar as the greater proportion of spare receptors for Ca^{++} signaling may favor activation by partial agonists over the K^+ channels, which require a greater signaling stimulus for actvitiy *(66)*. This has been proposed as a mechanism to reduce the sedative side effects of α2a receptor activation, while maintaining the therapeutic antihypertensive effect.

7.9 Therapeutic Implications of Adrenergic Receptor Functional Selectivity

The β-adrenergic receptors represent tremendous examples of the potential for ligand functional selectivity in therapeutics. As outlined earlier, there is much basic information about both mechanisms and pharmacology of functional selectivity at the β-adrenergic receptors. Also, β-adrenergic ligands are very important clinically. The role of β-blockers in many aspects of cardiovascular treatment and prevention is of specific interest because of evidence for ligand-specific effects rather than simple class-related actions. The COMET trial comparing carvedilol and metoprolol in congestive heart failure is one prominent example *(67)* with carvedilol showing lower mortality. There are many potential explanations for differences among β-blockers in these studies. Questions have been raised about dosages, and carvedilol blocks β2 and α1-receptors as well as β1 and has antioxidant actions as well. However, differential effects on different signaling outputs from the beta receptors are one exciting possibility. As more ligands are characterized for functional selectivity and both animal and human studies advance, the determinants of clinical actions and the appropriate balance of receptor actions for optimal therapeutic effect needs to be defined.

7.10 Outlook

The adrenergic receptors, in particular the β-receptors, will likely remain among the most important receptors in the development of concepts and uses of ligand functional selectivity. The availability of crystal structures for modeling and especially new structures solved with different ligands will contribute greatly to our understanding of the molecular basis of functional selectivity. It is clear that more information is needed about the degree to which simple conformational differences drive functional selectivity. Existing evidence that much of cellular functional selectivity is determined by receptor phosphorylation and trafficking brings substantial complexity to these systems. Indeed, direct evidence for robust differences in G protein outputs by ligands at a single receptor has been difficult to come by. Tools are now in place to directly test functional selectivity with isolated, reconstituted systems so hopefully those questions will be soon answered.

The larger questions about therapeutic utility will depend both on an enhanced basic understanding of functional selectivity at adrenergic receptors and creative animal and clinical studies of ligands with unique properties. It seems likely that much new and exciting information will be forthcoming in the very near future.

Acknowledgments Work in the author's lab described here was supported by NIH R01GM39561.

References

1. Black J. Drugs from emasculated hormones: the principle of syntopic antagonism. Science 1989;245:486–93.
2. Dixon RA, Kobilka BK, Strader DJ, et al. Cloning of the gene and cDNA for mammalian beta-adrenergic receptor and homology with rhodopsin. Nature 1986;321:75–9.
3. Ross EM, Gilman AG. Biochemical properties of hormone-sensitive adenylate cyclase. Annu Rev Biochem 1980;49:533–64.
4. Wettschureck N, Offermanns S. Mammalian G proteins and their cell type specific functions. Physiol Rev 2005;85:1159–204.
5. DeWire SM, Ahn S, Lefkowitz RJ, Shenoy SK. Beta-arrestins and cell signaling. Annu Rev Physiol 2007;69:483–510.
6. Cao W, Luttrell LM, Medvedev AV, et al. Direct binding of activated c-Src to the beta 3-adrenergic receptor is required for MAP kinase activation. J Biol Chem 2000;275:38131–4.
7. Smith NJ, Luttrell LM. Signal switching, crosstalk, and arrestin scaffolds: novel G protein-coupled receptor signaling in cardiovascular disease. Hypertension 2006;48:173–9.
8. Warne T, Serrano-Vega MJ, Baker JG, et al. Structure of a beta(1)-adrenergic G-protein-coupled receptor. Nature 2008;454:486–91.
9. Cherezov V, Rosenbaum DM, Hanson MA, et al. High-resolution crystal structure of an engineered human beta2-adrenergic G protein-coupled receptor. Science 2007;318:1258–65.
10. Rasmussen SG, Choi HJ, Rosenbaum DM, et al. Crystal structure of the human beta2 adrenergic G-protein-coupled receptor. Nature 2007;450:383–7.
11. Audet M, Bouvier M. Insights into signaling from the beta2-adrenergic receptor structure. Nat Chem Biol 2008;4:397–403.
12. Kobilka BK, Deupi X. Conformational complexity of G-protein-coupled receptors. Trends Pharmacol Sci 2007;28:397–406.
13. Xiao RP, Ji X, Lakatta EG. Functional coupling of the beta 2-adrenoceptor to a pertussis toxin-sensitive G protein in cardiac myocytes. Mol Pharmacol 1995;47:322–9.
14. Daaka Y, Luttrell LM, Lefkowitz RJ. Switching of the coupling of the beta2-adrenergic receptor to different G proteins by protein kinase A. Nature 1997;390:88–91.
15. Galandrin S, Oligny-Longpre G, Bonin H, Ogawa K, Gales C, Bouvier M. Conformational rearrangements and signaling cascades involved in ligand-biased mitogen-activated protein kinase signaling through the beta1-adrenergic receptor. Mol Pharmacol 2008;74:162–72.
16. Gutkind JS. Regulation of mitogen-activated protein kinase signaling networks by G protein-coupled receptors. Sci STKE 2000;2000:RE1.
17. Werry TD, Christopoulos A, Sexton PM. Mechanisms of ERK1/2 regulation by seven-transmembrane-domain receptors. Curr Pharm Des 2006;12:1683–702.

18. Luttrell LM. 'Location, location, location': activation and targeting of MAP kinases by G protein-coupled receptors. J Mol Endocrinol 2003;30:117–26.
19. Shenoy SK, Drake MT, Nelson CD, et al. beta-arrestin-dependent, G protein-independent ERK1/2 activation by the beta2 adrenergic receptor. J Biol Chem 2006;281:1261–73.
20. Katada T, Amano T, Ui M. Modulation by islet-activating protein of adenylate cyclase activity in C6 glioma cells. J Biol Chem 1982;257:3739–46.
21. Hazeki O, Ui M. Modification by islet-activating protein of receptor-mediated regulation of cyclic AMP accumulation in isolated rat heart cells. J Biol Chem 1981;256:2856–62.
22. Abramson SN, Martin MW, Hughes AR, et al. Interaction of beta-adrenergic receptors with the inhibitory guanine nucleotide-binding protein of adenylate cyclase in membranes prepared from cyc- S49 lymphoma cells. Biochem Pharmacol 1988;37:4289–97.
23. Xiao RP. Beta-adrenergic signaling in the heart: dual coupling of the beta2-adrenergic receptor to G(s) and G(i) proteins. Sci STKE 2001;2001:RE15.
24. Xiao RP, Avdonin P, Zhou YY, et al. Coupling of beta2-adrenoceptor to Gi proteins and its physiological relevance in murine cardiac myocytes. Circ Res 1999;84:43–52.
25. Xiang Y, Rybin VO, Steinberg SF, Kobilka B. Caveolar localization dictates physiologic signaling of beta 2-adrenoceptors in neonatal cardiac myocytes. J Biol Chem 2002;277:34280–6.
26. Foerster K, Groner F, Matthes J, Koch WJ, Birnbaumer L, Herzig S. Cardioprotection specific for the G protein Gi2 in chronic adrenergic signaling through beta 2-adrenoceptors. Proc Natl Acad Sci USA 2003;100:14475–80.
27. Fu Y, Huang X, Zhong H, Mortensen RM, D'Alecy LG, Neubig RR. Endogenous RGS proteins and Galpha subtypes differentially control muscarinic and adenosine-mediated chronotropic effects. Circ Res 2006;98:659–66.
28. He JQ, Balijepalli RC, Haworth RA, Kamp TJ. Crosstalk of beta-adrenergic receptor subtypes through Gi blunts beta-adrenergic stimulation of L-type Ca2+ channels in canine heart failure. Circ Res 2005;97:566–73.
29. Heubach JF, Ravens U, Kaumann AJ. Epinephrine activates both Gs and Gi pathways, but norepinephrine activates only the Gs pathway through human beta2-adrenoceptors overexpressed in mouse heart. Mol Pharmacol 2004;65:1313–22.
30. Wang Y, De Arcangelis V, Gao X, Ramani B, Jung YS, Xiang Y. Norepinephrine- and epinephrine-induced distinct beta2-adrenoceptor signaling is dictated by GRK2 phosphorylation in cardiomyocytes. J Biol Chem 2008;283:1799–807.
31. Xiao RP, Zhang SJ, Chakir K, et al. Enhanced G(i) signaling selectively negates beta2-adrenergic receptor (AR)–but not beta1-AR-mediated positive inotropic effect in myocytes from failing rat hearts. Circulation 2003;108:1633–9.
32. Ponicke K, Groner F, Heinroth-Hoffmann I, Brodde OE. Agonist-specific activation of the beta2-adrenoceptor/Gs-protein and beta2-adrenoceptor/Gi-protein pathway in adult rat ventricular cardiomyocytes. Br J Pharmacol 2006;147:714–9.
33. Zhu WZ, Zheng M, Koch WJ, Lefkowitz RJ, Kobilka BK, Xiao RP. Dual modulation of cell survival and cell death by beta(2)-adrenergic signaling in adult mouse cardiac myocytes. Proc Natl Acad Sci USA 2001;98:1607–12.
34. Bisognano JD, Weinberger HD, Bohlmeyer TJ, et al. Myocardial-directed overexpression of the human beta(1)-adrenergic receptor in transgenic mice. J Mol Cell Cardiol 2000;32:817–30.
35. Dorn GW, 2nd, Tepe NM, Lorenz JN, Koch WJ, Liggett SB. Low- and high-level transgenic expression of beta2-adrenergic receptors differentially affect cardiac hypertrophy and function in Galphaq-overexpressing mice. Proc Natl Acad Sci USA 1999;96:6400–5.
36. Luttrell LM, Daaka Y, Lefkowitz RJ. Regulation of tyrosine kinase cascades by G-protein-coupled receptors. Curr Opin Cell Biol 1999;11:177–83.
37. Gutkind JS. Cell growth control by G protein-coupled receptors: from signal transduction to signal integration. Oncogene 1998;17:1331–42.
38. Lefkowitz RJ, Pierce KL, Luttrell LM. Dancing with different partners: protein kinase a phosphorylation of seven membrane-spanning receptors regulates their G protein-coupling specificity. Mol Pharmacol 2002;62:971–4.

39. Azzi M, Charest PG, Angers S, et al. Beta-arrestin-mediated activation of MAPK by inverse agonists reveals distinct active conformations for G protein-coupled receptors. Proc Natl Acad Sci USA 2003;100:11406–11.
40. Wisler JW, DeWire SM, Whalen EJ, et al. A unique mechanism of beta-blocker action: carvedilol stimulates beta-arrestin signaling. Proc Natl Acad Sci USA 2007;104:16657–62.
41. Brzostowski JA, Kimmel AR. Signaling at zero G: G-protein-independent functions for 7-TM receptors. Trends Biochem Sci 2001;26:291–7.
42. Baker JG, Hall IP, Hill SJ. Agonist and inverse agonist actions of beta-blockers at the human beta 2-adrenoceptor provide evidence for agonist-directed signaling. Mol Pharmacol 2003;64:1357–69.
43. Galandrin S, Bouvier M. Distinct signaling profiles of beta1 and beta2 adrenergic receptor ligands toward adenylyl cyclase and mitogen-activated protein kinase reveals the pluridimensionality of efficacy. Mol Pharmacol 2006;70:1575–84.
44. Drake MT, Violin JD, Whalen EJ, Wisler JW, Shenoy SK, Lefkowitz RJ. beta-arrestin-biased agonism at the beta2-adrenergic receptor. J Biol Chem 2008;283:5669–76.
45. Baker JG, Hall IP, Hill SJ. Agonist actions of "beta-blockers" provide evidence for two agonist activation sites or conformations of the human beta1-adrenoceptor. Mol Pharmacol 2003;63: 1312–21.
46. Baker JG. Site of action of beta-ligands at the human beta1-adrenoceptor. J Pharmacol Exp Ther 2005;313:1163–71.
47. Baker JG. Evidence for a secondary state of the human beta3-adrenoceptor. Mol Pharmacol 2005;68:1645–55.
48. Gerhardt CC, Gros J, Strosberg AD, Issad T. Stimulation of the extracellular signal-regulated kinase 1/2 pathway by human beta-3 adrenergic receptor: new pharmacological profile and mechanism of activation. Mol Pharmacol 1999;55:255–62.
49. Sato M, Horinouchi T, Hutchinson DS, Evans BA, Summers RJ. Ligand-directed signaling at the beta3-adrenoceptor produced by 3-(2-Ethylphenoxy)-1-[(1,S)-1,2,3,4-tetrahydronapth-1-ylamino]-2S-2-propan ol oxalate (SR59230A) relative to receptor agonists. Mol Pharmacol 2007;72:1359–68.
50. Sato M, Hutchinson DS, Bengtsson T, et al. Functional domains of the mouse beta3-adrenoceptor associated with differential G protein coupling. J Pharmacol Exp Ther 2005;315:1354–61.
51. Perez DM, Karnik SS. Multiple signaling states of G-protein-coupled receptors. Pharmacol Rev 2005;57:147–61.
52. Perez DM, DeYoung MB, Graham RM. Coupling of expressed alpha 1B- and alpha 1D-adrenergic receptor to multiple signaling pathways is both G protein and cell type specific. Mol Pharmacol 1993;44:784–95.
53. Hu ZW, Shi XY, Lin RZ, Hoffman BB. Alpha1 adrenergic receptors activate phosphatidylinositol 3-kinase in human vascular smooth muscle cells. Role in mitogenesis. J Biol Chem 1996;271: 8977–82.
54. Nishio E, Nakata H, Arimura S, Watanabe Y. alpha-1-Adrenergic receptor stimulation causes arachidonic acid release through pertussis toxin-sensitive GTP-binding protein and JNK activation in rabbit aortic smooth muscle cells. Biochem Biophys Res Commun 1996;219:277–82.
55. Otani H, Oshiro A, Yagi M, Inagaki C. Pertussis toxin-sensitive and -insensitive mechanisms of alpha1-adrenoceptor-mediated inotropic responses in rat heart. Eur J Pharmacol 2001;419: 249–52.
56. O-Uchi J, Sasaki H, Morimoto S, et al. Interaction of alpha1-adrenoceptor subtypes with different G proteins induces opposite effects on cardiac L-type Ca2+ channel. Circ Res 2008;102: 1378–88.
57. Limbird LE. Receptors linked to inhibition of adenylate cyclase: additional signaling mechanisms. FASEB J 1988;2:2686–95.
58. Wang Q, Lu R, Zhao J, Limbird LE. Arrestin serves as a molecular switch, linking endogenous alpha2-adrenergic receptor to SRC-dependent, but not SRC-independent, ERK activation. J Biol Chem 2006;281:25948–55.

59. Kribben A, Herget-Rosenthal S, Lange B, Erdbrugger W, Philipp T, Michel MC. Alpha2-adrenoceptors in opossum kidney cells couple to stimulation of mitogen-activated protein kinase independently of adenylyl cyclase inhibition. Naunyn Schmiedebergs Arch Pharmacol 1997;356:225–32.
60. Eason MG, Kurose H, Holt BD, Raymond JR, Liggett SB. Simultaneous coupling of alpha 2-adrenergic receptors to two G-proteins with opposing effects. Subtype-selective coupling of alpha 2C10, alpha 2C4, and alpha 2C2 adrenergic receptors to Gi and Gs. J Biol Chem 1992;267:15795–801.
61. Eason MG, Liggett SB. Identification of a Gs coupling domain in the amino terminus of the third intracellular loop of the alpha 2A-adrenergic receptor. Evidence for distinct structural determinants that confer Gs versus Gi coupling. J Biol Chem 1995;270:24753–60.
62. Wade SM, Lim WK, Lan KL, Chung DA, Nanamori M, Neubig RR. G(i) activator region of alpha(2A)-adrenergic receptors: distinct basic residues mediate G(i) versus G(s) activation. Mol Pharmacol 1999;56:1005–13.
63. Brink CB, Wade SM, Neubig RR. Agonist-directed trafficking of porcine alpha(2A)-adrenergic receptor signaling in Chinese hamster ovary cells: l-isoproterenol selectively activates G(s). J Pharmacol Exp Ther 2000;294:539–47.
64. Eason MG, Jacinto MT, Liggett SB. Contribution of ligand structure to activation of alpha 2-adrenergic receptor subtype coupling to Gs. Mol Pharmacol 1994;45:696–702.
65. Surprenant A, Horstman DA, Akbarali H, Limbird LE. A point mutation of the alpha 2-adrenoceptor that blocks coupling to potassium but not calcium currents. Science 1992;257: 977–80.
66. Tan CM, Wilson MH, MacMillan LB, Kobilka BK, Limbird LE. Heterozygous alpha 2A-adrenergic receptor mice unveil unique therapeutic benefits of partial agonists. Proc Natl Acad Sci USA 2002;99:12471–6.
67. Poole-Wilson PA, Swedberg K, Cleland JG, et al. Comparison of carvedilol and metoprolol on clinical outcomes in patients with chronic heart failure in the Carvedilol Or Metoprolol European Trial (COMET): randomised controlled trial. Lancet 2003;362:7–13.

Chapter 8
Signaling Diversity Mediated by Muscarinic Acetylcholine Receptor Subtypes and Evidence for Functional Selectivity

R.A. John Challiss and Rachel L. Thomas

Abstract Muscarinic acetylcholine (mACh) receptor subtypes (M_1–M_5) mediate many of the central and peripheral actions of acetylcholine. Although two mACh receptor subgroups (M_1/M_3/M_5 and M_2/M_4) have been defined based on primary sequence similarities and G protein/effector coupling preferences, considerable overlap in the coupling of subtypes to different signaling pathways is evident from the literature. Here we summarize the available experimental evidence for single mACh receptor subtypes coupling via more than one G protein subtype, or independently of G proteins, to exert their cellular effects. We also critically analyze the current evidence for different ligands being able to activate selectively subsets of these downstream signaling readouts. We discuss why there is currently little available evidence for functional selectivity at mACh receptors and speculate whether this situation will change as more subtype-selective ligands, and ligands that act on mACh receptors at sites other than the acetylcholine binding site are developed and explored.

Keywords Muscarinic acetylcholine receptor, Acetylcholine, G protein-coupled receptor, G protein, Adenylyl cyclase, Phospholipase C, Receptor desensitization, Receptor internalization, Functional selectivity, Agonist-directed trafficking of receptor stimulus, Biased agonism

8.1 Introduction

Muscarinic acetylcholine (mACh) receptors have been one of the most widely studied G protein-coupled receptor (GPCR) families since their subclassification from nicotinic acetylcholine (nACh) receptors almost 100 years ago *(1)*. The existence

---

R.A.J. Challiss
Department of Cell Physiology and Pharmacology, University of Leicester, Leicester, LE1 9HN, UK
e-mail: jc36@leicester.ac.uk

K. Neve (ed.), *Functional Selectivity of G Protein-Coupled Receptor Ligands*,
DOI: 10.1007/978-1-60327-335-0_8,

of mACh receptor subtypes was much speculated on for at least two decades *(2, 3)* before definitive pharmacological experimental evidence was eventually published *(4)*. Despite the potential therapeutic uses of mACh receptor agonists and antagonists, the discovery of subtype-selective compounds has proven to be a slow and difficult process. The mACh receptors were initially pharmacologically divided into M_1 and M_2 subtypes, with an M_3 subtype subsequently being proposed and widely accepted; however, it was not until the advent of molecular cloning that the true diversity of the mACh receptor family was defined *(5)*. Today, it is known that mammalian species possess five mACh receptor subtypes, designated M_1–M_5 (or CHRM1–5), and these distinct gene products can be further subdivided into the M_1/M_3/M_5 and M_2/M_4 subclasses based on primary amino acid sequence similarities and the preferential coupling of the M_1/M_3/M_5 subclass via $G_{q/11}$ proteins to the stimulation of phospholipase C (PLC) activity, and the M_2/M_4 subclass via $G_{i/o}$ proteins to the inhibition of adenylyl cyclase (AC) activity *(6, 7)*. Although this $G_{q/11}$-phospholipase C-/ $G_{i/o}$-adenylyl cyclase-coupling discriminator provides a "shorthand" means of subdividing mACh receptors according to signaling output, as will be discussed later, mACh receptor subtypes are able to couple to a diverse array of intracellular signaling events that may or may not always be $G_{q/11}$- and/or $G_{i/o}$-dependent *(8, 9)*.

The stimulation of mACh receptors can account for many of the physiological actions of acetylcholine in the periphery, including smooth muscle contraction, glandular secretion, and the modulation of heart rate. Similarly, within the CNS, mACh receptors have roles in the central regulation of the autonomic nervous system (e.g., with respect to cardiovascular responses) and key modulatory roles within the motor control circuitry and a variety of learning and memory paradigms *(7,10)*. A variety of approaches have been applied to unraveling the localization and physiological/ pathophysiological roles of the different mACh receptor subtypes. One of the most powerful and revealing has been the use of knockout mice where one or more mACh receptor gene has been disrupted *(10)*. This approach has been especially important as, unlike for some other GPCR sub-families, categoric pharmacological discrimination between subtypes has only recently become possible using the highly subtype-selective muscarinic toxins, such as MT-3 (M_4-specific) and MT-7 (M_1-specific) *(11, 12)*, and "allosteric" agonists, such as AC-42, which appear to activate the receptor by binding to a site distinct from that occupied by acetylcholine *(13, 14)*. The knockout approach has also helped to clarify the respective roles of specific mACh receptor subtypes as these are often coexpressed within single cells; thus, categoric attribution of a particular effect to a single receptor subtype in native tissue was often a daunting task before the availability of transgenic technologies. Unsurprisingly therefore, over the past 20 years, heterologous expression of single mACh receptor subtypes, in a variety of model cell systems, has been a commonly used approach to study mACh receptor pharmacology, signal transduction, and regulation.

In this chapter, we will review the diversity of signaling mechanisms by which mACh receptors can determine cellular activities, consider the (presently limited) pharmacological evidence for functional selectivity, and speculate on how this potentially important aspect of mACh receptor pharmacology may develop in the immediate future.

8.2 mACh Receptor Signaling Pathways

Although it has been known for some time that mACh receptors can couple to both the inhibition of adenylyl cyclase and stimulation of phosphoinositide turnover, early studies presented experimental evidence both for and against these signaling outputs being regulated by one or more than one mACh receptor *(15)*. The receptor-G protein-effector assignments we recognize today were quickly established following the cloning of the mACh receptor subtypes in the late 1980s. Thus, heterologously expressed $M_1/M_3/M_5$ mACh receptor subtypes were shown to couple strongly to phospholipase C and the generation of inositol 1,4,5-trisphosphate and diacylglycerol *(16, 17)*; similarly, M_2/M_4 mACh receptor subtypes were shown to inhibit adenylyl cyclase activity via a pertussis toxin-sensitive coupling mechanism *(16)*.

[*Note*: in early papers from the Peralta/Ashkenazi/Capon groups mACh receptor subtypes designated "M_3" and "M_4" are now universally referred to according to the assignment of Bonner et al. *(18, 19)* as M_4 and M_3, respectively].

However, concurrent with these seminal studies, a number of groups were reporting alternative coupling mechanisms. For example, $M_1/M_3/M_5$ mACh receptors were also reported to positively or negatively modulate adenylyl cyclase *(20, 21)* and to regulate phospholipases A_2 and D via mechanisms apparently independent of phospholipase C *(22, 23)*. Similarly, it was reported that M_2/M_4 mACh receptors can stimulate phospholipase C activity via mechanisms involving either pertussis toxin-sensitive or insensitive G proteins *(24, 25)*. Although some of these early reports were subsequently shown to be brought about by indirect mechanisms (e.g., activation of adenylate cyclase might be brought about indirectly through the $M_1/M_3/M_5$ mACh receptor-$G_{q/11}$ protein-phospholipase C proximal signaling pathway resulting in increases in cytoplasmic Ca^{2+} and the modulation of Ca^{2+}-sensitive adenylyl cyclase isoenzymes and/or Ca^{2+}-sensitive phosphodiesterases *(26)*), others represent real examples of single mACh receptor subtypes initiating bifurcating signaling pathways.

8.2.1 mACh Receptor Coupling to Adenylyl Cyclase

Adenylyl cyclase isoenzymes (AC1–AC9) are ubiquitously and differentially expressed in mammalian cells and, like a number of other effector proteins, are controlled by multiple regulators, including $G\alpha_s$ and $G\alpha_{i/o}$ proteins, $G\beta\gamma$ subunits, Ca^{2+}, and protein kinases *(27, 28)*. Cyclic AMP concentration is also regulated by a plethora of phosphodiesterase (PDE) activities that too are subject to extrinsic regulation *(29)*. It is possible for M_1-M_5 mACh receptors either to increase or decrease cyclic AMP levels within cells via direct or indirect mechanisms. Crucial factors determining the observed effect of mACh receptor activation on cellular cyclic AMP concentration will be the subtype(s) of mACh receptor expressed and the complement of adenylyl cyclase/PDE isoforms present in a particular cell background.

Given the potential complexity of studying adenylyl cyclase regulation in native cells (which might express more that one mACh receptor subtype and multiple adenylyl cyclase/PDE isoenzymes), it is not surprising that recombinant systems have often been engineered for such studies to allow specific mACh receptor subtype/G protein/adenylyl cyclase isoform combinations to be assessed.

8.2.1.1 M_2/M_4 mACh Receptor Coupling to Adenylyl Cyclase

It is generally accepted that M_2 and M_4 mACh receptors preferentially couple to $G_{i/o}$ proteins to inhibit adenylyl cyclase activity and that the extent of this effect will depend, at least to some extent, on the complement of adenylyl cyclase (and PDE) isoenzymes expressed in the cell background. The Nathanson lab has reported a series of studies that aimed to pinpoint the G protein-coupling preference of M_2 and M_4 mACh receptors more precisely *(30–32)*. Thus, in a human chorioncarcinoma cell-line, JEG-3, the endogenous $G\alpha_{i1}$ and $G\alpha_{i3}$ proteins do not allow the M_4 mACh receptor subtype to couple to adenylyl cyclase, whereas recombinant expression of $G\alpha_{i2}$ *(30)* or $G\alpha_o$ *(31)* facilitated this inhibitory coupling. In the same cell background, the M_2 mACh receptor subtype was able to utilize $G\alpha_{i1-3}$ or $G\alpha_o$ proteins to couple to adenylyl cyclase *(32)*. Any M_2/M_4 mACh receptor-$G_{i/o}$ selectivity is, however, likely to be cell-type-dependent as in GH_4C_1 cells the M_4 mACh receptor couples to $G\alpha_{i1}/G\alpha_{i3}$ proteins to inhibit agonist-stimulated adenylyl cyclase activity *(33)*.

A number of studies have shown that at high occupancy (and high expression levels) M_2 and M_4 mACh receptors can increase cyclic AMP accumulation in a variety of cell backgrounds *(17,34–36)*. Although this might be brought about by direct or indirect actions of $G\beta\gamma$ subunits, a common finding is that M_2 and M_4 receptors can directly couple to G_s proteins to bring about this effect. M_2/M_4 mACh receptor-G_s coupling is more easily demonstrated following pertussis toxin inactivation of $G_{i/o}$ proteins, with the agonist concentration-dependencies for this coupling generally being 10- to 50-fold right-shifted relative to comparable IC_{50} values for M_2/M_4 mACh receptor-$G_{i/o}$ coupling *(36, 37)*. M_2 mACh receptor-G_s coupling and its functional consequences have also been shown by siRNA knockdown of the $G\alpha_s$ subunit *(38)*. A thorough study using stable M_2 and M_4 mACh receptor expressing Chinese hamster ovary (CHO) cell-lines to compare the abilities of methacholine, oxotremorine-M, oxotremorine, arecoline, pilocarpine, and bethanechol to inhibit or stimulate adenylyl cyclase via $G_{i/o}$ or G_s proteins, respectively, provided no clear evidence of functionally selective effects *(37)*.

8.2.1.2 $M_1/M_3/M_5$ mACh Receptor Coupling to Adenylyl Cyclase

Although adenylyl cyclase inhibitory effects of mACh receptor subtypes that stimulate phosphoinositide turnover have been observed *(21)*, more often an increase in cyclic AMP accumulation has been reported, especially where adenylyl cyclase activity is raised using forskolin and/or PDE activities are inhibited *(20, 39)*. As already mentioned, in many cases this can be shown to be PLC-dependent, through a Ca^{2+}- (and/or protein kinase C-) dependent modulation of adenylyl cyclase *(40, 41)* or PDE activity *(42)*. However, direct $M_1/M_3/M_5$ mACh receptor coupling to adenylyl cyclase via G_s proteins is likely to occur at least in some cell-types and may be of physiological significance *(8)*. Thus, a number of groups have reported Ca^{2+}/PKC and pertussis toxin-independent activation of adenylyl cyclase that is directly attributable to $M_1/M_3/M_5$ mACh receptor-G_s coupling *(43, 44)*. Where $M_1/M_3/M_5$ mACh receptor subfamily stimulation of phosphoinositide turnover and cyclic AMP accumulation has been observed, differences in coupling efficiencies have also been reported, with the concentration-dependency for coupling to the cyclic AMP response always lying to the right of that for phosphoinositide turnover, with the former response often overlying the ligand-receptor occupancy curve *(17, 39, 45)*. Interestingly, Schmidt and colleagues have provided evidence for a pathway linking the M_3 mACh receptor to phospholipase Cε (see later) activation that involves a G_s-dependent activation of adenylyl cyclase *(46)*.

A consistent finding, using a variety of methods to assess the ability of M_1 and M_3 mACh receptors to facilitate GTP-for-GDP exchange on Gα subunits, has been that significant exchange on $G_{i/o}$ (as well as $G_{q/11}$) proteins can be observed. Thus, using human embryonic kidney (HEK) cell-lines stably expressing recombinant human M_1 or M_3 mACh receptors, it was shown that pertussis toxin pretreatment differentially affects subsequent total Gα-[^{35}S]-GTPγS binding, with much more of the M_3 (but not the M_1) mACh receptor-stimulated [^{35}S]-GTPγS binding being lost in membranes prepared from pertussis toxin-treated cells *(47)*. The ability of M_3 (and to a lesser extent M_1) mACh receptors to couple to $G\alpha_{i1}$ and $G\alpha_{i3}$ proteins was confirmed by photo-labeling with [α-^{32}P]-GTP-azidoanilide and recovering specific Gα subunits by immunoprecipitation *(47)*. Comparable data, obtained using CHO cell-lines stably expressing recombinant human M_1 or M_3 mACh receptors, have been reported similar subtype differences for relative $G_{q/11}$- vs. $G_{i/o}$-[^{35}S]-GTPγS binding *(48)*. Intriguingly, Akam and colleagues also reported that agonist-stimulated $G_{q/11}$- vs. $G_{i/o}$-[^{35}S]-GTPγS binding in CHO-m3 cell membranes differed between the full agonist methacholine and partial agonist pilocarpine, with the latter agent appearing to be $G\alpha_i$- over $G\alpha_{q/11}$-selective *(48)*. Although these data indicate a potential for functionally selective effects, they require confirmation in other cell backgrounds and for a wider mACh receptor agonist repertoire to be explored.

8.2.2 mACh Receptor Coupling to Phospholipase C

The ability of acetylcholine to bring about the hydrolysis of membrane phospholipids has been recognized for more than 50 years *(49)*. A number of proteins, collectively called PLCs, possess the ability to hydrolyze membrane phospholipids to release diacylglycerol (DAG) and the polar phospho-head group. In mammalian systems, these activities are often selective for the hydrolysis of phosphatidylinositol 4,5-bisphosphate (PIP_2), and hence are referred to as phosphoinositide-specific PLCs *(50)*. The phosphoinositide-specific family is encoded by 13 genes and in humans consists of PLCβ1–4, PLCγ1,2, PLCδ1,3,4, PLCε, PLCζ, and PLCη1–2 *(51)*. The PLCβ family is most commonly associated with regulation by GPCR/G protein-dependent signaling; however, mACh receptors have also been reported to mediate their actions via other PLC isoenzymes, including PLCγ *(52, 53)*, although it is unclear what the inositol phospholipid substrate(s) is in this case *(54)*.

8.2.2.1 $M_1/M_3/M_5$ mACh Receptor Coupling to Phospholipase C

The $M_1/M_3/M_5$ mACh receptor subfamily preferentially couples to PLCβ isoenzymes to generate the second messengers inositol 1,4,5-trisphosphate (IP_3) and DAG. PLCβ activation can result in significant changes in PIP_2 concentration *(55)* and this phosphoinositide too can fulfill signaling roles *(56, 57)*, emphasizing the importance of PLC in regulating the cellular concentrations of three important mediator molecules. GPCRs can couple to PLCβ isoenzymes via $G\alpha_{q/11}$ and/or Gβγ subunits *(50, 58)*. $M_1/M_3/M_5$ mACh receptor-$G\alpha_{q/11}$-PLCβ coupling has been demonstrated *(59)* using a number of methods including the direct monitoring of receptor facilitation of GTP/GDP exchange on $G_{q/11}$ proteins using [^{35}S]-GTPγS and Gα protein-specific immunoprecipitation/immunocapture methods *(47, 48, 60)*. For the M_1 mACh receptor, a precise Gαβγ heterotrimer requirement (α_q/α_{11}, β_1/β_4, γ_4) for coupling to PLCβ has been reported *(61)*, although such Gαβγ stringency has not been observed by others. Other studies have highlighted the importance of the compartmentalization of the receptor as a determining factor in coupling to particular subsets of signaling pathway components *(62)*. Thus, in sympathetic neurons M_1 mACh receptors can activate PLCβ to generate IP_3 and DAG, but the receptors are not suitably compartmentalized to allow the IP_3 generated to mobilize Ca^{2+} from endoplasmic reticular stores *(63)*.

8.2.2.2 M_2/M_4 mACh Receptor Coupling to Phospholipase C

Initial characterizations of M_2 and M_4 mACh receptors expressed recombinantly reported that while these receptors couple preferentially to adenylyl cyclase inhibition via $G_{i/o}$ proteins, they are, nevertheless, also able to stimulate PLC activity, at least when expressed at high levels *(24, 64)*. This coupling may be mediated via the release of Gβγ subunits from $G_{i/o}$ proteins *(65)* in cell backgrounds expressing Gβγ-sensitive PLCβ isoenzymes; however, pertussis-toxin insensitive M_2/M_4-PLC coupling has also been reported *(66)*. Although a number of previous studies have failed to observe M_2/M_4 mACh receptor facilitation of GTP/GDP exchange on $G_{q/11}$ proteins using [^{35}S]-GTPγS and Gα protein-specific immunoprecipitation methods *(47, 48)*, a recent paper has reported M_2-$G_{q/11}$ coupling using an RNA interference approach in a CHO cell background *(38)*.

8.2.3 mACh Receptor Coupling to Other Phospholipases

In addition to phospholipase C activity, phospholipase A_2 (PLA_2, which cleaves the acyl-ester bond at the *sn*-2 position of the phospholipid glycerol backbone to release free fatty acid (arachidonic acid) and lyso-phospholipid) and phospholipase D (PLD, which cleaves the phospholipid polar head group to generate phosphatidic acid plus choline/ethanolamine/inositol) activities have also been shown to be regulated by mACh receptors *(22, 23, 67)*. PLA_2 activation appears to be a commonly observed consequence of mACh receptor activation in an array of tissues, including a variety of smooth muscle types *(6,8)*. Despite early debates on direct receptor-G protein-PLA_2 coupling, it is now generally accepted that PLA_2 activation is a more distal event, for example, being mediated through $M_1/M_3/M_5$ mACh receptor signaling via a $G_{q/11}$/PLCβ/DAG/PKC-dependent pathway *(68)*. mACh receptor activation of $cPLA_2$ activity can play a role in a variety of cell-types, including the neuronal processing of amyloid precursor protein *(69)*.

Cellular PLD activity is attributable to the expression of one or both isoenzymes, PLD1 and PLD2 *(70, 71)*. Contrary to some early reports of direct modulation of cellular PLD activity by heterotrimeric G proteins, it is now generally accepted that regulation is mediated primarily by PKCs and small GTPases (e.g., ARF and Rho family proteins). A key question regarding mACh receptor regulation of PLD activity has concerned G protein subtype involvement. Thus, in a HEK cell background, M_3 mACh receptor stimulation of PLD occurs via a $G_{12/13}$-, not $G_{q/11}$-dependent pathway *(72)*. These authors suggested that PLD activation is dependent on Rho-type small GTPase activation via $G\alpha_{12/13}$ protein scaffolding of guanine nucleotide exchange factors (GEFs), such as $p115^{RhoGEF}$. In contrast, others have reported that RhoA activation can also be brought about in a $G_{q/11}$-dependent manner through an alternative family of RhoGEFs (e.g., $p63^{RhoGEF}$; *(73)*). Indeed, $G_{q/11}$-dependent RhoA regulation has been linked to mACh receptor signaling in other cell backgrounds *(74, 75)*.

These data suggest that $M_1/M_3/M_5$ mACh receptors might be able to activate $G_{q/11}$ and/or $G_{12/13}$ proteins to regulate a variety of signaling outcomes, including divergent or convergent regulation of PLD activity and cytoskeletal rearrangements involving actin, myosin, and other cytoskeletal proteins *(71, 76)*. It is also worth noting that GPCRs, including mACh receptors, may differentially engage G_{12} and G_{13} proteins to transduce distinct cell signaling outcomes *(77)*.

8.2.4 mACh Receptor Coupling to Other Enzyme Effectors: The Mitogen-Activated Protein Kinases

In addition to the regulation of adenylyl cyclases and phospholipases, mACh receptor subtypes have also been demonstrated to regulate a variety of other cellular enzymatic activities, including phosphoinositide 3-kinase (PI3K) and its downstream effectors, e.g., protein kinase B (PKB/Akt), glycogen synthase kinase-3, nitric oxide synthase, non-receptor tyrosine kinases (e.g., Src), and mitogen-activated protein kinases (MAPKs). The vast majority of these can be considered distal readouts, being primarily dependent on initiating events more proximal to receptor activation. For example, in 1321N1 astrocytoma cells PKB/Akt activation is $G_{q/11}$/PLC-dependent and requires Ca^{2+}, the "transactivation" of a receptor tyrosine kinase (ErbB3), and PI3K *(78)*. It is likely with such distal readouts that cell background will play a greater role, with intermediary pathways between mACh receptor activation and enzyme modulation varying from cell-type to cell-type *(79)*, consequently only enzyme effectors relevant to subsequent discussions will be dealt with here.

The MAPK family of protein kinases includes extracellular signal-regulated kinases (ERK1/2, ERK5), c-Jun N-terminal kinases (JNK1–3), and p38 MAPKs (p38α-δ). The activity of each MAPK is controlled by dual threonine/tyrosine phosphorylation (within a T-X-Y motif) by parallel sets of MAPK kinase, and the activities of these upstream kinases are in turn regulated by MAPK kinase kinase activities *(80–82)*. Following the discovery of each MAPK subfamily, evidence quickly accrued for regulation by GPCRs, with mACh receptors (particularly M_1 and M_2 mACh receptors) often being used as prototypic $G_{q/11}$ and $G_{i/o}$-coupled GPCRs in these seminal studies. For example, it quickly became evident that the pathway linking a specific GPCR to ERK1/2 activation could vary greatly depending on the cellular system used in the investigation. Thus, for $M_1/M_3/M_5$ mACh receptors the pathway might show variable pertussis toxin sensitivity, PKC-, Ras-, and Ca^{2+}-dependency *(80, 81, 83–85)*. In contrast, studies on M_2/M_4 mACh receptor linkage to ERK1/2 suggested a lesser degree of cell background variability, with a Gβγ-, PI3K-, and Ras-dependent pathway being most often reported *(80, 86–88)*. The mACh receptors can also engage regulators of small GTPases (guanine nucleotide exchange factors (GEFs) or GTPase-activating proteins (GAPs)) to mediate Ras-Raf1- or Rap1-B-Raf-dependent effects on ERK1/2 *(89–91)*, and these mechanisms appear to be important in neuronal cell backgrounds. A major influence of cell background on GPCR coupling to JNK, p38 MAPK, and ERK5 signaling is also apparent *(85, 92–95)*.

An even greater diversity of mechanisms linking GPCRs to MAPK signaling has been reported in the last decade. One subset of mechanisms involves GPCR-stimulated tyrosine kinase activation. The ability of tyrosine kinase inhibitors to partially or fully inhibit GPCR-MAPK signaling has long been appreciated *(96)*, implicating non-receptor tyrosine kinases (non-RTKs) and/or receptor tyrosine kinases (RTKs) as intermediary proteins. The mACh receptors have been shown to engage non-RTKs to couple to ERK/JNK/p38 via a number of mechanisms, most commonly involving Gβγ subunits *(97, 98)*. Alternately, mACh receptors have been shown to stimulate the transactivation of ERK1/2 signaling using RTKs (e.g., the epidermal growth factor receptor). RTK recruitment can be brought about via cytoplasmic mechanisms involving the use of the intracellular kinase domains of the RTK as a scaffold, or through the shedding of a tethered RTK ligand from the surface of the cell *(99–101)*. It is possible that $M_1/M_3/M_5$ mACh receptors utilize both transactivation mechanisms, whereas M_2/M_4 mACh receptors utilize only the former *(102, 103)*. Irrespective of the mechanism(s) utilized by mACh receptors, the somewhat circuitous transactivation route of ERK regulation allows much scope for modulation by other signaling inputs and is increasingly being recognized as a major pathway linking GPCRs to ERK (and JNK/p38) in native tissues. For example, the regulation of fluid secretion by intestinal epithelia is regulated by mACh receptors that appear to utilize at least two distinct EGF receptor-dependent ERK signaling pathways *(104)*.

An alternative mechanism for coupling GPCRs to ERK/JNK/p38 MAPK activities involves the machinery of receptor desensitization/down-regulation (see later). Thus, it was shown for some GPCRs that ERK1/2 activation was inhibited if receptors were not allowed to internalize following activation, leading to the elucidation of a distinct mechanism of GPCR-ERK1/2 coupling involving the assembling of a MAPK signaling unit using GPCR-associated β-arrestin as a scaffold *(82, 105)*. β-arrestin scaffolding of JNK3 and p38 MAPK has been reported *(106, 107)*.

Whether mACh receptors utilize a β-arrestin-dependent pathway to link to MAPKs remains controversial. Thus, in HEK cells transiently expressing mouse M_1 mACh receptors, cotransfection of dominant-negative β-arrestin (V53D) or dynamin (K44A) constructs markedly inhibited ERK1/2 phosphorylation *(108)*, while activation of M_1 mACh receptors in COS cells caused β-arrestin-dependent ERK phosphorylation that was enhanced by the presence of the protein filamin-A *(109)*. In contrast, in CHO cells stably expressing human M_3 mACh receptors inhibiting receptor internalization had no effect on ERK1/2 activation *(110)*, which appears to proceed via β-arrestin-independent mechanisms *(85, 111)*.

At present it is not possible to draw up any general rules governing $M_1/M_3/M_5$ vs. M_2/M_4 mACh receptor-coupling to MAPK pathways. The mechanisms linking receptor activation to these signaling cascades appear to be highly dependent on cell background, with the same receptor (or coexpressed receptors) utilizing radically different pathways in different cells/tissues. Interestingly, it has recently been suggested that within the same cell different agonists might bring about ERK1/2 activation via distinct intermediary pathways *(112)*. Thus, in a human salivary cell-line (HSY), it was reported that while both carbachol and pilocarpine can activate ERK1/2, pilocarpine, but not carbachol utilizes an RTK-dependent route. In contrast, the carbachol, but not pilocarpine-stimulated ERK response was attenuated by PKC down-regulation *(112)*. HSY cells express both M_1 and M_3 mACh receptors, and it is conceivable that the partial agonist pilocarpine exhibits a functional selectivity for M_3 mACh receptor activation in these cells; nevertheless, the possibility that pilocarpine is functionally selective for a specific route of ERK1/2 activation deserves further investigation. The general issue of coexpression of GPCRs, or indeed the potential of GPCR subtypes to activate more than one subtype of G protein (see above for discussion with respect to mACh receptors), is also highly relevant to studies of GPCR-MAPK coupling as dual $G_{q/11}/G_{i/o}$ activation can markedly modify both the amplitude and duration of signaling *(113)*. Indeed, striking effects of M_2/M_3 mACh receptor coexpression, or manipulating the ability of M_3 mACh receptors to couple to both $G_{q/11}$ and $G_{i/o}$ proteins on the extent and concentration-dependency of ERK1/2 activation have been reported *(114)*.

8.2.5 mACh Receptor Coupling to Ion Channels

The regulation of ion channel function provides some of the most studied and physiologically important examples of mACh receptor signaling. Key examples include regulation of neuronal M-type K^+ current (KCNQ2/3/K_v7.2/7.3) by M_1 mACh receptors *(115, 116)*, and modulation of cardiac chronotropy by M_2 mACh receptors via G protein-activated, inward rectifier K^+(GIRK1–4/K_{ir}3.1–3.4) and hyperpolarization-activated, cyclic nucleotide-gated (HCN1–4) channels *(117, 118)*. It is beyond the scope of the present chapter to review systematically the considerable literature already available on Ca^{2+}/K^+/Na^+/Cl^- channel modulation by mACh receptor subtypes (for reviews see *(6, 9, 119)*).

8.3 mACh Receptor Phosphorylation: Desensitization and Alternate Signaling

The mACh receptors, like the vast majority of GPCRs, respond to continuous or repeated cycles of agonist activation by reducing their responsiveness. This occurs through a transient or longer-term adjustment in the level of cell surface expression of the receptor itself and sometimes other downstream signaling components. GPCRs are "desensitized" by a complex, multistage process involving receptor phosphorylation, the recruitment of arrestin proteins, receptor internalization, and an endosomal processing pathway leading to either reinsertion of the receptor into the plasma membrane (resensitization), or degradation of the receptor (down-regulation) *(120, 121)*. Similar to the emerging picture for MAPK regulation by mACh receptors, the precise route(s) leading to receptor desensitization also appears to be highly cell background-dependent, particularly with respect to the kinase(s) that initially phosphorylate the receptor resulting in receptor internalization, or equipping the receptor with alternate signaling potential. A number of recent reviews have directly addressed this topic *(122–124)*.

GPCRs are phosphorylated by a variety of protein kinases *(123)*; however, the greatest attention to date has focused on second messenger-regulated kinases (e.g., PKC), which can mediate heterologous desensitization (which is not specific to ligand-bound/activated GPCRs), and GPCR kinases (GRKs), which are specifically recruited to ligand-bound/activated receptors and therefore mediate homologous desensitization *(120, 125)*. There are seven GRKs (GRK1–7), of which GRKs 2 and 3, and GRKs 5 and 6 are considered to be ubiquitously expressed; similarly, of the arrestins (arrestin1–4), arrestin2 and 3 (β-arrestin1 and 2, respectively) are widely expressed *(120)*. Although receptor phosphorylation by second messenger-regulated kinases is considered to decrease the coupling efficiency between receptor and cognate G protein(s), GRK-mediated phosphorylation is considered necessary to recruit arrestin2/3 to the receptor and prevent further productive receptor-G protein coupling through steric hindrance *(120, 125)*.

A concept that has emerged in recent years is that receptor "desensitization" may serve alternate or additional purposes, allowing the receptor not simply to switch off, but to continue to signal, often via distinct sets of intermediary proteins *(126, 127)*. Such signaling most often occurs via the receptor-arrestin complex and its ability to scaffold the assembly of novel signaling components (e.g., proteins involved in ERK/JNK signaling pathways). A recent study has also reported that arrestins can form complexes with diacylglycerol kinase (DAGK) isoforms and this allows activated M_1 mACh receptors, expressed in COS7 and HEK cell backgrounds, to recruit arrestin-DAGK complexes and increase the rate of metabolism of diacylglycerol to phosphatidic acid *(128)*. These data echo previous studies that showed that arrestins also form complexes with phosphodiesterases allowing activated GPCRs selectively to modulate local cyclic AMP/cyclic GMP metabolism *(129)*. Thus, phosphorylation of activated GPCRs, particularly though not exclusively involving GRK-mediated phosphorylation, can be considered to fulfill a dual purpose: tagging receptors for arrestin capping and internalization, while allowing them to continue to signal while undergoing this process. This relatively recent realization has provided a mechanism by which GPCRs that undergo quite rapid phosphorylation/internalization can continue to influence cellular activity (e.g., at a transcriptional/translational level) independently of G proteins *(130)*. In addition, it is clearly possible for ligands differentially to affect immediate (G protein-dependent) and longer-term (G protein-independent, arrestin-dependent) signaling pathways, opening new possibilities for functional selectivity *(131, 132)*.

The mechanisms mediating mACh receptor desensitization have been extensively studied since early investigations identified the importance of phosphorylation in this process *(133)*. Initial studies focused on the effects of chronic agonist exposure on PKC-mediated heterologous desensitization *(134, 135)* and m1-m5 mACh receptor mRNA levels *(136)*. However, receptor phosphorylation and functional studies demonstrated that M_3 mACh receptors could undergo rapid desensitization *(137–139)*, and concurrent studies were also finding acute regulation of M_2 mACh receptors via GRK- and arrestin-dependent mechanisms *(140, 141)*. A thorough understanding of the protein kinases responsible for initial receptor phosphorylation has not yet been achieved as this appears to vary considerably between cell systems. An emerging consensus view is that the recombinant cell systems often used to study GPCR phosphorylation/regulation may not provide an accurate picture of the key kinases involved in receptor regulation in a particular cell-background in vivo *(142, 143)*. Nevertheless, in native cell systems where mACh receptor subtypes are endogenously expressed, a clearer picture is beginning to emerge regarding both GRK and arrestin involvement in mACh receptor regulation. For example, in rat hippocampal neurons M_1 mACh receptors are primarily regulated by GRK2-dependent mechanisms involving both phosphorylation-dependent and independent (limiting receptor access to $G_{q/11}$ proteins via GRK2 regulator of G protein signaling (RGS)-like (RH) domain interactions) uncoupling of receptor from $G_{q/11}$ *(144, 145)*. Interestingly, this regulation appears to be modulated by the level of network activity being experienced by the neuron *(146)*. In contrast, GRK6 has been shown to be preeminent in phosphorylating endogenously expressed M_3 mACh receptors in human neuroblastoma cells *(147)*, unlike the picture that has emerged from studies recombinantly expressing this receptor subtype *(148, 149)*. M_2 mACh receptors (but not M_3 mACh receptors) have been shown to be regulated by GRK5 in mouse airways smooth muscle *(150, 151)*; however, GRK5 may not be important in M_2 mACh receptor desensitization in all cell-backgrounds *(151)*, where other GRKs implicated by work in recombinant systems may play a role *(152, 153)*. Although the studies highlighted here represent real progress in understanding the physiologically-relevant initiation of mACh receptor subtype-specific desensitization, we are still a long way from being able to assess whether different agonists are able to generate ligand-specific patterns of phosphorylation and thus exhibit a functional selectivity toward particular regulatory outcomes (resensitization/down-regulation/alternate signaling/, etc.), as has been demonstrated for some other GPCRs *(154)*. The fact that mACh receptors (especially the M_1/M_3/M_5 subtypes) possess many (in some cases > 50) potential phosphorylation sites within their intracellular loop and *C*-terminal domains, and preliminary phospho-peptide mapping data suggest complex increases *and* decreases in M_3 mACh receptor phosphorylation on agonist stimulation *(155)*, make this a distant prospect.

Irrespective of the precise mechanisms bringing about receptor phosphorylation, the agonist-dependent internalization of mACh receptors has also been widely studied. The involvement of arrestins in mACh receptor internalization has not been fully resolved. A number of studies have demonstrated that dominant-negative arrestin constructs inhibit, while arrestin over-expression can increase, mACh receptor internalization *(108, 145, 152, 156, 157)*. However, a number of authors have found subtype-specific differences and arrestin-independent internalization has been reported, particularly, though not exclusively, for the M_2 mACh receptor subtype *(158–160)*. In addition, mACh receptor subtypes may internalize via distinct clathrin-dependent (M_1/M_3/M_4) and clathrin-independent (M_2) mechanisms *(159, 161, 162)*, although M_2 mACh receptors appear eventually to reach an endosomal compartment common to M_1, M_3, and M_4 mACh receptors *(163)*. Localization of M_2 (and possibly other subtype) mACh receptors to caveolae may also provide an alternate (arrestin-independent) route for internalization *(164)*. Current work is focusing on defining more precisely the cell biology of decision-making with respect to endocytosed mACh receptors, with recent studies defining the machinery of receptor trafficking *(165, 166)* and the receptor subtype-specific domains that define its endocytic fate *(167–169)*.

8.4 Critical Appraisal of the Evidence for Functional Selectivity at mACh Receptors

Experimental evidence for functional selectivity at GPCRs (also known as "agonist-directed trafficking of receptor stimulus' and 'biased agonism") has steadily accrued. Thus, different ranking orders have emerged for sets of ligands mediating two or more independent signaling readouts initiated by the same GPCR subtype (e.g., *(170–172)*; reviewed by *(154)*). Although the evidence for mACh receptor subtypes activating diverse signaling cascades is abundant, specific experimental evidence supporting true functional selectivity is presently limited. Here we critically examine the experimental evidence that presently exists in the literature, consider why more examples of this phenomenon involving mACh receptors have not been reported, and finally speculate on whether functionally-selective ligands are a realistic prospect and can provide new therapeutic opportunities for targeting mACh receptors in disease.

8.4.1 mACh Receptor Conformational States

As discussed earlier, the mACh receptors are able to interact with distinct subsets of G proteins to activate a range of cellular effectors. These and other data suggest that, like other GPCRs, mACh receptors are able to transition spontaneously between different conformational states, and this offers the opportunity for different ligands to stabilize one or a subset of the repertoire of conformations assumed by the receptor *(173, 174)*. Distinct receptor conformations may account for functional selectivity with respect to signaling cascades, as well as a number of other potential read-outs, including the mechanisms by which receptors become desensitized, internalized, and down-regulated, and interact with other cellular proteins in a ligand binding-dependent manner *(175, 176)*. mACh receptors are subject to regulation by different types of ligand, including classical orthosteric agonists and antagonists (inverse agonists) that bind to the acetylcholine-binding site *(177, 178)*, as well as allosteric ligands that bind to sites topographically distinct from the orthosteric-binding pocket *(179, 180)*. This latter group not only offers the potential to modulate the affinity and efficacy of orthosteric ligands, for example, as has recently been reported for VU10010 at the M_4 mACh receptor *(181)*, but also to cause subtype-selective activation of mACh receptors per se *(13, 14)*. These are particularly important developments as the synthesis of orthosteric mACh receptor agonists and antagonists that possess significant degrees of subtype selectivity has proved to be highly challenging, most likely due to the high degree of amino acid residue conservation (and presumably their 3D arrangement) within the orthosteric binding pocket of M_1-M_5 mACh receptors *(182, 183)*.

Investigation of conformational changes induced by different classes of orthosteric ligands at the molecular level using an in situ disulphide cross-linking strategy has demonstrated that, for the M_3 mACh receptor, orthosteric agonists (carbachol and oxotremorine-M) and inverse agonists (atropine and *N*-methylscopolamine) promote different intracellular loop movements *(175)*. These data show that at the molecular level, the opposite effects of agonists and inverse agonists can be explained by structural rearrangement. These distinct receptor conformations are not limited to different classes of ligands, whereby orthosteric agonists adopt one conformation, orthosteric antagonists adopt a second comformation, and allosteric ligands another. Indeed, conformational differences have been reported for different orthosteric agonists. In the rat cerebellar cortex, oxotremorine is thought to stabilize a different receptor conformation to acetylcholine or carbachol, despite activation through the same site *(184, 85)*. More recently, it has been reported that the M_1 mACh receptor is able to be activated in at least three distinct ways, dependent on the ligand used *(186)*.

An emerging concept over the past 10–15 years has been that G protein-coupled receptors have the capacity to dimerize and/or oligomerize into homo or hetero-multimers *(187, 188)*, and this can significantly affect the ligand binding, signaling, and regulatory properties of the receptor *(189, 190)*. Although it had been suspected for sometime that M_2 mACh receptors oligomerize *(191, 192)*, relatively few studies have so far addressed the potential importance of homo- or hetero-multimerization of mACh receptors to their pharmacological properties *(193, 194)*. Systematic investigation of this phenomenon is required as many cells express multiple mACh receptor subtypes, and therefore the possibility of hetero- and well as homo-multimers contributing to cell/tissue responses is considerable.

8.4.2 Promiscuous mACh Receptor Coupling to G Proteins

Direct measurement of receptor-G protein coupling presents an opportunity to investigate functional selectivity independent of possible "cross-talk" between downstream effector molecules. Work investigating the activation profiles of particular subsets of G proteins has used either a photolabile GTP analog method *(47)*, or the quasi-irreversible binding of [^{35}S]-GTPγS *(48,60,195)* combined with an immunoprecipitation/immunocapture step using a Gα subtype-specific antibody. This technique can therefore define the specific receptor-Gα subunit-coupling that an individual ligand is able to activate. Using this approach it has been shown that a number of mACh receptors have differential patterns of G protein activation that are dependent on not only the receptor subtype, but also the ligand used for receptor activation *(48)*. Thus, in CHO cells expressing M_1, M_2, M_3, or M_4 mACh receptors, Akam et al. *(48)* showed that, in contrast to methacholine, pilocarpine preferentially activated the $G\alpha_{i3/o}$ subpopulation of G proteins with no discernible activation of $G\alpha_{i1/2}$ proteins in CHO-m4 cell membranes. Further, when examining G protein activation profiles in CHO-m1 and CHO-m3 cell membranes, both methacholine and pilocarpine caused an increase in [^{35}S]-GTPγS binding to $G\alpha_{q/11}$ proteins as well as $G\alpha_{i1/2}$ and $G\alpha_{i3/o}$, demonstrating the promiscuous nature of G protein coupling by these receptor subtypes. Interestingly, at the M_3 mACh receptor, pilocarpine caused only a small increase in [^{35}S]-GTPγS binding to $G\alpha_{q/11}$ relative to methacholine, but a proportionately greater activation of $G\alpha_i$ proteins *(48)*. These data provided evidence that the orthosteric mACh receptor full agonist methacholine and partial agonist pilocarpine are able to activate different subsets of G proteins via the same receptor subtype. Further investigation of a wider range of mACh receptor ligand could unmask further examples of differential G protein activation. In addition, the functional consequences of preferential pilocarpine signaling via $G\alpha_i$ proteins need to be established.

8.4.3 Signaling Pathway Activation by mACh Receptor Subtypes

By far the most studies on mACh receptor signal transduction mechanisms have utilized more distal readouts to assess the consequences of receptor activation. Some early studies raised the possibility of functional selectivity, particularly with respect to multiple readouts (e.g., phosphoinositide turnover vs. cyclic AMP accumulation) stimulated by the M_1 and M_3 mACh receptors *(43, 45)*. Although these studies concluded that carbachol, oxotremorine, and McN-A-343 may differentially affect $G_{q/11}$ and G_s-linked signaling, the data reported were not conclusive and could be explained by the differing receptor reserves associated with the pathways and the differing efficacies of the agonists investigated. More recent studies of M_2 and M_4 mACh receptor coupling to $G_{i/o}$ and G_s-linked signaling have also unmasked interesting agonist-specific differences, but nothing that can unequivocally be defined as functional selectivity *(36–38)*.

The possibility that ligands can engage distinct signaling pathways to mediate the same end result has also been investigated. Agonists at M_3 mACh receptors, classified as full and partial agonists with respect to their abilities to stimulate phosphoinositide turnover/Ca^{2+} signaling in airway smooth muscle, are generally able to induce similar maximal contractions *(196)*. This is partly due to the receptor reserve between these two responses (a M_3 mACh receptor-mediated increase in phosphoinositide turnover that is ~5% of the maximum response is sufficient to induce a maximal contraction), but also involves a Rho-dependent Ca^{2+} sensitization of the contractile machinery *(197)*. Schaafsma and colleagues *(198)* investigated the abilities of methacholine, pilocarpine, and McN-A-343 to stimulate Ca^{2+} responses in airways smooth muscle cells in the absence and presence of the Rho-kinase inhibitor Y-27632. These experiments revealed that the partial agonists pilocarpine and McN-A-343 were dependent on Rho-kinase activity to induce their maximal ASM contraction; in contrast, the full agonist methacholine was only susceptible to Rho-kinase inhibition when the number of mACh receptors was diminished to remove the receptor reserve. These data provide us with evidence that while similar maximal responses can be induced by these ligands, the mechanism by which this is achieved is not necessarily the same.

For the mACh receptor subtypes, it is widely accepted that agonists tend to cause receptor internalization and down-regulation in a number of cell types. Moreover, this receptor redistribution appears to be affected by both the cell type and the ligand used to stimulate the receptor. Koenig and Edwardson *(199)* have shown that the rate of M_3 mACh receptor internalization in response to carbachol is greatly increased in SH-SY5Y cells when compared with CHO cells. In fact in CHO cells expressing the M_3 mACh receptor, the amount of receptor internalization was not shown to reach significance at time points up to 2 h. However, application of carbachol to M_3 mACh receptors expressed in HEK-293 cells reveals a different picture whereby there is an up-regulation of receptor expression as determined by [^{3}H]-QNB binding *(200)*. These contrasting data may be purely a result of the different cell backgrounds used in these experiments, or it may be that the ligands themselves are able to induce differential signaling in different cell backgrounds.

8.4.4 *Functional Selectivity Through Allosteric Interactions*

The mACh receptors are subject to allosteric modulation characterized by cooperativity that is dependent on the receptor subtype and on the orthosteric/allosteric ligand-pairs studied. Although much of the literature available to date concentrates on the quantitative analysis of allosteric interactions between ever increasing numbers of ligands, there is much less of an emphasis on the potential for functional selectivity. Nevertheless, if individual orthosteric ligands are able to induce functional selectivity, there is every reason to believe that orthosteric/allosteric ligand pairing will expand the potential for this phenomenon.

Thiochrome, an oxidation product and metabolite of thiamine, has been reported to act allosterically at mACh receptors *(201)*. Using a [^{3}H]radioligand binding approach, it was observed that thiochrome was able to enhance the binding of acetylcholine to the M_4, but not to M_1-M_3 or M_5 mACh receptors. In contrast, the allosteric ligands brucine, N-benzyl brucine, and strychnine all display negative cooperativity with respect to acetylcholine affinity at the M_4 mACh receptor *(202)*. Although neither of these studies has investigated further the potential that these modulations may result in differential signaling output, it is interesting to speculate that differential cooperativity could result in functional selectivity and that this would be driven by the choice of allosteric, rather than the orthosteric ligand. The M_2 mACh receptor subtype has been widely used to investigate the concept of allosterism. Thus, there is a wealth of information available reporting the interactions of a variety of orthosteric and allosteric ligands. Table 8.1 highlights the fact

Table 8.1 Cooperativity factors for the interaction of the allosteric modulator alcuronium with orthosteric ligands at the M_2 mACh receptor

Orthosteric ligand	α
Agonists	
Arecoline[a]	1.7
Acetylcholine[a]	10
Bethanecol[a]	10
Carbachol[a]	9.5
Furmethide[a]	8.4
Methylfurmethide[a]	7.3
Antagonists/inverse agonists	
Atropine[b]	0.26
Methyl-*N*-piperidinyl benzilate[b]	0.54
Methyl-*N*-quinuclidinyl benzilate[b]	63
Methyl-*N*-scopolamine[c]	0.24

The cooperativity factor (α) is the ratio of orthosteric agonist binding to the free receptor relative to the allosterically occupied receptor *(218)*
[a]Data are taken from Jakubik et al. *(219)*
[b]Data are taken from Hejnova et al. *(220)*
[c]Data are taken from Proska and Tucek *(221)*

that the interaction of the allosteric modulator alcuronium varies greatly depending on which orthosteric probe is used. Additionally, alcuronium allosterically modulates the efficacy of pilocarpine in both recombinant and native systems *(203)*. However, once again these studies do not report on the impact that these interactions may have on the signaling pathways activated, yet this could represent an area for the unmasking of functional selectivity at mACh receptors.

Finally, a group of compounds, exemplified by AC-42 *(13)*, have been reported that selectively activate signaling via the M_1 mACh receptor through binding to a locus distinct from the acetylcholine binding site. A very recent study has compared the activities of AC-42, and a homolog, 77-LH-28–1, that most likely also acts allosterically *(204)*, with a panel of orthosterically acting agonists to stimulate $G_{q/11}$-, G_s-, and $G_{i/o}$-mediated signaling events *(205)*. In the CHO cell background used in this study, agonists strongly stimulated M_1 mACh receptor-$G_{q/11}$ signaling events, but evidence for G_s- and $G_{i/o}$-coupling was also seen for full and partial orthosteric agonists. In contrast, allosterically-acting agonists significantly stimulated G_s- but not $G_{i/o}$-coupling, suggesting that at least AC-42 and 77-LH-28–1 are able to stimulate only a subset of responses activated by otherwise equi-efficacious orthosteric agonists. These new data suggest that allosterically-acting ligands that can stimulate signaling in the absence of an orthosteric agonist may do so in a functionally selective manner.

8.4.5 Why are there not more Examples of Functional Selectivity at mACh Receptors?

The roles of mACh receptors in an array of physiological and pathophysiological functions have for a long time made this GPCR family an attractive target for the development of new drugs to treat a variety of diseases, from gastrointestinal dysfunction to Alzheimer's disease *(206)*. Considering this pharmaceutical interest, it is perhaps surprising that so few examples of functional selectivity can be mustered here. One possible explanation for this might be the overall lack of progress in developing molecular entities, particularly orthosteric agonists that can distinguish mACh receptor subtypes. This is especially important as mACh receptor subtypes appear to be coexpressed in an array of cell-types. In the absence of sufficiently selective compounds, this always allows the possibility that conclusions regarding functionally selective effects might be wrongly drawn that are actually attributable to mACh receptor subtype-dependent differences. Nevertheless, in studies where only a single mACh receptor subtype is present (e.g., in recombinant cell systems), systematic studies have often failed to find evidence for this phenomenom *(37)*, suggesting that commonly used orthosteric agonists are not sufficiently conformationally-selective.

Put another way is there any reason for believing that functional selectivity at mACh receptors is somehow minimized? Although mACh receptor subtypes appear to be quite promiscuous (i.e., the activated receptor couples to multiple G protein subtypes), at least when expressed recombinantly, some evidence exists for a much more precise mACh receptor-G protein interaction requiring a precise Gα/Gβγ combination *(61)*. Furthermore, there is evidence that mACh receptor subtypes may be compartmentalized within the cell. This may be brought about either by specific membrane targeting of receptor subtypes in polarized cells *(124, 167)*, or by association with/exclusion from lipid rafts and/or caveolae *(164, 207,2 08)* or cytoskeletal elements *(76, 160)*. These means of compartmentalization may place receptors in micro-environments where signaling outcomes are constrained by the G proteins/effectors in the immediate vicinity.

Another potential constraint in vivo is the coexpression of regulators of G protein signaling (RGS) proteins in mACh receptor-expressing cell backgrounds. A number of members of this large family of proteins have been demonstrated to act as GTPase activating proteins (GAPs) for specific Gα protein subtypes, while others (possessing "RGS-like" domains) can fulfill other cellular functions, such as antagonizing receptor-G protein interactions *(209)*. RGS proteins have been shown to interact directly with mACh receptors. For example, RGS4 interactions with the M_3 *(210,211)* and RGS2 interaction with the M_1 *(212)* mACh receptor subtypes have been reported. These interactions can selectively influence receptor subtype-specific signaling outcomes *(213, 214)* and subtle changes in RGS protein structure (e.g., through alternative splicing) can influence the mACh receptor-RGS interaction and outcome *(215)*.

8.4.6 Potential Utility of Functionally-Selective mACh Receptor Agonists

At present the main focus of mACh receptor drug discovery programs is to achieve subtype-specificity, either at the acetylcholine binding site or via allosteric binding sites present within the receptor structure. Although efforts to achieve this objective at the orthosteric site continue *(216, 217)*, targeting allosteric sites shows the greatest promise presently *(13, 181)*, with genuine subtype specificity being shown by both allosterically-acting agonists and antagonists. Until a wider range of such compounds are generally available, it will continue to be challenging to map precisely signaling pathways activated by mACh receptor subtypes and to assess whether signaling promiscuity is a major limitation to the development of new drugs in a variety of disorders.

References

1. Dale HH. The action of certain esters and ethers of choline and their relation to muscarine. J Pharmacol Exp Ther 1914;6:147–190.
2. Roszkowski AP. An unusual type of sympathetic ganglionic stimulant. J Pharmacol Exp Ther 1961;132:156–170.

3. Burgen ASV, Spero L. The action of acetylcholine and other drugs on the efflux of potassium and rubidium from smooth muscle of the guinea-pig intestine. Br J Pharmacol 1968;34:99–115.
4. Hammer R, Berrie CP, Birdsall JNM, Burgen ASV, Hulme EC. Pirenzepine distinguishes between subclasses of muscarinic receptors. Nature 1980;283:90–92.
5. Hulme EC, Birdsall NJM, Buckley NJ. Muscarinic receptor subtypes. Ann Rev Pharmacol Toxicol 1990;30:633–673.
6. Caulfield MP. Muscarinic receptors – characterization, coupling and function. Pharmac Ther 1993;58:319–379
7. Caulfield MP, Birdsall NJM. International Union of Pharmacology. XVII. Classification of muscarinic acetylcholine receptors. Pharmacol Rev 1998;50:279–290.
8. Felder CC. Muscarinic acetylcholine receptors: signal transduction through multiple effectors. FASEB J 1995;9:619–625.
9. Lanzafame AA, Christopoulos A, Mitchelson F. Cellular signaling mechanisms for muscarinic acetylcholine receptors. Recep Channels 2003;9:241–260.
10. Wess J, Eglen RM, Gautam D. Muscarinic acetylcholine receptors: mutant mice provide new insights for drug development. Nat Rev Drug Discov 2007;6:721–733.
11. Potter LT. Snake toxins that bind specifically to individual subtypes of muscarinic receptors. Life Sci 2001;68:2541–2547.
12. Onali P, Adem A, Karlsson E, Olianas MC. The pharmacological action of MT-7. Life Sci 2005;76:1547–1552.
13. Spalding TA, Trotter C, Skjaerbaek N, et al. Discovery of an ectopic activation site on the M_1 muscarinic receptor. Mol Pharmacol 2002;61:1297–1302.
14. Langmead CJ, Fry VA, Forbes IT, et al. Probing the molecular mechanism of interaction between 4-n-butyl-1-[4-(2-methylphenyl)-4-oxo-1-butyl]-piperidine (AC-42) and the muscarinic M1 receptor: direct pharmacological evidence that AC-42 is an allosteric agonist. Mol Pharmacol 2006;69:236–246.
15. Brown JH, Brown SL. Agonists differentiate muscarinic receptors that inhibit cyclic AMP formation from those that stimulate phosphoinositide metabolism. J Biol Chem 1984;259:3777–3781.
16. Peralta EG, Ashkenazi A, Winslow JW, Ramachandran J, Capon DJ. Differential regulation of PI hydrolysis and adenylyl cyclase by muscarinic receptor subtypes. Nature 1988;334:434–437.
17. Jones SV, Heilman CJ, Brann MR. Functional responses of cloned muscarinic receptors expressed in CHO-K1 cells. Mol Pharmacol 1991;40:242–47.
18. Bonner TI, Buckley NJ, Young AC, Brann MR. Identification of a family of muscarinic acetylcholine receptor genes. Science 1987;237:527–532.
19. Bonner TI, Young AC, Brann MR, Buckley NJ. Cloning and expression of the human and rat m5 muscarinic acetylcholine receptor genes. Neuron 1988;1:403–410.
20. Baumgold J, Fishman PH. Muscarinic receptor-mediated increase in cAMP levels in SK-N-SH human neuroblastoma cells. Biochem Biophys Res Commun 1988;154:1137–1143.
21. Stein R, Pinkas-Kramarski R, Sokolovsky M. Cloned M_1 muscarinic receptors mediate both adenylate cyclase inhibition and phosphoinositide turnover. EMBO J 1988;7:3031–3035.
22. Conklin BR, Brann MR, Buckley NJ, Ma AL, Bonner TI, Axelrod J. Stimulation of arachidonic acid release and inhibition of mitogenesis by cloned genes for muscarinic receptor subtypes stably expressed in A9 L cells. Proc Natl Acad Sci USA 1988;85:8698–8702.
23. Sandmann J, Peralta EG, Wurtmann RJ. Coupling of transfected muscarinic acetylcholine receptor subtypes to phospholipase D. J Biol Chem 1991;266:6031–6034.
24. Ashkenazi A, Peralta EG, Winslow JW, Ramachandran J, Capon DJ. Functionally distinct G proteins selectively couple different receptors to PI hydrolysis in the same cell. Cell 1989;56:487–493.
25. Schmidt M, Bienek C, van Koppen CJ, Michel MC, Jakobs KH. Differential calcium signalling by m2 and m3 muscarinic acetylcholine receptors in a single cell type. Naunyn-Schmiedeberg's Arch Pharmacol 1995;352:469–476.

26. Cooper DM, Mons N, Karpen JW. Adenylyl cyclases and the interaction between calcium and cAMP signalling. Nature 1995;374:421–424.
27. Hanoune J, Defer N. Regulation and role of adenylyl cyclase isoforms. Ann Rev Pharmacol Toxicol 2001;41:145–174.
28. Sunahara RK, Taussig R. Isoforms of mammalian adenylyl cyclase: multiplicities of signaling. Mol Interv 2002;2:168–184.
29. Bender AT, Beavo JA. Cyclic nucleotide phosphodiesterases: molecular regulation to clinical use. Pharmacol Rev 2006;58:488–520.
30. Migeon JC, Nathanson NM. Differential regulation of cAMP-mediated gene transcription by m1 and m4 muscarinic acetylcholine receptors. Preferential coupling of m4 receptors to Giα-2. J Biol Chem 1994;269:9767–9773.
31. Migeon JC, Thomas SL, Nathanson NM. Regulation of cAMP-mediated gene transcription by wild type and mutated G-protein α-subunits. J Biol Chem 1994;269:29146–29152.
32. Migeon JC, Thomas SL, Nathanson NM. Differential coupling of m2 and m4 muscarinic receptors to inhibition of adenylyl cyclase by $G_i\alpha$ and $G_o\alpha$ subunits. J Biol Chem 1995;270:16070–16074.
33. Liu YF, Ghahremani MH, Rasenick MM, Jakobs KH, Albert PR. Stimulation of cAMP synthesis by G_i-coupled receptors upon ablation of distinct $G\alpha_i$ protein expression. J Biol Chem 1999;274:16444–16450.
34. Dittman AH, Weber JP, Hinds TR, et al. A novel mechanism for coupling of m4 muscarinic acetylcholine receptors to calmodulin-sensitive adenylyl cyclases: crossover from G protein-coupled inhibition to stimulation. Biochemistry 1994;33:943–951.
35. Vogel WK, Mosser VA, Bulseco DA, Schimerlik MI. Porcine m2 muscarinic acetylcholine receptor-effector coupling in Chinese hamster ovary cells. J Biol Chem 1995;270:15485–15493.
36. Michal P, Lysíková M, Tucek S. Dual effects of muscarinic M_2 acetylcholine receptors on the synthesis of cyclic AMP in CHO cells: dependence on time, receptor density and receptor agonists. Br J Pharmacol 2001;132:1217–1228.
37. Mistry R, Dowling MR, Challiss RAJ. An investigation of whether agonist-selective receptor conformations occur with respect to M_2 and M_4 muscarinic acetylcholine receptor signalling via $G_{i/o}$ and G_s proteins. Br J Pharmacol 2005;144:566–575.
38. Michal P, El-Fakahany EE, Doležal V. Muscarinic M_2 receptors directly activate $G_{q/11}$ and G_s G-proteins. J Pharmacol Exp Ther 2007;320:607–614.
39. Burford NT, Tobin AB, Nahorski SR. Differential coupling of m1, m2 and m3 muscarinic receptor subtypes to inositol 1,4,5-trisphosphate and adenosine 3′,5′-cyclic monophosphate accumulation in Chinese hamster ovary cells. J Pharmacol Exp Ther 1995;274:134–142.
40. Felder CC, Kanterman RY, Ma AL, Axelrod J. A transfected m1 muscarinic acetylcholine receptor stimulates adenylate cyclase via phosphatidylinositol hydrolysis. J Biol Chem 1989;264:20356–20362.
41. Zhou XM, Curran P, Baumgold J, Fishman PH. Modulation of adenylylcyclase by protein kinase C in human neurotumor SK-N-MC cells: evidence that the α isozyme mediates both potentiation and desensitization. J Neurochem 1994;63:1361–1370.
42. Meeker RB, Harden TK. Muscarinic cholinergic receptor-mediated control of cyclic AMP metabolism. Agonist-induced changes in nucleotide synthesis and degradation. Mol Pharmacol 1983;23:384–392.
43. Gurwitz D, Haring R, Heldman E, Fraser CM, Manor D, Fisher A. Discrete activation of transduction pathways associated with acetylcholine m1 receptor by several muscarinic ligands. Eur J Pharmacol 1994;267:21–31.
44. Burford NT, Nahorski SR. Muscarinic m1 receptor-stimulated adenylate cyclase activity in Chinese hamster ovary cells is mediated by $G_s\alpha$ and is not a consequence of phosphoinositidase C activation. Biochem J 1996;315:883–888.
45. Heldman E, Barg J, Fisher A, et al. Pharmacological basis for functional selectivity of partial muscarinic receptor agonists. Eur J Pharmacol 1996;297:283–291.
46. Evellin S, Nolte J, Tysack K, et al. Stimulation of phospholipase C-ε by the M_3 muscarinic acetylcholine receptor mediated by cyclic AMP and the GTPase Rap2B. J Biol Chem 2002;277:16805–16813.

47. Offermanns S, Wieland T, Homann D, et al. Transfected muscarinic acetylcholine receptors selectively couple to G_i-type G proteins and $G_{q/11}$. Mol Pharmacol 1994;45:890–898.
48. Akam EC, Challiss RA, Nahorski SR. $G_{q/11}$ and $G_{i/o}$ activation profiles in CHO cells expressing human muscarinic acetylcholine receptors: dependence on agonist as well as receptor-subtype. Br J Pharmacol 2001;132:950–958.
49. Hokin MR, Hokin LE. Effects of acetylcholine on phospholipides in the pancreas. J Biol Chem 1954;209:549–558.
50. Rhee SG. Regulation of phosphoinositide-specific phospholipase C. Annu Rev Biochem 2001;70:281–312.
51. Suh PG, Park JI, Manzoli L, et al. Multiple roles of phosphoinositide-specific phospholipase C isozymes. BMB Rep 2008;41:415–434.
52. Gusovsky F, Lueders JE, Kohn EC, Felder CC. Muscarinic receptor-mediated tyrosine phosphorylation of phospholipase C-γ. An alternative mechanism for cholinergic-induced phosphoinositide breakdown. J Biol Chem 1993;268:7768–7772.
53. Nishida M, Sugimoto K, Hara Y, et al. Amplification of receptor signalling by Ca^{2+} entry-mediated translocation and activation of PLCγ2 in B lymphocytes. EMBO J 2003;22:4677–4688.
54. Mitchell CJ, Kelly MM, Blewitt M, Wilson JR, Biden TJ. Phospholipase C-γ mediates the hydrolysis of phosphatidylinositol, but not of phosphatidylinositol 4,5-bisphoshate, in carbamylcholine-stimulated islets of Langerhans. J Biol Chem 2001;276:19072–19077.
55. Willars GB, Nahorski SR, Challiss RAJ. Differential regulation of muscarinic acetylcholine receptor-sensitive polyphosphoinositide pools and consequences for signaling in human neuroblastoma cells. J Biol Chem 1998;273:5037–5046.
56. Suh BC, Hille B. Regulation of ion channels by phosphatidylinositol 4,5-bisphosphate. Curr Opin Neurobiol 2005;15:370–378.
57. Halstead JR, Jalink K, Divecha N. An emerging role for PtdIns(4,5)P_2-mediated signalling in human disease. Trends Pharmacol Sci 2005;26:654–660.
58. Sternweis PC, Smrcka AV. Regulation of phospholipase C by G proteins. Trends Biochem Sci 1992;17:502–506.
59. Berstein G, Blank JL, Smrcka AV, et al. Reconstitution of agonist-stimulated phosphatidylinositol 4,5-bisphosphate hydrolysis using purified m1 muscarinic receptor, $G_{q/11}$, and phospholipase C-β1. J Biol Chem 1992;267:8081–8088.
60. DeLapp NW, McKinzie JH, Sawyer BD, et al. Determination of [^{35S}]guanosine-5'-O-(3-thio) triphosphate binding mediated by cholinergic muscarinic receptors in membranes from Chinese hamster ovary cells and rat striatum using an anti-G protein scintillation proximity assay. J Pharmacol Exp Ther 1999;289:946–955.
61. Dippel E, Kalkbrenner F, Wittig B, Schultz G. A heterotrimeric G protein complex couples the muscarinic m1 receptor to phospholipase C-β. Proc Natl Acad Sci USA 1996;93:1391–1396.
62. Delmas P, Crest M, Brown DA. Functional organization of PLC signaling microdomains in neurons. Trends Neurosci 2004;27:41–47.
63. Delmas P, Wanaverbecq N, Abogadie FC, Mistry M, Brown DA. Signaling microdomains define the specificity of receptor-mediated $InsP_3$ pathways in neurons. Neuron 2002;34:209–220.
64. Vogel WK, Mosser VA, Bulseco DA, Schimerlik MI. Porcine m2 muscarinic acetylcholine receptor-effector coupling in Chinese hamster ovary cells. J Biol Chem 1995;270:15485–15493.
65. Katz A, Wu D, Simon MI. Subunits βγ of heterotrimeric G protein activate β2 isoform of phospholipase C. Nature 1992;360:686–689.
66. Schmidt M, Bienek C, van Koppen CJ, Michel MC, Jakobs KH. Differential calcium signalling by m2 and m3 muscarinic acetylcholine receptors in a single cell type. Naunyn Schmiedeberg's Arch Pharmacol 1995;352:469–476.
67. Martinson EA, Goldstein D, Brown JH. Muscarinic receptor activation of phosphatidylcholine hydrolysis. Relationship to phosphoinositide hydrolysis and diacylglycerol metabolism. J Biol Chem 1989;264:14748–14754.
68. Bayon Y, Hernandez M, Alonso A, et al. Cytosolic phospholipase A_2 is coupled to muscarinic receptors in the human astrocytoma cell line 1321N1: characterization of the transducing mechanism. Biochem J 1997;323:281–287.

69. Cho HW, Kim JH, Choi S, Kim HJ. Phospholipase A_2 is involved in muscarinic receptor-mediated sAPPα release independently of cyclooxygenase or lypoxygenase activity in SH-SY5Y cells. Neurosci Lett 2006;397:214–218.
70. Cockcroft S. Signalling roles of mammalian phospholipase D1 and D2. Cell Mol Life Sci 2001;58:1674–1687.
71. Oude Weernink PA, López de Jesús M, Schmidt M. Phospholipase D signaling: orchestration by PIP_2 and small GTPases. Naunyn Schmiedeberg's Arch Pharmacol 2007;374:399–411.
72. Rümenapp U, Asmus M, Schablowski H, et al. The M_3 muscarinic acetylcholine receptor expressed in HEK-293 cells signals to phospholipase D via G_{12}, but not G_q-type G proteins. J Biol Chem 2001;276:2474–2479.
73. Lutz S, Shankaranarayanan A, Coco C, et al. Structure of $G\alpha_q$-p63^{RhoGEF}-RhoA complex reveals a pathway for the activation of RhoA by GPCRs. Science 2007;318:1923–1927.
74. Strassheim D, May LG, Varker KA, et al. M_3 muscarinic acetylcholine receptors regulate cytoplasmic myosin by a process involving RhoA and requiring conventional protein kinase C isoforms. J Biol Chem 1999;274:18675–18685.
75. Vogt S, Grosse R, Schultz G, Offermanns S. Receptor-dependent RhoA activation in G_{12}/G_{13}-deficient cells. J Biol Chem 2003;278:28743–28749.
76. Street M, Marsh SJ, Stabach PR, Morrow JS, Brown DA, Buckley NJ. Stimulation of $G\alpha_q$-coupled M_1 muscarinic receptor causes reversible spectrin redistribution mediated by PLC, PKC and ROCK. J Cell Sci 2006;119:1528–1536.
77. Gohla A, Offermanns S, Wilkie TM, Schultz G. Differential involvement of $G\alpha_{12}$ and $G\alpha_{13}$ in receptor-mediated stress fiber formation. J Biol Chem 1999;274:17901–17907.
78. Tang X, Batty IH, Downes CP. Muscarinic receptors mediate phospholipase C-dependent activation of protein kinase B via Ca^{2+}, ErbB3, and phosphoinositide 3-kinase in 1321N1 astrocytoma cells. J Biol Chem 2002;277:338–344.
79. Nelson CP, Challiss RAJ. "Phenotypic" pharmacology: the influence of cellular environment on G protein-coupled receptor antagonist and inverse agonist pharmacology. Biochem Pharmacol 2007;73:737–751.
80. Gutkind JS. The pathways connecting G protein-coupled receptors to the nucleus through divergent mitogen-activated protein kinase cascades. J Biol Chem 1998;273:1839–1842.
81. Marinissen MJ, Gutkind JS. G-protein-coupled receptors and signaling networks: emerging paradigms. Trends Pharmacol Sci 2001;22:368–376.
82. Pierce KL, Luttrell LM, Lefkowitz RJ. New mechanisms in heptahelical receptor signaling to mitogen activated protein kinase cascades. Oncogene 2001;20:1532–1539.
83. Offermanns S, Bombien E, Schultz G. Stimulation of tyrosine phosphorylation and mitogen-activated-protein (MAP) kinase activity in human SH-SY5Y neuroblastoma cells by carbachol. Biochem J 1993;294:545–550.
84. Crespo P, Xu N, Simonds WF, Gutkind JS. Ras-dependent activation of MAP kinase pathway mediated by G protein βγ-subunits. Nature 1994;369:418–420.
85. Wylie PG, Challiss RA, Blank JL. Regulation of extracellular-signal regulated kinase and c-Jun N-terminal kinase by G protein-linked muscarinic acetylcholine receptors. Biochem J 1999;338:619–628.
86. Winitz S, Russell M, Qian NX, Gardner A, Dwyer L, Johnson GL. Involvement of Ras and Raf in the G_i-coupled acetylcholine muscarinic m2 receptor activation of mitogen-activated protein (MAP) kinase kinase and MAP kinase. J Biol Chem 1993;268:19196–19199.
87. Koch WJ, Hawes BE, Allen LF, Lefkowitz RJ. Direct evidence that G_i-coupled receptor stimulation of mitogen-activated protein kinase is mediated by Gβγ activation of p21^{ras}. Proc Natl Acad Sci USA 1994;91:12706–12710.
88. Lopez-Ilasaca M, Crespo P, Pellici PG, Gutkind JS, Wetzker R. Linkage of G protein-coupled receptors to the MAPK signaling pathway through PI3-kinaseg. Science 1997; 275:394–397.
89. Mochizuki N, Ohba Y, Kiyokawa E, et al. Activation of the ERK/MAPK pathway by an isoform of rap1GAP associated with $G\alpha_i$. Nature 1999;400:891–894.

90. Guo FF, Kumahara E, Saffen D. A CalDAG-GEFI/Rap1/B-Raf cassette couples M1 muscarinic acetylcholine receptors to the activation of ERK1/2. J Biol Chem 2001; 276:25568–25581.
91. Yang H, Cooley D, Legakis JE, Ge Q, Andrade R, Mattingly RR. Phosphorylation of the Ras-GRF1 exchange factor at Ser916/898 reveals activation of Ras signaling in the cerebral cortex. J Biol Chem 2003;278:13278–13285.
92. Coso OA, Teramoto H, Simonds WF, Gutkind JS. Signaling from G protein-coupled receptors to c-Jun kinase involves bg subunits of heterotrimeric G proteins acting on a Ras and Rac1-dependent pathway. J Biol Chem 1996;271:3963–3966.
93. Yamauchi J, Nagao M, Kaziro Y, Itoh H. Activation of p38 mitogen-activated protein kinase by signaling through G protein-coupled receptors. Involvement of Gbg and Gaq/11 subunits. J Biol Chem 1997;272:27771–27777.
94. Marinissen MJ, Chiariello M, Pallante M, Gutkind JS. A network of mitogen-activated protein kinases links G protein-coupled receptors to the c-jun promoter: a role for c-Jun NH2-terminal kinase, p38s, and extracellular signal-regulated kinase 5. Mol Cell Biol 1999;19:4289–4301.
95. Fukuhara S, Marinissen MJ, Chiariello M, Gutkind JS. Signaling from G protein-coupled receptors to ERK5/Big MAPK 1 involves Gaq and Ga12/13 families of heterotrimeric G proteins. J Biol Chem 2000;275:21730–21736.
96. van Corven EJ, Hordijk PL, Medema RH, Bos JL, Moolenaar WH. Pertussis toxin-sensitive activation of p21ras by G protein-coupled receptor agonists in fibroblasts. Proc Natl Acad Sci USA 1993;90:1257–1261.
97. Wan Y, Kurosaki T, Huang XY. Tyrosine kinases in activation of the MAP kinase cascade by G-protein-coupled receptors. Nature 1996;380:541–544.
98. Nagao M, Yamauchi J, Kaziro Y, Itoh H. Involvement of protein kinase C and Src family tyrosine kinase in Gaq/11-induced activation of c-Jun N-terminal kinase and p38 mitogen-activated protein kinase. J Biol Chem 1998;273:22892–22898.
99. Daub H, Weiss FU, Wallasch C, Ullrich A. Role of transactivation of the EGF receptor in signalling by G-protein-coupled receptors. Nature 1996;379:557–560.
100. Keely SJ, Uribe JM, Barrett KE. Carbachol stimulates transactivation of epidermal growth factor receptor and mitogen-activated protein kinase in T84 cells. Implications for carbachol-stimulated chloride secretion. J Biol Chem 1998;273:27111–27117.
101. Prenzel N, Zwick E, Daub H, et al. EGF receptor transactivation by G-protein-coupled receptors requires metalloproteinase cleavage of proHB-EGF. Nature 1999;402:884–888.
102. Slack BE. The m3 muscarinic acetylcholine receptor is coupled to mitogen-activated protein kinase via protein kinase C and epidermal growth factor receptor kinase. Biochem J 2000;348:381–387.
103. Stirnweiss J, Valkova C, Ziesché E, Drube S, Liebmann C. Muscarinic M_2 receptors mediate transactivation of EGF receptor through Fyn kinase and without matrix metalloproteases. Cell Signal 2006;18:1338–1349.
104. McCole DF, Truong A, Bunz M, Barrett KE. Consequences of direct versus indirect activation of epidermal growth factor receptor in intestinal epithelial cells are dictated by protein-tyrosine phosphatase 1B. J Biol Chem 2007;282:13303–13315.
105. Luttrell LM, Daaka Y, Della Rocca GJ, Lefkowitz RJ. G protein-coupled receptors mediate two functionally distinct pathways of tyrosine phosphorylation in rat 1a fibroblasts. Shc phosphorylation and receptor endocytosis correlate with activation of ERK kinases. J Biol Chem 1997;272:31648–31656.
106. McDonald PH, Chow CW, Miller WE, et al. β-arrestin2: a receptor-regulated MAPK scaffold for the activation of JNK3. Science 2000;290:1574–1577.
107. Sun Y, Cheng Z, Ma L, Pei G. β-arrestin2 is critically involved in CXCR4-mediated chemotaxis, and this is mediated by its enhancement of p38 MAPK activation. J Biol Chem 2002;277:49212–49219.
108. Vögler O, Nolte B, Voss M, Schmidt M, Jakobs KH, van Koppen CJ. Regulation of muscarinic acetylcholine receptor sequestration and function by β-arrestin. J Biol Chem 1999;274:12333–12338.

109. Scott MG, Pierotti V, Storez H, et al. Cooperative regulation of extracellular signal-regulated kinase activation and cell shape change by filamin A and β-arrestins. Mol Cell Biol 2006;26:3432–445.
110. Budd DC, Rae A, Tobin AB. Activation of the mitogen-activated protein kinase pathway by a $G_{q/11}$-coupled muscarinic receptor is independent of receptor internalization. J Biol Chem 1999;274:12355–12360.
111. Budd DC, Willars GB, McDonald JE, Tobin AB. Phosphorylation of the $G_{q/11}$-coupled m3-muscarinic receptor is involved in receptor activation of the ERK-1/2 mitogen-activated protein kinase pathway. J Biol Chem 2001;276:4581–4587.
112. Lin AL, Zhu B, Zhang W, et al. Distinct pathways of ERK activation by the muscarinic agonists pilocarpine and carbachol in a human salivary cell line. Am J Physiol 2008;294:C1454-C1464.
113. Blaukat A, Barac A, Cross MJ, Offermanns S, Dikic I. G protein-coupled receptor-mediated mitogen-activated protein kinase activation through cooperation of $G\alpha_q$ and $G\alpha_i$ signals. Mol Cell Biol 2000;20:6837–6848.
114. Hornigold DC, Mistry R, Raymond PD, Blank JL, Challiss RA. Evidence for cross-talk between M_2 and M_3 muscarinic acetylcholine receptors in the regulation of second messenger and extracellular signal-regulated kinase signalling pathways in Chinese hamster ovary cells. Br J Pharmacol 2003;138:1340–1350.
115. Delmas P, Brown DA. Pathways modulating neural KCNQ/M (Kv7) potassium channels. Nat Rev Neurosci 2005;6:850–862.
116. Hernandez CC, Zaika O, Tolstykh GP, Shapiro MS. Regulation of neural KCNQ channels: signalling pathways, structural motifs and functional implications. J Physiol 2008;586:1811–1821.
117. Mark MD, Herlitze S. G-protein mediated gating of inward-rectifier K^+ channels. Eur J Biochem 2000;267:5830–5836.
118. Baruscotti M, Bucchi A, Difrancesco D. Physiology and pharmacology of the cardiac pacemaker ("funny") current. Pharmacol Ther 2005;107:59–79.
119. Harvey RD, Belevych AE. Muscarinic regulation of cardiac ion channels. Br J Pharmacol 2003;139:1074–1084.
120. Ferguson SS. Evolving concepts in G protein-coupled receptor endocytosis: the role in receptor desensitization and signaling. Pharmacol Rev 2001;53:1–24.
121. Hanyaloglu AC, von Zastrow M. Regulation of GPCRs by endocytic membrane trafficking and its potential implications. Annu Rev Pharmacol Toxicol 2008;48:537–568.
122. van Koppen CJ, Kaiser B. Regulation of muscarinic acetylcholine receptor signaling. Pharmacol Ther 2003;98:197–220.
123. Torrecilla I, Tobin AB. Co-ordinated covalent modification of G protein-coupled receptors. Curr Pharm Des 2006;12:1797–1808.
124. Nathanson NM. Synthesis, trafficking, and localization of muscarinic acetylcholine receptors. Pharmacol Ther 2008;119:33–43.
125. Hausdorff WP, Caron MG, Lefkowitz RJ. Turning off the signal: desensitization of β-adrenergic receptor function. FASEB J 1990;4:2881–2889.
126. Pierce KL, Lefkowitz RJ. Classical and new roles of β-arrestins in the regulation of G protein-coupled receptors. Nat Rev Neurosci 2001;2:727–733.
127. Lefkowitz RJ, Shenoy SK. Transduction of receptor signals by β-arrestins. Science 2005;308:512–517.
128. Nelson CD, Perry SJ, Regier DS, Prescott SM, Topham MK, Lefkowitz RJ. Targeting of diacylglycerol degradation to M_1 muscarinic receptors by β-arrestins. Science 2007;315:663–666.
129. Perry SJ, Baillie GS, Kohout TA, et al. Targeting of cyclic AMP degradation to β_2-adrenergic receptors by β-arrestins. Science 2002;298:834–836.
130. Ma L, Pei G. β-arrestin signaling and regulation of transcription. J Cell Sci 2007;120:213–218.
131. Kenakin T. New concepts in drug discovery: collateral efficacy and permissive antagonism. Nat Rev Drug Discov 2005;4:919–927.
132. Violin JD, Lefkowitz RJ. β-arrestin-biased ligands at seven-transmembrane receptors. Trends Pharmacol Sci 2007;28:416–422.

133. Kwatra MM, Leung E, Maan AC, et al. Correlation of agonist-induced phosphorylation of chick heart muscarinic receptors with receptor desensitization. J Biol Chem 1987;262:16314–16321.
134. Haga K, Haga T, Ichiyama A. Phosphorylation by protein kinase C of the muscarinic acetylcholine receptor. J Neurochem 1990;54:1639–1644.
135. Richardson RM, Hosey MM. Agonist-independent phosphorylation of purified cardiac muscarinic cholinergic receptors by protein kinase C. Biochemistry 1990;29:8555–8561.
136. Habecker BA, Nathanson NM. Regulation of muscarinic acetylcholine receptor mRNA expression by activation of homologous and heterologous receptors. Proc Natl Acad Sci USA 1992;89:5035–5038.
137. Tobin AB, Lambert DG, Nahorski SR. Rapid desensitization of muscarinic m3 receptor-stimulated polyphosphoinositide responses. Mol Pharmacol 1992;42:1042–1048.
138. Tobin AB, Nahorski SR. Rapid agonist-mediated phosphorylation of m3-muscarinic receptors revealed by immunoprecipitation. J Biol Chem 1993;268:9817–9823.
139. Wojcikiewicz RJ, Tobin AB, Nahorski SR. Muscarinic receptor-mediated inositol 1,4,5-trisphosphate formation in SH-SY5Y neuroblastoma cells is regulated acutely by cytosolic Ca^{2+} and by rapid desensitization. J Neurochem 1994;63:177–185.
140. Gurevich VV, Richardson RM, Kim CM, Hosey MM, Benovic JL. Binding of wild type and chimeric arrestins to the m2 muscarinic cholinergic receptor. J Biol Chem 1993;268:16879–16882.
141. Richardson RM, Kim C, Benovic JL, Hosey MM. Phosphorylation and desensitization of human m2 muscarinic cholinergic receptors by two isoforms of the β-adrenergic receptor kinase. J Biol Chem 1993;268:13650–13656.
142. Willets JM, Challiss RA, Nahorski SR. Non-visual GRKs: are we seeing the whole picture? Trends Pharmacol Sci 2003;24:626–633.
143. Gurevich VV, Gurevich EV. Rich tapestry of G protein-coupled receptor signaling and regulatory mechanisms. Mol Pharmacol 2008;74:312–316.
144. Willets JM, Nash MS, Challiss RA, Nahorski SR. Imaging of muscarinic acetylcholine receptor signaling in hippocampal neurons: evidence for phosphorylation-dependent and -independent regulation by G-protein-coupled receptor kinases. J Neurosci 2004;24:4157–4162.
145. Willets JM, Nahorski SR, Challiss RA. Roles of phosphorylation-dependent and -independent mechanisms in the regulation of M_1 muscarinic acetylcholine receptors by G protein-coupled receptor kinase 2 in hippocampal neurons. J Biol Chem 2005;280:18950–18958.
146. Willets JM, Nelson CP, Nahorski SR, Challiss RA. The regulation of M_1 muscarinic acetylcholine receptor desensitization by synaptic activity in cultured hippocampal neurons. J Neurochem 2007;103:2268–2280.
147. Willets JM, Mistry R, Nahorski SR, Challiss RA. Specificity of G protein-coupled receptor kinase 6-mediated phosphorylation and regulation of single-cell M_3 muscarinic acetylcholine receptor signaling. Mol Pharmacol 2003;64:1059–1068.
148. DebBurman SK, Kunapuli P, Benovic JL, Hosey MM. Agonist-dependent phosphorylation of human muscarinic receptors in *Spodoptera frugiperda* insect cell membranes by G protein-coupled receptor kinases. Mol Pharmacol 1995;47:224–233.
149. Wu G, Bogatkevich GS, Mukhin YV, Benovic JL, Hildebrandt JD, Lanier SM. Identification of Gβγ binding sites in the third intracellular loop of the M_3-muscarinic receptor and their role in receptor regulation. J Biol Chem 2000;275:9026–9034.
150. Gainetdinov RR, Bohn LM, Walker JK, et al. Muscarinic super-sensitivity and impaired receptor desensitization in G. Protien-coupled receptor kinase 5-deficient mice. Neuron 1999:1029-1036.
151. Walker JK, Gainetdinov RR, Feldman DS, et al. G protein-coupled receptor kinase 5 regulates airway responses induced by muscarinic receptor activation. Am J Physiol 2004;286:L312-L319.
152. Schlador ML, Nathanson NM. Synergistic regulation of m2 muscarinic acetylcholine receptor desensitization and sequestration by G protein-coupled receptor kinase-2 and β-arrestin-1. J Biol Chem 1997;272:18882–18890.
153. Tsuga H, Okuno E, Kameyama K, Haga T. Sequestration of human muscarinic acetylcholine receptor hm1-hm5 subtypes: effect of G protein-coupled receptor kinases GRK2, GRK4, GRK5 and GRK6. J Pharmacol Exp Ther 1998;284:1218–1226.

154. Urban JD, Clarke WP, von Zastrow M, et al. Functional selectivity and classical concepts of quantitative pharmacology. J Pharmacol Exp Ther 2007;320:1–13.
155. Torrecilla I, Spragg EJ, Poulin B, et al. Phosphorylation and regulation of a G protein-coupled receptor by protein kinase CK2. J Cell Biol 2007;177:127–137.
156. Pals-Rylaarsdam R, Gurevich VV, Lee KB, Ptasienski JA, Benovic JL, Hosey MM. Internalization of the m2 muscarinic acetylcholine receptor. Arrestin-independent and -dependent pathways. J Biol Chem 1997;272:23682–23689.
157. Claing A, Perry SJ, Achiriloaie M, et al. Multiple endocytic pathways of G protein-coupled receptors delineated by GIT1 sensitivity. Proc Natl Acad Sci USA 2000;97:1119–1124.
158. Wu G, Krupnick JG, Benovic JL, Lanier SM. Interaction of arrestins with intracellular domains of muscarinic and α_2-adrenergic receptors. J Biol Chem 1997;272:17836–17842.
159. Lee KB, Pals-Rylaarsdam R, Benovic JL, Hosey MM. Arrestin-independent internalization of the m1, m3, and m4 subtypes of muscarinic cholinergic receptors. J Biol Chem 1998;273:12967–12972.
160. Popova JS, Rasenick MM. Clathrin-mediated endocytosis of m3 muscarinic receptors. Roles for Gβγ and tubulin. J Biol Chem 2004;279:30410–30418.
161. Pals-Rylaarsdam R, Xu Y, Witt-Enderby P, Benovic JL, Hosey MM. Desensitization and internalization of the m2 muscarinic acetylcholine receptor are directed by independent mechanisms. J Biol Chem 1995;270:29004–29011.
162. Vögler O, Bogatkewitsch GS, Wriske C, Krummenerl P, Jakobs KH, van Koppen CJ. Receptor subtype-specific regulation of muscarinic acetylcholine receptor sequestration by dynamin. Distinct sequestration of m2 receptors. J Biol Chem 1998;273:12155–12160.
163. Delaney KA, Murphy MM, Brown LM, Radhakrishna H. Transfer of M_2 muscarinic acetylcholine receptors to clathrin-derived early endosomes following clathrin-independent endocytosis. J Biol Chem 2002;277:33439–33446.
164. Dessy C, Kelly RA, Balligand JL, Feron O. Dynamin mediates caveolar sequestration of muscarinic cholinergic receptors and alteration in NO signaling. EMBO J 2000;19:4272–4280.
165. Volpicelli LA, Lah JJ, Fang G, Goldenring JR, Levey AI. Rab11a and myosin Vb regulate recycling of the M_4 muscarinic acetylcholine receptor. J Neurosci 2002;22:9776–9784.
166. Reiner C, Nathanson NM. The internalization of the M_2 and M_4 muscarinic acetylcholine receptors involves distinct subsets of small G-proteins. Life Sci 2008;82:718–27.
167. Iverson HA, Fox D, Nadler LS, Klevit RE, Nathanson NM. Identification and structural determination of the M_3 muscarinic acetylcholine receptor basolateral sorting signal. J Biol Chem 2005;280:24568–24575.
168. Hashimoto Y, Morisawa K, Saito H, Jojima E, Yoshida N, Haga T. Muscarinic M_4 receptor recycling requires a motif in the third intracellular loop. J Pharmacol Exp Ther 2008;325:947–953.
169. Sawyer GW, Ehlert FJ, Shults CA. Cysteine pairs in the third intracellular loop of the muscarinic m1 acetylcholine receptor play a role in agonist-induced internalization. J Pharmacol Exp Ther 2008;324:196–205.
170. Berg KA, Maayani S, Goldfarb J, Scaramellini C, Leff P, Clarke WP. Effector pathway-dependent relative efficacy at serotonin type 2A and 2C receptors: evidence for agonist-directed trafficking of receptor stimulus. Mol Pharmacol 1998;54:94–104.
171. Gay EA, Urban JD, Nichols DE, Oxford GS, Mailman RB. Functional selectivity of D_2 receptor ligands in a Chinese hamster ovary hD2L cell line: evidence for induction of ligand-specific receptor states. Mol Pharmacol 2004;66:97–105.
172. Galandrin S, Bouvier M. Distinct signaling profiles of beta1 and β_2 adrenergic receptor ligands toward adenylyl cyclase and mitogen-activated protein kinase reveals the pluridimensionality of efficacy. Mol Pharmacol 2006;70:1575–1584.
173. Kobilka BK, Deupi X. Conformational complexity of G-protein-coupled receptors. Trends Pharmacol Sci 2007;28:397–406.
174. Yeagle PL, Albert AD. G-protein coupled receptor structure. Biochim Biophys Acta 2007;1768:808–824.
175. Li JH, Han SJ, Hamdan FF, et al. Distinct structural changes in a G protein-coupled receptor caused by different classes of agonist ligands. J Biol Chem 2007;282:26284–26293.

176. Li JH, Hamdan FF, Kim SK, et al. Ligand-specific changes in M_3 muscarinic acetylcholine receptor structure detected by a disulfide scanning strategy. Biochemistry 2008;47:2776–2788.
177. Wess J, Liu J, Blin N, Yun J, Lerche C, Kostenis E. Structural basis of receptor/G protein coupling selectivity studied with muscarinic receptors as model systems. Life Sci 1997;60:1007–1014.
178. Hulme EC, Lu ZL, Ward SD, Allman K, Curtis CA. The conformational switch in 7-transmembrane receptors: the muscarinic receptor paradigm. Eur J Pharmacol 1999;375:247–260.
179. Birdsall NJ, Lazareno S. Allosterism at muscarinic receptors: ligands and mechanisms. Mini Rev Med Chem 2005;5:523–543.
180. Gregory KJ, Sexton PM, Christopoulos A. Allosteric modulation of muscarinic acetylcholine receptors. Curr Neuropharmacol 2007;5:157–167.
181. Shirey JK, Xiang Z, Orton D, et al. An allosteric potentiator of M_4 mAChR modulates hippocampal synaptic transmission. Nat Chem Biol 2008;4:42–50.
182. Lu ZL, Saldanha JW, Hulme EC. Seven-transmembrane receptors: crystals clarify. Trends Pharmacol Sci 2002;23:140–146.
183. Vistoli G, Pedretti A, Dei S, Scapecchi S, Marconi C, Romanelli MN. Docking analyses on human muscarinic receptors: unveiling the subtypes peculiarities in agonists binding. Bioorg Med Chem 2008;16:3049–3058.
184. Gurwitz D, Sokolovsky M. Rat brain and heart muscarinic receptors: modification with tetranitromethane. Biochem Biophys Res Commun 1985;131:1124–1131.
185. van Koppen CJ, Sokolovsky M. Chemical modification of rat cerebral cortex M_1 muscarinic receptors: role of histidyl residues in antagonist and agonist binding. Biochem Biophys Res Commun 1988;151:1069–1073.
186. Spalding TA, Ma JN, Ott TR, et al. Structural requirements of transmembrane domain 3 for activation by the M1 muscarinic receptor agonists AC-42, AC-260584, clozapine, and *N*-desmethylclozapine: evidence for three distinct modes of receptor activation. Mol Pharmacol 2006;70:1974–1983.
187. Bouvier M. Oligomerization of G-protein-coupled transmitter receptors. Nat Rev Neurosci 2001;2:274–286.
188. Milligan G. G protein-coupled receptor dimerization: function and ligand pharmacology. Mol Pharmacol 2004;66:1–7.
189. Devi LA. Heterodimerization of G-protein-coupled receptors: pharmacology, signaling and trafficking. Trends Pharmacol Sci 2001;22:532–537.
190. Springael JY, Urizar E, Costagliola S, Vassart G, Parmentier M. Allosteric properties of G protein-coupled receptor oligomers. Pharmacol Ther 2007;115:410–418.
191. Avissar S, Amitai G, Sokolovsky M. Oligomeric structure of muscarinic receptors is shown by photoaffinity labeling: subunit assembly may explain high- and low-affinity agonist states. Proc Natl Acad Sci USA 1983;80:156–159.
192 Wreggett KA, Wells JW. Cooperativity manifest in the binding properties of purified cardiac muscarinic receptors. J Biol Chem 1995;270:22488–22499.
193 Zeng FY, Wess J. Identification and molecular characterization of m3 muscarinic receptor dimers. J Biol Chem 1999;274:19487–19497.
194 Goin JC, Nathanson NM. Quantitative analysis of muscarinic acetylcholine receptor homo- and heterodimerization in live cells: regulation of receptor down-regulation by heterodimerization. J Biol Chem 2006;281:5416–5425.
195 Okamoto T, Ikezu T, Murayama Y, Ogata E, Nishimoto I. Measurement of GTPγS binding to specific G proteins in membranes using G-protein antibodies. FEBS Lett 1992;305:125–128.
196. Meurs H, Roffel AF, Postema JB, et al. Evidence for a direct relationship between phosphoinositide metabolism and airway smooth muscle contraction induced by muscarinic agonists. Eur J Pharmacol 1988;156:271–274.
197. Somlyo AP, Somlyo AV. Ca^{2+} sensitivity of smooth muscle and nonmuscle myosin II: modulated by G proteins, kinases, and myosin phosphatase. Physiol Rev 2003;83:1325–1358.
198. Schaafsma D, Boterman M, de Jong AM, et al. Differential Rho-kinase dependency of full and partial muscarinic receptor agonists in airway smooth muscle contraction. Br J Pharmacol 2006;147:737–743.

199. Koenig JA, Edwardson JM. Intracellular trafficking of the muscarinic acetylcholine receptor: importance of subtype and cell type. Mol Pharmacol 1996;49:351–359.
200. Stope MB, Kunkel C, Kories C, Schmidt M, Michel MC. Differential agonist-induced regulation of human M_2 and M_3 muscarinic receptors. Biochem Pharmacol 2003;66:2099–2105.
201. Lazareno S, Dolezal V, Popham A, Birdsall NJ. Thiochrome enhances acetylcholine affinity at muscarinic M_4 receptors: receptor subtype selectivity via cooperativity rather than affinity. Mol Pharmacol 2004;65:257–266.
202. Lazareno S, Gharagozloo P, Kuonen D, Popham A, Birdsall NJ. Subtype-selective positive cooperative interactions between brucine analogues and acetylcholine at muscarinic receptors: radioligand binding studies. Mol Pharmacol 1998;53:573–589.
203. Zahn K, Eckstein N, Tränkle C, Sadée W, Mohr K. Allosteric modulation of muscarinic receptor signaling: alcuronium-induced conversion of pilocarpine from an agonist into an antagonist. J Pharmacol Exp Ther 2002;301:720–728.
204. Langmead CJ, Austin NE, Branch CL, et al. Characterization of a CNS penetrant, selective M_1 muscarinic receptor agonist, 77-LH-28–1. Br J Pharmacol 2008;154:1104–1115.
205. Thomas RL, Mistry R, Langmead CJ, Wood MD, Challiss RAJ. G protein-coupling and signaling pathway activation by M_1 muscarinic acetylcholine receptor orthosteric and allosteric agonists. J Pharmacol Exp Ther 2008;327:365–74.
206. Eglen RM. Muscarinic receptor subtype pharmacology and physiology. Prog Med Chem 2005;43:105–136.
207. Gosens R, Stelmack GL, Dueck G, et al. Caveolae facilitate muscarinic receptor-mediated intracellular Ca^{2+} mobilization and contraction in airway smooth muscle. Am J Physiol 2007;293:L1406–L1418.
208. Shmuel M, Nodel-Berner E, Hyman T, Rouvinski A, Altschuler Y. Caveolin 2 regulates endocytosis and trafficking of the M_1 muscarinic receptor in MDCK epithelial cells. Mol Biol Cell 2007;18:1570–1585.
209. Hollinger S, Hepler JR. Cellular regulation of RGS proteins: modulators and integrators of G protein signaling. Pharmacol Rev 2002;54:527–559.
210. Zeng W, Xu X, Popov S, et al. The N-terminal domain of RGS4 confers receptor-selective inhibition of G protein signaling. J Biol Chem 1998;273:34687–34690.
211. Xu X, Zeng W, Popov S, et al. RGS proteins determine signaling specificity of G_q-coupled receptors. J Biol Chem 1999;274:3549–3556.
212. Bernstein LS, Ramineni S, Hague C, et al. RGS2 binds directly and selectively to the M_1 muscarinic acetylcholine receptor third intracellular loop to modulate $G_{q/11}\alpha$ signaling. J Biol Chem 2004;279:21248–21256.
213. Melliti K, Meza U, Adams B. Muscarinic stimulation of α_{1E} Ca channels is selectively blocked by the effector antagonist function of RGS2 and phospholipase C-β1. J Neurosci 2000;20:7167–7173.
214. Melliti K, Meza U, Adams BA. RGS2 blocks slow muscarinic inhibition of N-type Ca^{2+} channels reconstituted in a human cell line. J Physiol 2001;532:337–347.
215. Itoh M, Nagatomo K, Kubo Y, Saitoh O. Alternative splicing of RGS8 gene changes the binding property to the M_1 muscarinic receptor to confer receptor type-specific G_q regulation. J Neurochem 2006;99:1505–1516.
216. Dallanoce C, De Amici M, Barocelli E, et al. Novel oxotremorine-related heterocyclic derivatives: Synthesis and *in vitro* pharmacology at the muscarinic receptor subtypes. Bioorg Med Chem 2007;15:7626–7637.
217. Lewis LM, Sheffler D, Williams R, et al. Synthesis and SAR of selective muscarinic acetylcholine receptor subtype 1 (M_1 mAChR) antagonists. Bioorg Med Chem Lett 2008;18:885–890.
218. Christopoulos A, Kenakin T. G protein-coupled receptor allosterism and complexing. Pharmacol Rev 2002;54:323–374.
219. Jakubík J, Bacáková L, El-Fakahany EE, Tucek S. Positive cooperativity of acetylcholine and other agonists with allosteric ligands on muscarinic acetylcholine receptors. Mol Pharmacol 1997;52:172–179.

220. Hejnová L, Tucek S, el-Fakahany EE. Positive and negative allosteric interactions on muscarinic receptors. Eur J Pharmacol 1995;291:427–430.
221. Proska J, Tucek S. Mechanisms of steric and cooperative actions of alcuronium on cardiac muscarinic acetylcholine receptors. Mol Pharmacol 1994;45:709–717.

Chapter 9
Functional Selectivity at Serotonin Receptors

Kelly A. Berg and William P. Clarke

Abstract Drug selectivity has been attributed solely to differential affinity for different receptor subtypes. Recent experimental evidence suggests that in addition to receptor subtype selectivity, drugs are selective for individual signaling pathways coupled to a single receptor subtype. In this chapter, experimental evidence for functional selectivity of ligands acting at serotonin receptor subtypes is reviewed. The existence of functional selectivity greatly expands the diversity of action of drugs and holds the promise of improved therapeutic selectivity and reduced adverse effects.

Keywords Serotonin; G protein-coupled receptors; Functional selectivity; Efficacy; Receptor theory

9.1 Introduction

Drug selectivity is typically associated with differential affinity for different receptor subtypes. In drug development, selectivity is considered desirable since adverse effects are usually attributed to the action of a drug at nontarget receptors. In this framework, the perfect drug would be one that has perfect selectivity (i.e., it binds only to the target), although a strong case has been made for therapeutic agents that are selectively nonselective (magic shotguns), as opposed to magic bullets *(1)*. Recent advances in our understanding of receptor function indicate that drugs have more selectivity than that afforded by differential affinity for different receptor subtypes. Experimental evidence that has accumulated over the

K.A. Berg and W.P. Clarke (✉)
Department of Pharmacology, University of Texas Health Science Center,
San Antonio, TX 78229-3900, USA
e-mail: clarkew@uthscsa.edu

K. Neve (ed.), *Functional Selectivity of G Protein-Coupled Receptor Ligands*,
DOI: 10.1007/978-1-60327-335-0_9,

past decade reveals that drugs can selectively activate different cellular signaling cascades coupled to a single receptor subtype. Figures 9.1 and 9.2 illustrate the concepts of receptor selectivity and signaling specificity. Although several terminologies have been used to describe this signal transduction pathway selectivity (e.g., agonist-directed trafficking of receptor stimulus, biased agonism, stimulus trafficking, collateral efficacy), it appears that the field is settling on the term "functional selectivity" *(2)*.

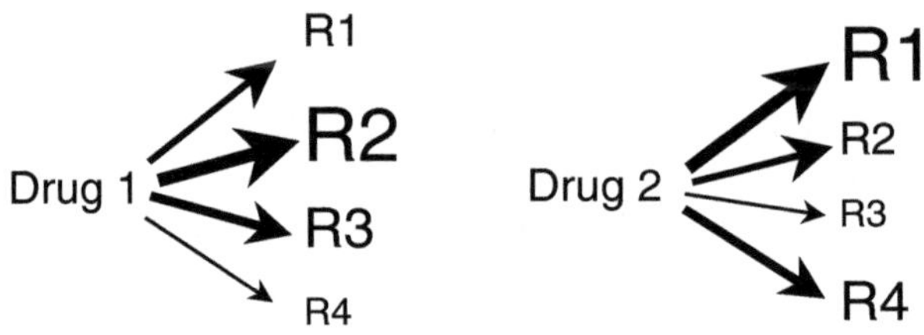

Fig. 9.1 Conceptual illustration of receptor selectivity of drugs. Both Drug 1 and Drug 2 interact with 4 receptors (R1, R2, R3, R4), but with differential affinity as illustrated by the thickness of the *arrows* and the size of the receptor label. Based upon this differential affinity, Drug 1 is selective for receptor 2, whereas Drug 2 is selective for receptor 1

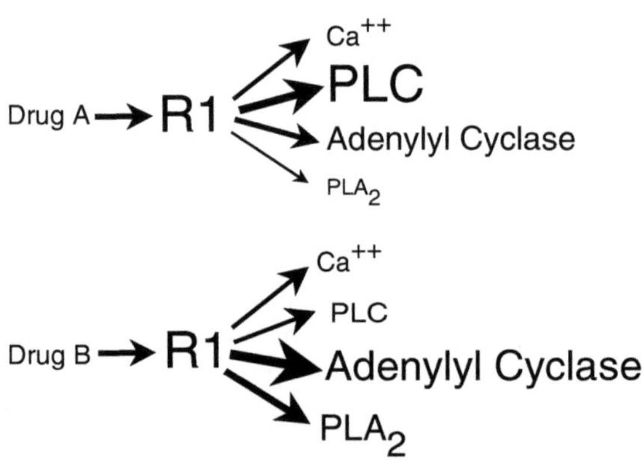

Fig. 9.2 Conceptual illustration of functional selectivity of drugs. Both Drugs A and B have affinity for receptor 1, but promote differential activation of signaling pathways coupled to the receptor. Drug A is more selective for PLC and adenylyl cyclase than for mobilization of intracellular calcium and activation of phospholipase A_2 (PLA_2). Drug B, on the other hand, is selective for the adenylyl cyclase and PLA_2 responses and produces less activation of PLC and calcium, even though Drug B interacts with the same receptor as Drug A. Notice also that both drugs produce approximately the same level of calcium pathway activation. If these two drugs were characterized using high throughput FLIPR methods, they would be considered to have the same efficacy, although their efficacy profiles differ markedly. The efficiency of receptor–effector coupling is illustrated by the thickness of the *arrows* and the size of the font

Although functional selectivity has been demonstrated in a wide variety of receptor systems, this chapter will review what is known about functional selectivity with respect to serotonin receptor subtypes.

9.2 Serotonin Receptors

Serotonin is an ubiquitous chemical messenger in the body, present in both the central nervous system and the periphery *(3)*. It would be difficult to find a physiological function or behavior that is not influenced by serotonin in some way. Some of the notable functions regulated by serotonin include affective state, body temperature, sleep, food intake, sexual behavior, cardiovascular function, endocrine secretions, gastrointestinal function, and hemostasis. Given this wide array of actions, it may not be surprising that serotonin's effects are mediated by a large variety of cell surface receptors.

Serotonin receptors are subdivided into 7 distinct families (5-HT_1–5-HT_7; see Fig. 9.3). With the exception of 5-HT_3 receptors, which are ligand-gated ion channels, all other 5-HT receptor subtypes are G protein-coupled receptors (now referred to

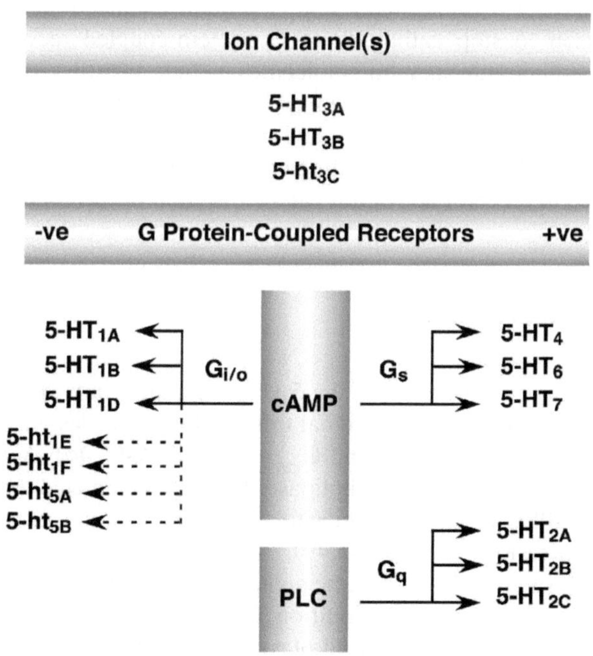

Fig. 9.3 Graphical representation of the current classification of 5-HT receptors. Receptor subtypes associated with *dashed lines* and lower case designate receptors that have not been demonstrated to definitively function in native systems. Abbreviations: 3′-5′ cyclic adenosine monophosphate (cAMP); phospholipase C (PLC); negative (–ve); positive (+ve). Modified from *(4)*

as 7 transmembrane spanning [7-TMS] receptors). There are 13 distinct genes that encode for serotonin G protein-coupled receptors. The current classification scheme is based on sequence homology, ligand pharmacology, and (common) signal transduction pathway coupling *(see 4–6, and further)*. The 5-HT_2 subfamily, which encompasses the 5-HT_{2A}, 5-HT_{2B} and 5-HT_{2C} (formerly 5-HT_{1C}) receptor subtypes, couple to Gαq proteins leading to increases in phospholipase C (PLC) activity. Interestingly, the remaining receptor subtypes all couple to the adenylyl cyclase signaling cascade either via Gαi proteins leading to inhibition of adenylyl cyclase activity or Gαs proteins leading to stimulation of adenylyl cyclase activity. In addition to the 13 genes for 5-HT G protein-coupled receptors, the diversity of 5-HT receptors is greatly increased as a result of alternative splicing and RNA-editing (of the 5-HT_{2C} receptor), which brings the total number of receptors expressed to over 30, not including single nucleotide polymorphisms *(7)*. The range of actions of drugs acting at this diverse group of receptors is further enhanced by functional selectivity of 5-HT ligands *(8,9)*.

Although Fig. 9.3 presents a single signaling pathway coupled to each G protein-coupled receptor subtype, virtually all receptors can regulate the activity of multiple intracellular signaling cascades *(10–14)*. The pathways shown in Fig. 9.3 are simply the "best studied pathways." There are numerous examples of 7-TMS receptor coupling to multiple G proteins *(15–22)*. Moreover, it is also clear that 7-TMS receptors can regulate many non-G protein mediated signaling cascades (hence the reason to stop calling them G protein-coupled receptors) *(23–27)*. Examples of non-G protein signaling molecules include arrestin *(28–31)*, Jak-STAT *(32,33)*, calmodulin *(34,35)*, and NHERF *(36,37)*. Functional selectivity is based on the formation of drug-specific receptor conformations *(38–43)* that have differential capacity to regulate the multiple signaling cascades coupled to a receptor. Thus the cellular response to receptor activation can be different for different drugs.

9.3 5-HT_{2C} Receptors

The 5-HT_{2C} receptor is a subtype of the 5-HT_2 receptor family that is best known for activation of second messenger signal transduction cascades via pertussis toxin insensitive G proteins of the $G_{q/11}$ and $G_{12/13}$ families *(6,44–46)*. Activation of the 5-HT_{2C} receptor leads to both phospholipase A_2 (PLA_2)-mediated arachidonic acid (AA) release and PLC-phosphatidylinositol (PI) hydrolysis in brain *(47,48)* and in heterologous expression systems *(45,49)*. Although it has been established that coupling to PLC is via Gαq/11 *(44)*, the signal transduction molecule(s) mediating PLA_2-AA signaling is unknown. In addition, PLD is activated via G_{13} by the 5-HT_{2C} receptor expressed naturally in choroid plexus epithelial cells and when expressed heterologously in NIH-3T3 cells *(50)*. This receptor can also couple to $G\alpha_i$ *(51,52)* ß-arrestin *(53)*, and MAPK signaling *(54,55)*, and can activate a variety of desensitization mechanisms *(56–58)*.

Studies with the 5-HT_{2C} receptor provided the first unequivocal evidence that agonists have selectivity in the signaling pathways they regulate. These early studies demonstrated reversal of efficacy order for 5-HT_{2C} agonists to stimulate PLC-mediated hydrolysis of inositol phosphates (PLC-IP) vs. PLA_2-mediated arachidonic acid release (PLA_2-AA) *(45)*. In Chinese hamster ovary (CHO) cells, stably expressing low levels of the human 5-HT_{2C} receptor, the relative efficacies of a series of 5-HT_{2C} agonists were measured by conducting full concentration-response curves to test agonists and to 5-HT as a reference agonist. Lack of receptor reserve for 5-HT in the cells was established using irreversible receptor alkylation with phenoxybenzamine. Importantly, both the PLC-IP and PLA_2-AA responses were independent of each other under the conditions of the experiment and were measured simultaneously from the same cells thereby obviating possible differences in experimental conditions in interpretation of the results. Figure 9.4 shows concentration-response curves of 5-HT_{2C} agonists expressed as the percentage of the maximal response of 5-HT (curves for 5-HT were run in each experiment). When expressed in this way, the plateau of each curve represents the relative efficacy of the test agonist. Traditional receptor theory predicts that the curves for both responses for each agonist should be superimposable – there should be no difference in relative efficacy between responses. It is clear, however, that agonist relative efficacy differed depending upon whether the PLC-IP or the PLA_2-AA response was measured, thus supporting functional selectivity for these ligands. More striking, however, is that there was a *reversal* of efficacy order. Some agonists (DOI, LSD, bufotenin) preferentially activated the PLA_2-AA response whereas others (TFMPP and quipazine) favored the PLC-IP response. The rank order of agonist relative efficacy was dependent upon the response measured. For the PLC-IP response, rank order of relative efficacy was TFMPP = quipazine > bufotenin = DOI > LSD, whereas when the measure was AA release rank order of relative efficacy was bufotenin = DOI > quipazine = TFMPP > LSD. As expected for responses elicited from activation of a single receptor in the absence of receptor reserve, potency of the agonists did not differ between responses. Data such as these indicate that, upon receptor activation, agonists are capable of providing multiple stimuli of different intensity that are differentially distributed to the signaling pathways coupled to the receptor.

It should be noted that the measurement of ligand *relative* efficacy is important to assess functional selectivity. Because system-dependent parameters (quantity and type of G protein expression, levels of effector expression, etc.) influence drug efficacy, simple comparison of maximal responses of a drug for different signaling pathways does not provide a measure of the receptor stimulus that is delivered by the drug-receptor complex to each of the signaling systems. For example, a drug may produce a receptor conformation that weakly couples to a particular G protein, but if that G protein is expressed at high levels in the cell under study, a strong response can occur. Consequently, to assess drug action at a receptor-signaling pathway pair, relative measures that nullify cellular system-dependent factors must be used.

Relative efficacy is calculated as the ratio of the maximal response of a test drug to that of a reference drug $\left(E_{\max(\text{test})} / E_{\max(\text{ref})}\right)$. Under some circumstances

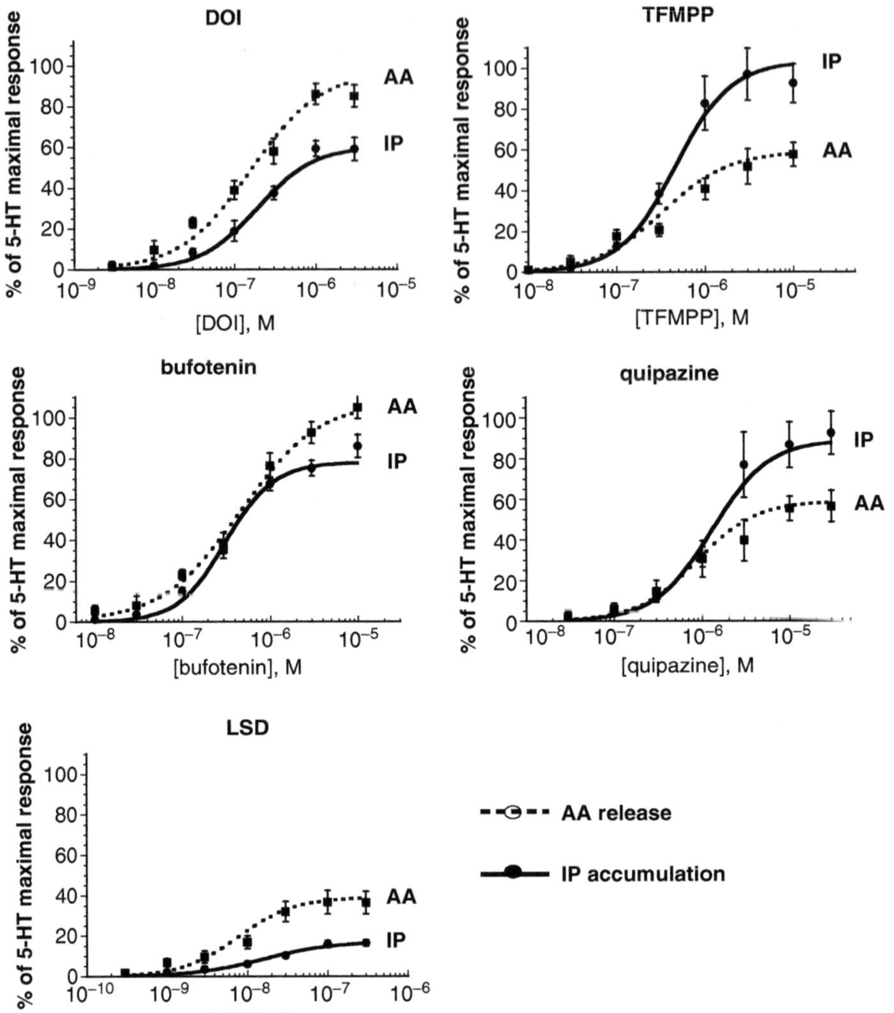

Fig. 9.4 Functional selectivity of 5-HT$_{2C}$ agonists. Concentration response curves for 5-HT$_{2C}$ agonists measuring arachidonic acid (AA) release (PLA$_2$-AA) and inositol phosphate (IP) accumulation (PLC-IP) in CHO-1C19 cells expressing the human 5-HT$_{2C}$ receptor (200 fmol/mg protein). Cells, in serum free medium, were labeled with 1 μCi/ml [3H]-*myo*-inositol (10–25 Ci/mmol) for 24 h and with 0.1 μCi/ml [14C]-arachidonic acid (57 mCi/mmol) for 4 h at 37°C. Measurements of PLC-mediated IP accumulation and PLA2-AA release were made from the same multiwell, simultaneously, after 10 min of agonist exposure. Data are expressed as the percent of the maximal response to 5-HT, concentration-response curves for which were run in each experiment. Data were fit to a three parameter logistic equation using nonlinear regression analysis to obtain estimates of Emax, EC50, and slope parameters. Because of the absence of receptor reserve, the maximal agonist response represents agonist relative efficacy for each effector. Notice that the drugs do not differ in potency between responses (AA vs. IP) as expected for drugs acting at a single receptor in the absence of receptor reserve. Reproduced with permission from *(45)*

(presence of receptor reserve), it is necessary to incorporate receptor occupancy into the calculation

$$\left(\frac{K_A / EC_{50}(\text{test})}{K_A / EC_{50}(\text{ref})}\right).$$

Because the responses to the test drugs and the reference drug are made in the same cell system (measuring the same responses at the same time and under the same experimental conditions), system-dependent variables (e.g. receptor density, efficiency of receptor-effector coupling, etc.) that influence efficacy (maximal response) cancel out. As a result, relative efficacy provides the best measure of the ability of the test drug to alter receptor activation for a particular response. Differences in potency of drugs (especially reversal of potency order) have sometimes been used as support for functional selectivity. The underlying basis for this is that efficacy can influence potency under conditions where there is high efficiency of receptor–effector coupling (receptor reserve). Therefore, response-dependent intrinsic efficacy (i.e., functional selectivity) can lead to response-dependent potency of drugs. However, caution must be used in interpreting potency differences as support for functional selectivity since potency is influenced by both affinity and efficacy. A response-dependent potency difference, not due to functional selectivity, would result from differences in the efficiency of receptor–effector coupling between responses (e.g., receptor reserve present for one response but not the other).

Using site-directed mutagenesis and molecular modeling techniques, Shapiro et al. *(59)* reported that even small changes in the structure of a ligand or receptor can lead to dramatic and unpredictable changes in receptor activation. Figure 9.5 shows that small differences in the structure of 5-HT_{2C} receptor ligands can result in significant changes in functional selectivity *(60)*. For PLC-IP accumulation response, DOM and 2,5-DMA are partial agonists, with relative efficacies (with respect to

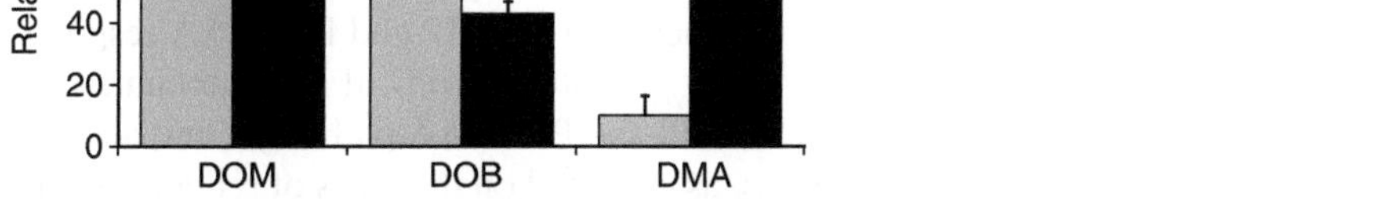

Fig. 9.5 Functional selectivity of DOM, DOB, and DMA at 5-HT_{2C} receptors. PLC-IP and PLA_2-AA were measured as described above in Fig. 9.4. The *y*-axis represents the relative efficacy of the drugs with respect to 5-HT. Data are reproduced from *(60)* in which the functional selectivity of a series of phenethylamines and phenylisopropylamines at 5-HT_{2A} and 5-HT_{2C} receptors was studied

5-HT) of 85% and 65%, respectively. However, DOM is a full agonist for PLA_2-AA release, whereas 2,5-DMA does not elicit any significant response. It should be noted that DOM and 2,5-DMA differ only in the presence or absence of a methyl group at C4. Thus, even in a closely related series of compounds, subtle structural modifications have a profound impact on drug efficacy, and importantly, affect the different signaling pathways in a different (and at this time unpredictable) manner.

There have been other reports that also suggest that ligands may differentially signal via the 5-HT_{2C} receptor. Werry et al. *(54)* found that the relative efficacy of DOI and quipazine was reversed when measured for the PLC-IP and calcium responses vs. phosphorylation of ERK1/2 via the 5-HT_{2C} receptor expressed in CHO cells. ERK1/2 phosphorylation by 5-HT was dependent upon a PLD-mediated signaling cascade and independent of PLC. Lysergic acid diethylamide (LSD) also may elicit unique signaling via the 5-HT_{2C} receptor *(61)*. In NIH-3T3 cells expressing the 5-HT_{2C} receptor, it was reported that although LSD was equally efficacious as 5-HT in stimulating PLC-IP, it does not elicit increases in intracellular calcium, or cause phosphorylation of the 5-HT_{2C} receptor. Although LSD is known to have weaker efficacy than 5-HT, it does not appear that this differential signaling could be the result of differences in receptor–effector coupling efficiency or response amplification ("strength of stimulus", see *42)* as the relative efficacies of other weak agonists (DOI, DOB, and m-CPP) were similar to that of 5-HT. However, strength of stimulus does appear to underlie ligand-specific coupling of the 5-HT_{2C} receptor ($5\text{-HT}_{2C\text{-VSV}}$ isoform, see below) to $G\alpha q/11$ vs $G\alpha i$ *(62)*.

The 5-HT_{2C} receptor is unique among 7-TMS receptors in that mRNA transcripts of the rat and human 5-HT_{2C} receptor have been found to undergo adenosine-to-inosine editing events at five sites, which encompass amino acids 156–160 within the putative second intracellular loop of the encoded human receptor, resulting in production of 14 receptor isoforms *(63,64)*. In human brain, the nonedited receptor contains the amino acids isoleucine, asparagine, and isoleucine (i.e., INI) at positions 156, 158, and 160, respectively, while two of the principle edited isoforms express valine, serine, and valine (i.e., VSV) or valine, glycine, and valine (i.e., VGV) corresponding to these amino acid positions. Several groups have reported differences in function of some edited receptor isoforms *(63–67)*, and it has been suggested that alterations in RNA editing efficiency may be involved in the etiology of schizophrenia *(68)* and affective disorders *(69,70)*.

RNA-editing of 5-HT_{2C} mRNA alters the capacity of the receptors to subserve functional selectivity. In contrast to selective signaling at the nonedited receptor, agonist relative efficacy was not different between PLC-IP and PLA_2-AA responses for the fully edited $5\text{-HT}_{2C\text{-VSV}}$ and $5\text{-HT}_{2C\text{-VGV}}$ isoforms *(65)*. It is important to note that 5-HT was able to stimulate both PLC-PI and PLA_2-AA pathways, and although potency was reduced for both responses, as expected on the basis of reduced affinity of 5-HT *(63,66,67)*, the potency did not differ between the two responses. The lack of effector pathway-dependence of agonist relative efficacy for the $5\text{-HT}_{2C\text{-VSV}}$ and $5\text{-HT}_{2C\text{-VGV}}$ isoforms indicates that the potential for agonists to traffic receptor stimuli differently to effector mechanisms is missing from these edited 5-HT_{2C} receptor isoforms. Since the $5\text{-HT}_{2C\text{-VSV}}$ and $5\text{-HT}_{2C\text{-VGV}}$ receptor isoforms differ from the nonedited isoform, 5-HT$_{2C-INI}$, in 3 amino acids located in the second intracellular loop, the lack of agonist-directed signaling by these edited isoforms suggests that the second intracellular loop plays an important role in transmitting agonist-specific information to the cellular signaling systems. The loss of functional selectivity suggests that either (a) these fully edited receptor isoforms are not capable of adopting ligand-specific conformations or (b) that the capacity of ligand-specific conformations to differentially interact with signaling molecules is impaired.

It appears that even a single, conservative amino acid change in the second intracellular loop that occurs as a result of mRNA-editing (valine for isoleucine at position 156 of the human 5-HT$_{2C}$ receptor; I156 V; 5-HT$_{2C-VNI}$), alters, but does not abolish, stimulus trafficking *(71)*. Ligands with selectivity for the PLC-IP response in 5-HT$_{2C-INI}$ cells retained this selectivity in 5-HT$_{2C-VNI}$ expressing cells. However, ligands with selectivity for PLA$_2$-AA pathway in 5-HT$_{2C-INI}$ cells lost the capacity for preferential PLA$_2$ activation in 5-HT$_{2C-VNI}$ cells. The reduction in agonist relative efficacy toward PLA$_2$-AA as a result of the I156 V substitution was not simply due to reduced general capacity of the edited isoforms to signal to PLA$_2$-AA as the capacity for serotonin to stimulate PLC-PI and PLA$_2$-AA was not different for the 5-HT$_{2C-VNI}$ vs. the 5-HT$_{2C-INI}$ isoforms. These data suggest that the I156 V substitution that occurs naturally as a result of mRNA editing can dramatically alter the signaling profile of the 5-HT$_{2C}$ receptor and underscores the need to study the other signaling pathways as well as the signaling capacity of each of the 5-HT$_{2C}$ receptor isoforms.

Functional selectivity also extends to the capacity of 5-HT$_{2C}$ agonists to activate cellular desensitization mechanisms *(57)*. Application of 5-HT to CHO cells expressing low levels of the 5-HT$_{2C}$ receptor leads to a rapid, time-dependent loss of responsiveness for subsequent stimulation of the PLC-IP and PLA$_2$-AA signaling pathways (Fig. 9.6). The magnitude of the reduction in responsiveness of both pathways was similar (≈60%). However, desensitization elicited by pretreatment with other ligands was quite varied and was not related to their relative efficacy to activate PLC-IP or PLA$_2$-AA. For example, the magnitude of desensitization of the AA and IP responses elicited by TFMPP, an agonist with relative efficacy values of 1 and 0.6 for PLC and PLA$_2$, respectively, was similar to that produced by 5-HT. However, desensitization elicited by quipazine, a drug with relative efficacy for PLC and PLA$_2$ similar to that of TFMPP (0.9 and 0.6, respectively) was markedly different from that produced by TFMPP. Pretreatment with quipazine produced less desensitization of the PLC response and enhanced the responsiveness of the PLA$_2$ pathway. Bufotenin, an agonist with efficacies similar to that of 5-HT to activate PLC and PLA$_2$ (0.8 and 1, respectively) produced much less desensitization of both responses than did 5-HT. Interestingly, pretreatment with LSD produced a small reduction in responsiveness of the PLC-PI pathway, but increased responsiveness of the PLA$_2$-AA cascade.

Differences in the capacity of agonists to promote 5-HT$_{2C}$-receptor system desensitization may be important to consider with respect to physiological activity of 5-HT$_{2C}$ drugs in vivo. The 5-HT$_{2C}$ receptor may be an important target for pharmacotherapy related to control of food intake and body weight *(72)*. Differential degrees of tolerance developed to the hypophagic effects of prolonged treatment

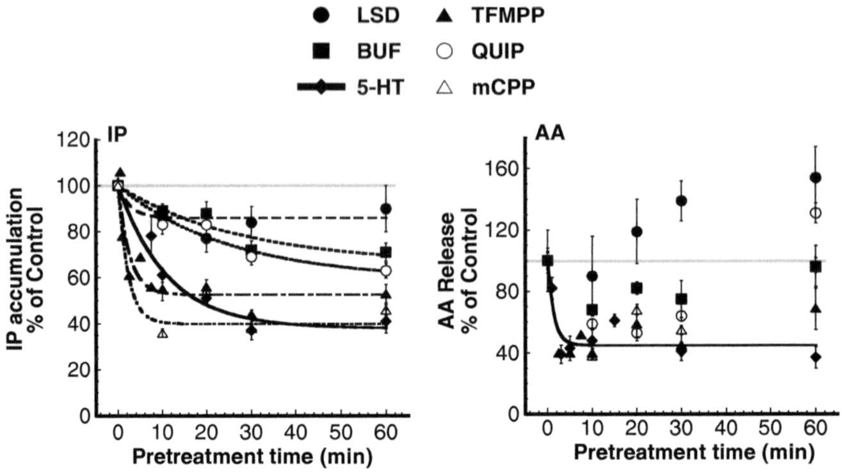

Fig. 9.6 Agonist-dependence of desensitization of 5-HT_{2C} receptor-mediated inositol phosphate (IP) accumulation and arachidonic acid (AA) release. Cells were pretreated for 0–60 min with maximal concentrations (at least 20× K_A) of the agonists 5-HT (10 μM), quipazine (30 μM), LSD (100 nM), TFMPP (10 μM), mCPP (20 μM), or bufotenin (10 μM), washed, and AA release and IP accumulation measured in response to 10 μM 5-HT for 10 min. AA release and IP accumulation were measured simultaneously, from the same cells. Data are expressed as a percent of control 5-HT stimulation and are reproduced from *(57)* with permission

(14 days) with three 5-HT_{2C} agonists, YM348, mCPP, and Ro 60–0175 *(73)*. Acute administration of all three drugs was equally efficacious at reducing food intake. However, after 14 days, the hypophagic response to YM348 was absent and the response to mCPP was reduced, whereas Ro 60–0175 remained fully efficacious. Importantly, tolerance did not develop to the hyperthermic effects of the drugs. These data suggest differences in the capacity of 5-HT_{2C} agonists to promote desensitization in vivo. Agonists that are able to activate signaling mechanisms that lead to a therapeutic response without activating signaling mechanisms that promote desensitization and tolerance would be highly desirable therapeutic agents.

Ligand-dependent differences in relative efficacy to promote desensitization of different responses coupled to the same receptor could be due to differences in the ability of the agonists to *activate* desensitization mechanisms for each response. Alternatively, ligand-specific receptor conformations could serve differentially as *targets* for desensitization mechanisms that target the receptor. It is known that some desensitization mechanisms, such as GRK and arrestin, appear to be sensitive to receptor conformation because they preferentially interact with agonist-activated receptors *(74)*. When occupied by morphine, but not DAMGO, the μ opioid receptor is not a target for arrestin-mediated uncoupling from G protein *(75)*. Moreover, such effects could be cell type-dependent. Gray et al. *(76)* found different mechanisms for 5-HT_{2A} receptor desensitization and resensitization in HEK 293 vs. C6 glioma cells. Thus, ligands may promote receptor conformations with differential capacity to activate, or to serve as targets for, desensitization mechanisms, in a cell-dependent manner.

Current models of receptor function provide receptor conformations that interconvert between inactive (R) and active states (R*), which are capable of regulating signal transduction pathways in the absence of an activating ligand (constitutive receptor activity). Such models also incorporate the existence of ligands, which reduce constitutive activity (inverse agonists). Functional selectivity is based on the notion that receptors can adopt multiple active conformations (R*, R**, R***, etc.), each of which differs in its capacity to activate effector pathways. As such, constitutive receptor activity should also be response-dependent and inverse agonists may display functional selectivity. Furthermore, it is possible for a ligand to enrich an active conformation (R*) at the expense of depleting a different active conformation (R**). Such ligands would be agonists for one response and inverse agonists for the other response, at the same time. This type of ligand behavior has been observed by several groups *(77–81)* and the ligands that act in this way have been termed protean ligands *(12,41,82,83)*.

Figure 9.7 shows that, indeed, the relative efficacy of 5-HT_{2C} inverse agonists differs markedly depending upon the response measured *(45,84,85)*. For four different responses coupled to the 5-HT_{2C} receptor in CHO cells, relative efficacy not only differed for different responses, but the response profile of relative efficacy differed between ligands. Notice especially the markedly different relative efficacies of SB 242,084 and SB 243,213. SB 243,213 is a relatively strong

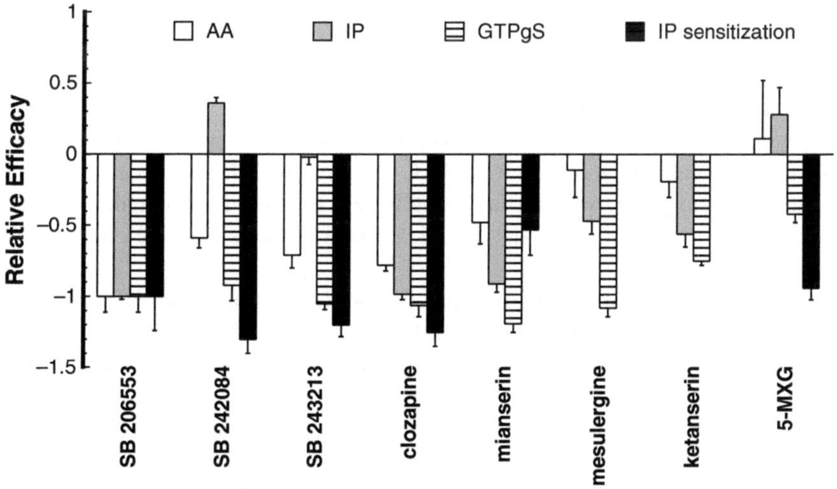

Fig. 9.7 Relative efficacy of 5-HT_{2C} ligands measured from different effector systems illustrates that inverse agonism is response-dependent. Relative efficacy for inverse agonists was calculated with SB 206,553 as the reference ligand for PLC, PLA2, and GTP[γ^{35}S] binding responses in CHO-1C7 cells (high levels of 5-HT_{2C} receptor expression, optimized to detect inverse agonism). For IP sensitization, CHO-1C19 cells (low levels of receptor expression) were treated for 24 h with the indicated ligands, washed to remove the inverse agonist, and the subsequent enhanced response to the agonist DOI was measured. Concentrations of drugs were used to produce maximal receptor occupancy (>100 × K_d). Some of these data appear in *(45,84,85)*

inverse agonist for all responses except the basal PLC response, where it has 0 relative efficacy (i.e., it is an antagonist). SB 242,084 also displays relatively strong inverse agonism toward PLA_2, Gαi activation, and the PLC sensitization response; however, for the basal PLC response, SB 242,084 is an agonist. Thus, SB 242,084 is a protean ligand. Protean ligand behavior can result from actions of ligands at receptors with different levels of constitutive receptor activity for different responses *(41,86)*. Protean ligand behavior can also occur as a result of functional selectivity *(87)*. A number of examples of protean behavior by ligands have been reported *(78,88–92)*.

9.4 5-HT_{2A} Receptors

The 5-HT_{2A} receptor is highly homologous to the 5-HT_{2C} receptor in both sequence and cellular signaling function. The receptors share about 50% overall sequence homology and about 80% homology within the transmembrane spanning domains. It is not surprising therefore that the pharmacological characteristics of these receptors are also quite similar with relatively few selective ligands available. Although there are many commonalities in the signaling cascades regulated by both receptors, important differences in cellular signaling have been observed *(49,58,93–95)*.

As with the 5-HT_{2C} receptor, many 5-HT_{2A} ligands are also functionally selective with respect to PLC and PLA_2 *(45,60,96)*. Figure 9.8 shows marked differences in the relative efficacies of a series of 5-HT_{2A} ligands for stimulation of the PLC-IP and PLA_2-AA pathways in CHO cells expressing a low density of the human 5-HT_{2A} receptor (absence of receptor reserve for 5-HT). For example, tryptamine was a very strong agonist for the PLA_2-AA cascade (relative efficacy of 1.3 with respect to 5-HT), but was much weaker than 5-HT for the PLC-IP response (relative efficacy of 0.6). On the other hand, quipazine was not functionally selective. It was an equally efficacious partial agonist for both the PLC-IP and PLA_2-AA responses (relative efficacy of 0.4 for both responses).

As a result of activation of PLC and the production of inositol trisphosphate, 5-HT_{2A} agonists also increase intracellular calcium levels ($[Ca^{2+}]_i$). It would be expected that agonists would not be functionally selective between PLC and $[Ca^{2+}]_i$ since the calcium response is derived from the PLC response. As shown in Fig. 9.8, indeed, there was no difference in agonist relative efficacy between the two responses. Thus, the lack of functional selectivity of ligands for sequential signaling responses reinforces confidence in the differences in relative efficacy between the PLC-IP and PLA_2-AA responses and in the capacity of ligands to traffic receptor stimulus differentially to independent responses.

In a similar study, the Nichols group *(96)* demonstrated functional selectivity for 5-HT_{2A} ligands in NIH-3T3 cells expressing a high density of rat 5-HT_{2A} receptors (5.5 pmol/mg protein). Although receptor reserve was present for 5-HT, comparison of the maximal responses of partial agonists revealed strong evidence of functional selectivity for some, but not all ligands. For example, the relative efficacy of LSD

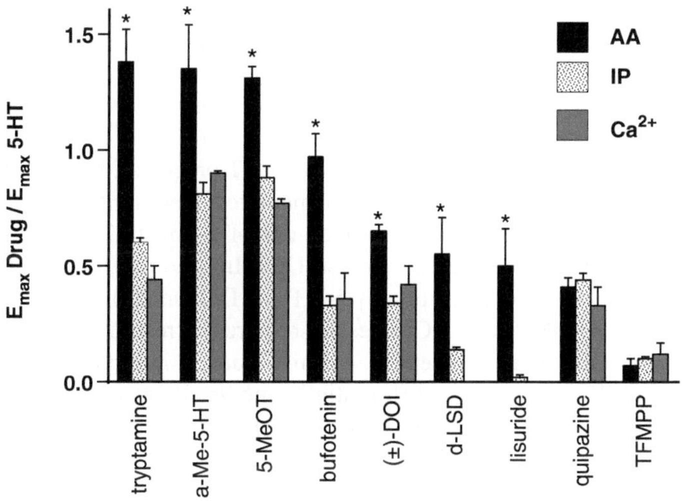

Fig. 9.8 Functional selectivity of 5-HT_{2A} receptor agonists. CHO cells expressing a low level of the 5-HT_{2A} receptor (≈200 fmol/mg protein) were incubated with maximal concentration of the indicated ligands and with 5-HT. Accumulation of inositol phosphate (IP) and release of arachidonic acid (AA) was measured from the same cells, simultaneously after 10 min of agonist exposure. Control experiments indicated that the PLC-PI and the PLA_2-AA responses were independent. Increases in intracellular calcium (Ca^{2+}) were measured using FURA-2 spectrofluorimetry. *Bars* represent the ratio of the responses to maximal concentrations of the agonists with respect to a maximal concentration of 5-HT (10 μM). Agonist concentrations used were 20 × EC50or *K*ivalues as follows (in μM, ±DOI, 1; d-LSD, 0.3; lisuride, 0.3; TFMPP, 20; quipazine, 30; bufotenin, 3; α-me-5-HT, 3; 5-MeOT, 3; tryptamine, 20). Lisuride and LSD did not elicit a measurable increase in $[Ca^{2+}]i$. *Asterisks* denote significant differences in agonist relative efficacy between agonist-elicited AA vs. IP accumulation ($p < 0.05$). Response-dependent differences in relative efficacy are consistent with agonist-directed trafficking of receptor stimulus as traditional receptor theory requires that agonist relative efficacy be the same for all responses coupled to the receptor. Notice that for sequential responses (PLC-PI and Ca^{2+}), which ligands cannot differentially regulate, relative efficacy is not different. Reproduced with permission from *(45)*

for PLC-IP differed from that for PLA_2-AA (0.22 vs. 0.56, respectively) whereas that of DOB was not different (0.79 vs. 0.75, respectfully). The largest difference in relative efficacy between responses was for tryptamine (0.91 vs. 0.41 for PLC-IP vs. PLA_2-AA, respectfully). This reversal of relative efficacy between tryptamine and LSD is strong evidence for functional selectivity. Interestingly, however, this difference in the relative efficacy of tryptamine was opposite that reported by Berg et al. (see above, *45)* for the human 5-HT_{2A} receptor expressed in CHO cells where tryptamine strongly favored the PLA_2-AA over the PLC-IP response. This difference could be the result of species differences in structure of the rat vs. human receptors, or perhaps this is an example of the impact of cell phenotype on the pharmacological properties of ligands. It is possible that the same 5-HT_{2A} receptor conformation(s) promoted by tryptamine have different activities in CHO cells vs. NIH-3T3 cells, by virtue of different types and quantities of signaling molecules expressed by the cell types. This difference in the pharmacological characteristics of tryptamine highlights the importance of the choice of assay system(s) for the lead optimization phase of drug discovery to optimize therapeutic relevance.

Recent work from Bohn's group provides strong evidence to suggest that functional selectivity of 5-HT_{2A} agonists is physiologically relevant *(97)*. In ß-arrestin2 knockout mice, 5-HT, produced by administration of the 5-HT precursor, 5-hydroxytrytophan (5-HTP), does not elicit 5-HT_{2A} receptor-mediated head-twitches as it does in wild-type mice (Fig. 9.9). However, knock-out of ß-arrestin2 does not alter the head-twitch response to the 5-HT_{2A} agonist, DOI. In in vitro experiments, receptor internalization and ERK1/2 activation by 5-HT and DOI are differentially dependent upon ß-arrestin2. Unlike 5-HT, DOI does not require ß-arrestin2 to activate ERK1/2 or to internalize the 5-HT_{2A} receptor. Agonist-induced internalization of 5-HT_{2A} receptors in HEK cells is also differentially mediated by PKC *(98)*. Although caution must be used in interpreting drug action in vivo due to the possibility of non-target receptor-mediated effects, nevertheless, these data are consistent with functional selectivity of 5-HT_{2A} agonists in vivo and emphasize the potential importance of functional selectivity with respect to drug discovery efforts.

It is intriguing to consider that agonist functional selectivity may underlie the differential hallucinogenic action of 5-HT_{2A} agonists. It has long been known that some (e.g., LSD, DOI), but not all (e.g. 5-HT, lisuride), 5-HT_{2A} agonists produce hallucinations in humans *(99)*. In fact it has been suggested that 5-HT_{2A} agonists with selectivity for PLA_2 over PLC may be associated with hallucinogenic efficacy *(96,100)*. As described earlier with respect to the Bohn study, DOI and 5-HT differ in their requirement for ß-arrestin2 for behavioral and cellular signaling activities.

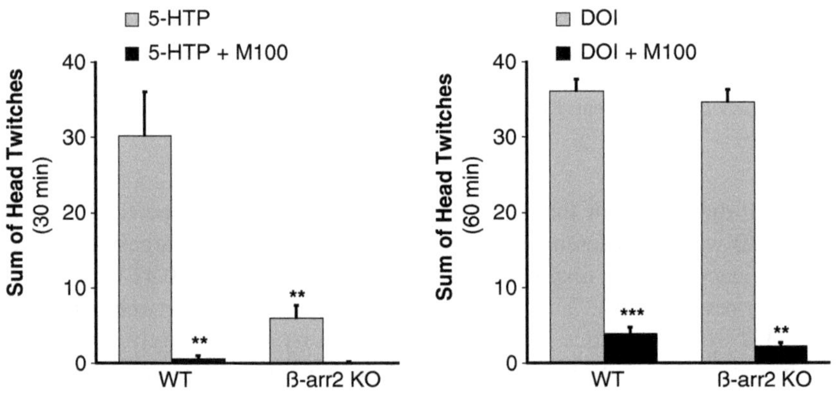

Fig. 9.9 Differential dependence upon ß-arrestin2 for head twitch behavior for 5-HT and DOI acting via the 5-HT_{2A} receptor. Wild type (WT) and ß-arrestin2 knockout (ß-arr KO) mice were treated with vehicle (0.9% saline) or drug (1 mg/kg DOI or 100 mg/kg 5-HTP) given i.p. at a volume of 10 µl per gram of body weight. Immediately after the injection, each mouse was placed individually into a Plexiglas box, and the number of head twitches was counted by two observers in 5 or 10-min increments. The 5-HT_{2A} receptor antagonist M100907 was injected i.p. at a dose of 0.05 mg/kg, 10 min before 5-HTP or DOI. Reproduced from *(97)* with permission

Sealfon's group has shown that hallucinogenic compounds such as DOI and LSD elicit similar patterns of gene expression via activation of the 5-HT$_{2A}$ receptor in mouse somatosensory cortex that are different from the gene expression profile produced by the nonhallucinogenic compound lisuride *(101,102)*. Moreover, signaling by hallucinogenic compounds mediated by cortical 5-HT$_{2A}$ receptors was distinct from nonhallucinogenic drugs and predictive of behavioral responses. Although lisuride and LSD signal via the well-established PLC cascade in primary cortical neurons, LSD signaling also occurred via Gαi/o proteins and Src because responses were blocked by pretreatment with pertussis toxin and the Src kinase inhibitor PP2 *(101)*.

9.5 5-HT$_{1A}$ Receptors

The 5-HT$_{1A}$ receptor is a member of the 5-HT$_1$ family of serotonin receptors, which are best known for their capacity to inhibit adenylyl cyclase activity via pertussis toxin-sensitive G proteins. These receptors have also been shown to couple to K^+ channels, Ca^{++} channels, PLC, intracellular calcium, the mitogen-activated protein kinase cascade, and Na + /K + ATPase *(103,104)*. This receptor is an important target for the management of anxiety, depression, and schizophrenia *(105–108)*.

The first evidence that 5-HT$_{1A}$ ligands might be functionally selective came from the work of Gettys et al. *(109)*. Using a coimmunoprecipitation strategy with a photoreactive radiolabeled GTP analog in CHO cells expressing the human 5-HT$_{1A}$ receptor, the efficacy of ipsapirone to promote coupling of the 5-HT$_{1A}$ receptor to Gαi3 was greater than that for Giα2, whereas the full agonist 5-HT and the partial agonist rauwolscine did not distinguish between the two G proteins. Newman-Tancredi et al. *(110)* also examined agonist-mediated coupling of the human 5-HT$_{1A}$ receptor expressed in CHO cells to Gαi3 using antibody-capture with a scintillation proximity assay. Some ligands reported to have high efficacy, such as 5-HT and 8-OH-DPAT, produced bell-shaped concentration-response curves for GTP[γ^{35}S] binding to Gαi3, whereas curves for weaker agonists (e.g., pindolol) were monophasic. Although these differences in ligand efficacy to activate Gαi3 could be evidence of stimulus trafficking, more experiments are necessary to rule out "strength of stimulus" mechanisms, as the authors suggest.

Pauwels and Colpaert *(111)* investigated the action of a variety of 5-HT$_{1A}$ ligands on 5-HT$_{1A}$-mediated increases in intracellular calcium and GTP[γ^{35}S] binding in CHO cells and in C6-glial cells. Both responses were sensitive to prior treatment with pertussis toxin suggesting both were mediated by Gαi/o proteins. The relative efficacy of some ligands differed between the two responses. Many ligands were not able to mobilize intracellular calcium, even though their relative efficacy to stimulate GTP[γ^{35}S] binding (with respect to 5-HT) was reasonably high. The pharmacological profiles for F 14,679 and flesinoxan were especially interesting. Although both drugs had similar relative efficacy for GTP[γ^{35}S] binding (0.93 and 0.81, respectively), the relative efficacy to promote intracellular calcium release was markedly different (0.87 and 0.05, respectively). These data suggest that some drugs can strongly differentiate between two responses coupled to the 5-HT_{1A} receptor.

9.6 5-HT_{1B} Receptors

Using rabbit common carotid artery rings in vitro, Akin et al. *(112)* found that sumatriptan and eletriptan produced 5-HT_{1B}-mediated contraction, whereas naratriptan behaved as an antagonist. However, all three drugs were equally efficacious at inhibition of forskolin-stimulated adenylyl cyclase activity mediated by the 5-HT_{1B} receptor. The response-dependent efficacies of the drugs cannot be explained on the basis of "strength of stimulus." The potencies of all three agonists to inhibit adenylyl cyclase activity were similar to their affinities for 5-HT_{1B} receptors suggesting an absence of receptor reserve. Consequently, it would appear that naratriptan, acting via the 5-HT_{1B} receptor, is functionally selective for the inhibition of adenylyl cyclase activity vs. vasocontraction in the rabbit common carotid artery.

9.7 Conclusions

It is difficult to overestimate the importance of ligand functional selectivity with respect to the experimental use of drugs to understand physiological processes and for drug discovery efforts. Traditional receptor theory that has guided pharmacological thinking and drug development for the past 50 years held that drugs, acting at a single receptor subtype, could differ in the magnitude of effect they produced but not in the quality of effect. Consequently, qualitative differences in effects of drugs on physiological functions or behaviors in vivo have generally been attributed to action at different receptor subtypes. Functional selectivity means that the qualitative actions of drugs mediated by a single receptor can differ, and suggests that drugs have more selectivity than that afforded by differential affinity for different receptor subtypes. This extended level of selectivity greatly enhances the range of actions that serotonergic drugs have beyond that provided by the large array of serotonergic receptor subtypes.

Although it now appears that pharmacological actions of drugs are considerably more complex than previously thought, with this complexity comes opportunity for better drug development and improved therapeutics. To take advantage of this opportunity, we need to develop a greater understanding of the dynamics of receptor systems and signaling mechanisms and to develop appropriate assay systems that will be predictive of therapeutic drug action. The challenge now for pharmacologists is to learn to exploit, for therapeutic advantage, the newly discovered diversity found in the pharmacological treasure chest of drugs.

References

1. Roth BL, Sheffler DJ, Kroeze WK. Magic shotguns versus magic bullets: selectively non-selective drugs for mood disorders and schizophrenia. Nat Rev Drug Discov 2004;3(4):353–9.
2. Urban JD, Clarke WP, von Zastrow M, et al. Functional selectivity and classical concepts of quantitative pharmacology. J Pharmacol Exp Ther 2006;320(1):1–13.
3. Green RA. Neuropharmacology of 5-hydroxytryptamine. Br J Pharmacol 2006;147 Suppl 1:S145–52.
4. Hoyer D, Hannon JP, Martin GR. Molecular, pharmacological and functional diversity of 5-HT receptors. Pharmacol Biochem Behav 2002;71(4):533–54.
5. Kroeze W, Roth B. Molecular biology and genomic organization of G protein-coupled serotonin receptors. In: Roth B, ed. The Serotonin Receptors: From Molecular Pharmacology to Human Therapeutics. Totowa, NJ: Humana Press, Inc; 2006:1–38.
6. Raymond JR, Turner JH, Gelasco AK, et al. 5-HT receptor signal transduction pathways. In: Roth B, ed. The Serotonin Receptors: From Molecular Pharmacology to Human Therapeutics. Totowa, NJ: Humana Press, Inc; 2006:143–206.
7. Davies M, Chang C-Y, Roth B. Polymorphic and posttranslational modifications of 5-HT receptor structure. In: Roth B, ed. The Serotonin Receptors: From Molecular Pharmacology to Human Therapeutics. Totowa, NJ: Humana Press, Inc; 2006:59–90.
8. Berg KA, Clarke WP. Development of functionally selective agonists as novel therapeutic agents. Drug Discov Today: Ther Strat 2006;3(4):421–8.
9. Berg KA, Clarke WP. Agonist-directed trafficking of 5-HT receptor-mediated signal transduction. In: Roth B, ed. The Serotonin Receptors: From Molecular Pharmacology to Human Therapeutics. Totowa, NJ: Humana Press, Inc; 2006:207–36.
10. Hermans E. Biochemical and pharmacological control of the multiplicity of coupling at G-protein-coupled receptors. Pharmacol Ther 2003;99(1):25–44.
11. Maudsley S, Martin B, Luttrell LM. The origins of diversity and specificity in G protein-coupled receptor signaling. J Pharmacol Exp Ther 2005;314(2):485–94.
12. Perez DM, Karnik SS. Multiple signaling states of G-protein-coupled receptors. Pharmacol Rev 2005;57(2):147–61.
13. Raymond JR, Mukhin YV, Gelasco A, et al. Multiplicity of mechanisms of serotonin receptor signal transduction. Pharmacol Ther 2001;92(2–3):179–212.
14. Patel TB. Single transmembrane spanning heterotrimeric G protein-coupled receptors and their signaling cascades. Pharmacol Rev 2004;56(3):371–85.
15. Kenakin T. Collateral efficacy in drug discovery: taking advantage of the good (allosteric) nature of 7TM receptors. Trends Pharmacol Sci 2007;28(8):407–15.
16. Kukkonen JP. Regulation of receptor-coupling to (multiple) G proteins. A challenge for basic research and drug discovery. Receptors Channels 2004;10(5–6):167–83.
17. Zamah AM, Delahunty M, Luttrell LM, Lefkowitz RJ. Protein kinase A-mediated phosphorylation of the β_2-adrenergic receptor regulates its coupling to G_s and G_i. Demonstration in a reconstituted system. J Biol Chem 2002;277(34):31249–56.
18. Cordeaux Y, Briddon SJ, Megson AE, McDonnell J, Dickenson JM, Hill SJ. Influence of receptor number on functional responses elicited by agonists acting at the human adenosine A_1 receptor: evidence for signaling pathway-dependent changes in agonist potency and relative intrinsic activity. Mol Pharmacol 2000;58(5):1075–84.
19. Jin LQ, Wang HY, Friedman E. Stimulated D_1 dopamine receptors couple to multiple $G\alpha$ proteins in different brain regions. J Neurochem 2001;78(5):981–90.
20. Laugwitz KL, Allgeier A, Offermanns S, et al. The human thyrotropin receptor: a heptahelical receptor capable of stimulating members of all four G protein families. Proc Natl Acad Sci USA 1996;93(1):116–20.
21. Offermanns S, Wieland T, Homann D, et al. Transfected muscarinic acetylcholine receptors selectively couple to G_i-type G proteins and $G_{q/11}$. Mol Pharmacol 1994;45(5):890–8.

22. Zhu X, Gilbert S, Birnbaumer M, Birnbaumer L. Dual signaling potential is common among G_s-coupled receptors and dependent on receptor density. Mol Pharmacol 1994;46(3):460–9.
23. Bockaert J, Fagni L, Dumuis A, Marin P. GPCR interacting proteins (GIP). Pharmacol Ther 2004;103(3):203–21.
24. Rajagopal K, Lefkowitz RJ, Rockman HA. When 7 transmembrane receptors are not G protein-coupled receptors. J Clin Invest 2005;115(11):2971–4.
25. Luttrell LM, Lefkowitz RJ. The role of beta-arrestins in the termination and transduction of G-protein-coupled receptor signals. J Cell Sci 2002;115(Pt 3):455–65.
26. Brzostowski JA, Kimmel AR. Signaling at zero G: G-protein-independent functions for 7-TM receptors. Trends Biochem Sci 2001;26(5):291–7.
27. Hall RA, Premont RT, Lefkowitz RJ. Heptahelical receptor signaling: beyond the G protein paradigm. J Cell Biol 1999;145(5):927–32.
28. Shenoy SK, Lefkowitz RJ. Seven-transmembrane receptor signaling through ß-arrestin. Sci STKE 2005: cm10.
29. Drake MT, Violin JD, Whalen EJ, Wisler JW, Shenoy SK, Lefkowitz RJ. ß-Arrestin-biased agonism at the $ß_2$-adrenergic receptor. J Biol Chem 2008;283(9):5669–76.
30. Violin JD, DiPilato LM, Yildirim N, Elston TC, Zhang J, Lefkowitz RJ. $ß_2$-Adrenergic receptor signaling and desensitization elucidated by quantitative modeling of real time cAMP dynamics. J Biol Chem 2008;283(5):2949–61.
31. Violin JD, Lefkowitz RJ. ß-Arrestin-biased ligands at seven-transmembrane receptors. Trends Pharmacol Sci 2007;28(8):416–22.
32. Ali MS, Sayeski PP, Dirksen LB, Hayzer DJ, Marrero MB, Bernstein KE. Dependence on the motif YIPP for the physical association of Jak2 kinase with the intracellular carboxyl tail of the angiotensin II AT_1 receptor. J Biol Chem 1997;272(37):23382–8.
33. Guillet-Deniau I, Burnol AF, Girard J. Identification and localization of a skeletal muscle secrotonin $5\text{-}HT_{2A}$ receptor coupled to the Jak/STAT pathway. J Biol Chem 1997;272(23):14825–9.
34. Turner JH, Gelasco AK, Raymond JR. Calmodulin interacts with the third intracellular loop of the serotonin $5\text{-hydroxytryptamine}_{1A}$ receptor at two distinct sites: putative role in receptor phosphorylation by protein kinase C. J Biol Chem 2004;279(17):17027–37.
35. Turner JH, Raymond JR. Interaction of calmodulin with the serotonin $5\text{-hydroxytryptamine}_{2A}$ receptor. A putative regulator of G protein coupling and receptor phosphorylation by protein kinase C. J Biol Chem 2005;280(35):30741–50.
36. Weinman EJ, Hall RA, Friedman PA, Liu-Chen LY, Shenolikar S. The association of NHERF adaptor proteins with G protein-coupled receptors and receptor tyrosine kinases. Annu Rev Physiol 2006;68:491–505.
37. Hall RA, Premont RT, Chow CW, et al. The ß2-adrenergic receptor interacts with the Na^+/H^+-exchanger regulatory factor to control Na^+/H^+ exchange. Nature 1998;392(6676):626–30.
38. Hoffmann D, Zürn A, Bünemann M, Lohse M. Conformational changes in G-protein-coupled receptors - the quest for functionally selective conformations is open. Br J Pharmacol 2008; 153:S358–S66.
39. Vauquelin G, Van Liefde I. G protein-coupled receptors: a count of 1001 conformations. Fund Clin Pharmacol 2005;19(1):45–56.
40. Kenakin T. Agonist-receptor efficacy II: agonist trafficking of receptor signals. Trends Pharmacol Sci 1995;16(7):232–8.
41. Kenakin T. Inverse, protean, and ligand-selective agonism: matters of receptor conformation. FASEB J 2001;15(3):598–611.
42. Kenakin T. Functional selectivity through protean and biased agonism: who steers the ship? Mol Pharmacol 2007;72(6):1393–401.
43. Kenakin T. Drug efficacy at G protein-coupled receptors. Ann Rev Pharmacol Toxicol 2002;42:349–79.
44. Chang M, Zhang LS, Tam JP, Sanders-Bush E. Dissecting G protein-coupled receptor signaling pathways with membrane-permeable blocking peptides - endogenous 5-HT2C receptors in choroid plexus epithelial cells. J Biol Chem 2000;275(10):7021–9.

45. Berg KA, Maayani S, Goldfarb J, Scaramellini C, Leff P, Clarke WP. Effector pathway-dependent relative efficacy at serotonin type 2A and 2C receptors: evidence for agonist-directed trafficking of receptor stimulus. Mol Pharmacol 1998;54:94–104.
46. Gohla A, Offermanns S, Wilkie TM, Schultz G. Differential involvement of $G\alpha_{12}$ and $G\alpha_{13}$ in receptor-mediated stress fiber formation. J Biol Chem 1999;274(25):17901–7.
47. Kaufman MJ, Hartig PR, Hoffman BJ. Serotonin 5-HT_{2C} receptor stimulates cyclic GMP formation in choroid plexus. J Neurochem 1995;64:199–205.
48. Conn PJ, Sanders-Bush E, Hoffman BJ, Hartig PR. A unique serotonin receptor in choroid plexus is linked to phosphatidylinositol turnover. Proc Natl Acad Sci USA 1986;83(11):4086–8.
49. Berg KA, Maayani S, Clarke WP. 5-hydroxytryptamine$_{2C}$ receptor activation inhibits 5- hydroxytryptamine$_{1B}$-like receptor function via arachidonic acid metabolism. Mol Pharmacol 1996;50(4):1017–23.
50. McGrew L, Chang MS, Sanders-Bush E. Phospholipase D activation by endogenous 5-hydroxytryptamine 2C receptors is mediated by $G\alpha_{13}$ and pertussis toxin-insensitive G ßγ subunits. Mol Pharmacol 2002;62(6):1339–43.
51. Alberts GL, Pregenzer JF, Im WB, Zaworski PG, Gill GS. Agonist-induced GTPγ^{35}S binding mediated by human 5-HT_{2C} receptors expressed in human embryonic kidney 293 cells. Eur J Pharmacol 1999;383(3):311–9.
52. Lucaites VL, Nelson DL, Wainscott DB, Baez M. Receptor subtype and density determine the coupling repertoire of the 5-HT_2 receptor subfamily. Life Sci 1996;59(13):1081–95.
53. Marion S, Oakley RH, Kim KM, Caron MG, Barak LS. A ß-arrestin binding determinant common to the second intracellular loops of rhodopsin family G protein-coupled receptors. J Biol Chem 2006;281(5):2932–8.
54. Werry T, Gregory K, Sexton P, Christopoulos A. Characterization of serotonin 5-HT_{2C} signaling to extracellular signal related kinases 1 and 2. J Neurochem 2005;93(6):1603–15.
55. Fitzgerald LW, Burn TC, Brown BS, et al. Possible role of valvular serotonin 5-HT_{2B} receptors in the cardiopathy associated with fenfluramine. Mol Pharmacol 2000;57(1):75–81.
56. Westphal RS, Backstrom JR, Sanders-Bush E. Increased basal phosphorylation of the constitutively active serotonin 2C receptor accompanies agonist-mediated desensitization. Mol Pharmacol 1995;48(2):200–5.
57. Stout BD, Clarke WP, Berg KA. Rapid desensitization of the serotonin$_{2C}$ receptor system: Effector pathway and agonist dependence. J Pharmacol Exp Therap 2002;302(3):957–62.
58. Berg KA, Stout BD, Maayani S, Clarke WP. Differences in rapid desensitization of 5-hydroxytryptamine$_{2A}$ and 5-hydroxytryptamine$_{2C}$ receptor-mediated phospholipase C activation. J Pharmacol Exp Therap 2001;299(2):593–602.
59. Shapiro DA, Kristiansen K, Kroeze WK, Roth BL. Differential modes of agonist binding to 5-hydroxytryptamine$_{2A}$ serotonin receptors revealed by mutation and molecular modeling of conserved residues in transmembrane region 5. Mol Pharmacol 2000;58(5):877–86.
60. Moya PR, Berg KA, Gutierrez-Hernandez MA, et al. Functional selectivity of hallucinogenic phenethylamine and phenylisopropylamine derivatives at human 5-hydroxytryptamine (5-HT)$_{2A}$ and 5-HT_{2C} receptors. J Pharmacol Exp Ther 2007;321(3):1054–61.
61. Backstrom JR, Chang MS, Chu H, Niswender CM, Sanders-Bush E. Agonist-directed signaling of serotonin 5-HT_{2C} receptors: Differences between serotonin and lysergic acid diethylamide (LSD). Neuropsychopharmacology 1999;21(2):S77–S81.
62. Cussac D, Newman-Tancredi A, Duqueyroix D, Pasteau V, Millan MJ. Differential activation of Gq/11 and Gi_3 proteins at 5-hydroxytryptamine$_{2C}$ receptors revealed by antibody capture assays: Influence of receptor reserve and relationship to agonist-directed trafficking. Mol Pharmacol 2002;62(3):578–89.
63. Niswender CM, Copeland SC, Herrick-Davis K, Emeson RB, Sanders-Bush E. RNA editing of the human serotonin 5-hydroxytryptamine 2C receptor silences constitutive activity. J Biol Chem 1999;274(14):9472–8.
64. Burns CM, Chu H, Rueter SM, et al. Regulation of serotonin-2C receptor G-protein coupling by RNA editing. Nature 1997;387(6630):303–8.

65. Berg KA, Cropper JD, Niswender CM, Sanders-Bush E, Emeson RB, Clarke WP. RNA-editing of the 5-HT_{2C} receptor alters agonist-receptor-effector coupling specificity. Brit J Pharmacol 2001;134(2):386–92.
66. Fitzgerald LW, Iyer G, Conklin DS, et al. Messenger RNA editing of the human serotonin 5-HT_{2C} receptor. Neuropsychopharmacology 1999;21(2):S82-S90.
67. Herrick-Davis K, Grinde E, Niswander CM. Serotonin 5-HT_{2C} receptor RNA editing alters receptor basal activity: implications for serotonergic signal transduction. J Neurochem 1999;73(4):1711–7.
68. Sodhi MS, Burnet PW, Makoff AJ, Kerwin RW, Harrison PJ. RNA editing of the 5-HT_{2C} receptor is reduced in schizophrenia. Mol Psychiatry 2001;6(4):373–9.
69. Gurevich I, Tamir H, Arango V, Dwork AJ, Mann JJ, Schmauss C. Altered editing of serotonin 2C receptor pre-mRNA in the prefrontal cortex of depressed suicide victims. Neuron 2002; 34(3):349–56.
70. Niswender CM, Herrick-Davis K, Dilley GE, et al. RNA editing of the human serotonin 5-HT_{2C} receptor. alterations in suicide and implications for serotonergic pharmacotherapy. Neuropsychopharmacology 2001;24(5):478–91.
71. Berg KA, Dunlop J, Sanchez T, Silva M, Clarke WP. A conservative, single-amino acid substitution in the second cytoplasmic domain of the human $serotonin_{2C}$ receptor alters both ligand-dependent and -independent receptor signaling. J Pharmacol Exp Ther 2008;324(3):1084–92.
72. Miller KJ. Serotonin 5-HT_{2C} receptor agonists: potential for the treatment of obesity. Mol Interv 2005;5(5):282–91.
73. Hayashi A, Suzuki M, Sasamata M, Miyata K. Agonist diversity in 5-HT_{2C} receptor-mediated weight control in rats. Psychopharmacology (Berl) 2005;178(2–3):241 9.
74. Benovic JL, Strasser RH, Caron MG, Lefkowitz RJ. ß-adrenergic receptor kinase: identification of a novel protein kinase that phosphorylates the agonist-occupied form of the receptor. Proc Natl Acad Sci USA 1986;83(9):2797–801.
75. Whistler JL, von Zastrow M. Morphine-activated opioid receptors elude desensitization by beta-arrestin. Proc Natl Acad Sci USA 1998;95(17):9914–9.
76. Gray JA, Sheffler DJ, Bhatnagar A, et al. Cell-type specific effects of endocytosis inhibitors on $5\text{-hydroxytryptamine}_{2A}$ receptor desensitization and resensitization reveal an arrestin-, GRK2-, and GRK5-independent mode of regulation in human embryonic kidney 293 cells. Mol Pharmacol 2001;60(5):1020–30.
77. De Deurwaerdére P, Navailles S, Berg KA, Clarke WP, Spampinato U. Constitutive activity of the serotonin2C receptor inhibits in vivo dopamine release in the rat striatum and nucleus accumbens. J Neurosci 2004;24(13):3235–41.
78. Ganguli SC, Park CG, Holtmann MH, Hadac EM, Kenakin TP, Miller LJ. Protean effects of a natural peptide agonist of the G protein-coupled secretin receptor demonstrated by receptor mutagenesis. J Pharmacol Exp Ther 1998;286(2):593–8.
79. Lane JR, Powney B, Wise A, Rees S, Milligan G. Protean agonism at the dopamine D2 receptor: (S)-3-(3-hydroxyphenyl)-*N*-propylpiperidine is an agonist for activation of G_{o1} but an antagonist/ inverse agonist for G_{i1},G_{i2}, and G_{i3}. Mol Pharmacol 2007;71(5):1349–59.
80. Newman-Tancredi A, Cussac D, Marini L, Touzard M, Millan MJ. h5-HT_{1B} receptor-mediated constitutive Galphai3-protein activation in stably transfected Chinese hamster ovary cells: an antibody capture assay reveals protean efficacy of 5-HT. Br J Pharmacol 2003;138(6):1077–84.
81. Pauwels PJ, Rauly I, Wurch T, Colpaert FC. Evidence for protean agonism of RX 831003 at α_{2A}-adrenoceptors by co-expression with different Gα protein subunits. Neuropharmacology 2002;42(6):855–63.
82. Neubig RR. Missing links: mechanisms of protean agonism. Mol Pharmacol 2007;71(5):1200–2.
83. Chidiac P. Considerations in the evaluation of inverse agonism and protean agonism at G protein-coupled receptors. Methods Enzymol 2002;343:3–16.
84. Berg KA, Navailles S, Sanchez TA, et al. Differential effects of 5-methyl-1-[[2-[(2-methyl-3-pyridyl) oxyl]-5-pyridyl]carbamoyl]-6-trifluoro methylindone (SB 243213) on $5\text{-hydroxytryptamine}_{2C}$ receptor-mediated responses. J Pharmacol Exp Ther 2006;319(1):260–8.

85. Berg KA, Stout BD, Cropper JD, Maayani S, Clarke WP. Novel actions of inverse agonists on 5-HT_{2C} receptor systems. Mol Pharmacol 1999;55(5):863–72.
86. Kenakin T. Pharmacological proteus? Trends Pharmacol Sci 1995;16(8):256–8.
87. Brink CB. Protean behavior by agonists: agonist-directed trafficking of receptor signaling. Trends Pharmacol Sci 2002;23(10):454–5.
88. Chidiac P, Hebert TE, Valiquette M, Dennis M, Bouvier M. Inverse agonist activity of beta-adrenergic antagonists. Mol Pharmacol 1994;45(3):490–9.
89. Chidiac P, Nouet S, Bouvier M. Agonist-induced modulation of inverse agonist efficacy at the $ß_2$-adrenergic receptor. Mol Pharmacol 1996;50(3):662–9.
90. Gbahou F, Rouleau A, Morisset S, et al. Protean agonism at histamine H_3 receptors in vitro and in vivo. Proc Natl Acad Sci USA 2003;100(19):11086–91.
91. Jansson CC, Kukkonen JP, Nasman J, et al. Protean agonism at α_{2A}-adrenoceptors. Mol Pharmacol 1998;53(5):963–8.
92. Pauwels PJ, Rauly I, Wurch T, Colpaert FC. Evidence for protean agonism of RX 831003 at α_{2A}-adrenoceptors by co-expression with different Gα protein subunits. Neuropharmacology 2002;42(6):855–63.
93. Roth BL, Willins DL, Kristiansen K, Kroeze WK. 5-Hydroxytryptamine$_2$-family receptors (5-hydroxytryptamine$_{2A}$, 5- hydroxytryptamine$_{2B}$, 5-hydroxytryptamine$_{2C}$): where structure meets function. Pharmacol Ther 1998;79(3):231–57.
94. Grotewiel MS, Sanders-Bush E. Differences in agonist-independent activity of 5-HT_{2A} and 5-HT_{2C} receptors revealed by heterologous expression. Naunyn Schmiedebergs Arch Pharmacol 1999;359(1):21–7.
95. Berg KA, Clarke WP, Sailstad C, Saltzman A, Maayani S. Signal transduction differences between 5-hydroxytryptamine type 2A and type 2C receptor systems. Mol Pharmacol 1994;46(3):477–84.
96. Kurrasch-Orbaugh DM, Watts VJ, Barker EL, Nichols DE. Serotonin 5-hydroxytryptamine$_{2A}$ receptor-coupled phospholipase C and phospholipase A_2 signaling pathways have different receptor reserves. J Pharmacol Exp Therap 2003;304(1):229–37.
97. Schmid CL, Raehal KM, Bohn LM. Agonist-directed signaling of the serotonin 2A receptor depends on ß-arrestin2 interactions in vivo. Proc Natl Acad Sci USA 2008;105(3):1079–84.
98. Bhattacharyya S, Raote I, Bhattacharya A, Miledi R, Panicker MM. Activation, internalization, and recycling of the serotonin 2A receptor by dopamine. Proc Natl Acad Sci USA 2006;103(41):15248–53.
99. Nichols DE. Hallucinogens. Pharmacol Therap 2004;101(2):131–81.
100. McLean TH, Parrish JC, Braden MR, Marona-Lewicka D, Gallardo-Godoy A, Nichols DE. 1-Aminomethylbenzocycloalkanes: conformationally restricted hallucinogenic phenethylamine analogues as functionally selective 5-HT_{2A} receptor agonists. J Med Chem 2006;49(19):5794–803.
101. Gonzalez-Maeso J, Weisstaub NV, Zhou M, et al. Hallucinogens recruit specific cortical 5-HT_{2A} receptor-mediated signaling pathways to affect behavior. Neuron 2007;53(3):439–52.
102. Gonzalez-Maeso J, Yuen T, Ebersole BJ, et al. Transcriptome fingerprints distinguish hallucinogenic and nonhallucinogenic 5-hydroxytryptamine 2A receptor agonist effects in mouse somatosensory cortex. J Neurosci 2003;23(26):8836–43.
103. Raymond JR, Mukhin YV, Gettys TW, Garnovskaya MN. The recombinant 5-HT_{1A} receptor: G protein coupling and signalling pathways. Br J Pharmacol 1999;127(8):1751–64.
104. Lanfumey L, Hamon M. 5-HT_1 receptors. Curr Drug Targets CNS Neurol Disord 2004; 3(1):1–10.
105. Celada P, Puig M, Amargos-Bosch M, Adell A, Artigas F. The therapeutic role of 5-HT_{1A} and 5-HT_{2A} receptors in depression. J Psychiatry Neurosci 2004;29(4):252–65.
106. Kroeze WK, Roth BL. The molecular biology of serotonin receptors: therapeutic implications for the interface of mood and psychosis. Biol Psychiatry 1998;44(11):1128–42.
107. Millan MJ. Improving the treatment of schizophrenia: focus on serotonin $(5\text{-HT})_{1A}$ receptors. J Pharmacol Exp Ther 2000;295(3):853–61.

108. Roth BL, Hanizavareh SM, Blum AE. Serotonin receptors represent highly favorable molecular targets for cognitive enhancement in schizophrenia and other disorders. Psychopharmacology (Berl) 2004;174(1):17–24.
109. Gettys TW, Fields TA, Raymond JR. Selective activation of inhibitory G-protein α-subunits by partial agonists of the human 5-HT_{1A} receptor. Biochemistry 1994;33(14):4283–90.
110. Newman-Tancredi A, Cussac D, Marini L, Millan MJ. Antibody capture assay reveals bell-shaped concentration-response isotherms for h5-HT_{1A} receptor-mediated $G\alpha_{i3}$ activation: conformational selection by high-efficacy agonists, and relationship to trafficking of receptor signaling. Mol Pharmacol 2002;62(3):590–601.
111. Pauwels PJ, Colpaert FC. Ca^{2+} responses in Chinese hamster ovary-K1 cells demonstrate an atypical pattern of ligand-induced 5-HT_{1A} receptor activation. J Pharmacol Exp Therap 2003;307(2):608–14.
112. Akin D, Onaran HO, Gurdal H. Agonist-directed trafficking explaining the difference between response pattern of naratriptan and sumatriptan in rabbit common carotid artery. Br J Pharmacol 2002;136(2):171–6.

Chapter 10
Functional Selectivity at Dopamine Receptors

Richard B. Mailman, Yan-Min Wang, Andrew Kant, and Justin Brown

Abstract This chapter reviews functional selectivity at dopamine receptors with emphasis on ligands for both D_2 and D_1 receptors. The signaling mechanisms for both the D_1 and D_2 receptors are reviewed, and we illustrate an early example of functional selectivity of the drug dihydrexidine and a related analog propylDHX. In traditional binding and functional assays, both compounds have affinity for D_{2L} receptors with typical shallow, GTP-sensitive curves. Both compounds are full agonists at inhibiting adenylate cyclase in a variety of heterologous and physiological systems. Yet the hypothesis from these data that these compounds are full D_{2L} agonists was challenged by evidence of pure antagonism at other endpoints, both in heterologous systems and physiological preparations. The fact that the behavioral effects of both compounds are atypical suggests that functionally selective drugs can cause pharmacological responses in vivo that may be therapeutically useful. This hypothesis is supported by data with the third generation antipsychotic drug aripiprazole (Abilify). Although thought to be a simple partial agonist, it is actually a functionally selective ligand whose intrinsic activity and potency are markedly affected by the milieu of the D_{2L} receptor. We conclude by a review of the current controversy regarding the hypothesized functional selectivity of certain D_1 ligands and its implications for novel therapy for Parkinson's disease and cognition. These examples strongly support the hypothesis that functional selectivity is a major mechanism that can affect drug action in vivo. For dopamine and other receptors, this offers research challenges, but also great opportunities for future drug design.

Keywords 6-hydroxydopamine, A77636, Adenylate cyclase, Antipsychotic drugs, Arachidonic acid, Aripriprazole, Behavior, Cognition, REB, Cyclooxygenase, D_1, D_1-like, D_{2L}, D_2-like, DARPP-32, Dihydrexidine, Dinapsoline, Dopamine,

R.B. Mailman (✉)
Department of Pharmacology, Penn State University College of Medicine, R130500 University Dr., Hershey, PA17033-0850
e-mail: rmailman@psu.edu

K. Neve (ed.), *Functional Selectivity of G Protein-Coupled Receptor Ligands*,
DOI: 10.1007/978-1-60327-335-0_10,

Dopamine autoreceptors, Dopamine hypothesis of schizophrenia, Dopamine stabilizer, Extracellular signal-regulated kinase (ERK), G protein-regulated inwardly rectifying potassium channel (GIRK or Kir3), $G\alpha_{OLF}$, $G\alpha_S$, GSK3, $G\alpha_{i2}$, $G\alpha_{i3}$, $G\alpha_o$, $G\alpha_z$, Heterodimers, Homodimers, Lactotrophs, Lipoxygenase, Locomotor activity, Mitogen- and stress-activated protein kinase 1 (MSK1) MN9D cells, Neurological disorders, *N*-n-propyldihydrexidine (PrDHX), Parkinson's disease, Partial agonist, Phospholipase C, PKB(AKT), Protein kinase A, Protein kinase C, Psychiatric disorders, Receptor internalization, Receptor trafficking, Schizophrenia, ß-arrestin2, Stress-activated protein kinase/Jun amino-terminal kinase (SAPK/JNK), Working memory,

10.1 Introduction

The publication of this book coincides with the 20-year anniversary of our lab stumbling into the arena of functional selectivity. When we discovered the first full D_1 agonist *(1)*, we were surprised to find that it also had D_2-like affinity. Even more surprising was our discovery that this same drug had both agonist and antagonist effects at D_2 receptors. Like many other labs entering this arena, we recognized the discordance between such observations and classical pharmacological dogma. Yet we soon became confident enough in both the findings and their ramifications to present these data *(2)*. Moreover, we believe that the first public to use of the term functional selectivity to describe ligand-induced differential signaling *(3)* was when we described the functional properties of the drug now known as aripiprazole. It is noteworthy that although all of these early data were eventually published, they were rejected several times because of the commonly-expressed view that the data "could not be right" and that some uncontrolled off-site action had not been controlled experimentally. It is therefore a particular pleasure to be a part of a volume that focuses on an arena that has finally reached "prime-time."

10.1.1 Dopamine Receptor Background

Except for the adrenergic receptors, there have probably been more drugs developed that target dopamine receptor function than for any other member of the heptahelical G protein-coupled receptor (GPCR) superfamily. This rich availability of clinical and research ligands facilitates basic research, but also allows facile clinical translation. The dopamine receptors are divided into two pharmacological families called "D_1-like" and "D_2-like" *(4,5)*. The molecular biology of the dopamine receptors has been the subject of numerous recent chapters and books *(6–9)*. Dopamine receptors are encoded by five genes. Both of the "D_1-like" receptor genes are intron-less, and include the D_1 (also called the D_{1A} in murine species) and the D_5 (also termed D_{1B}) *(10–13)*. The "D_2-like" receptors include two major splice variants of the D_2 gene,

D_{2L} (long) and D_{2S} (short), that together are the most highly expressed of the D_2-like receptors *(14–17)*. The other D_2-like receptors are D_3 and D_4 receptors, of which multiple splice variants of the D_3 receptor have been isolated, but which seem to be expressed in mice but not in humans *(15,18)*. The D_4 receptor has variable 16-nucleotide repeats in the region coding for the third intracellular loop that can vary between individuals in periodicity from 4 to 16 repeats *(19,20)*. Some of these alleles have been suggested to have major effects on behavioral phenotype *(21–23)*, although this has been controversial *(24–26)*. Throughout this chapter, the use of D_1 will refer to the primate D_1 or murine D_{1A}, whereas D_5 will refer to the primate D_5 and murine D_{1B}.

10.1.2 Clinical Impact of Dopamine Receptors: Importance to Functional Selectivity

Despite a half-century of research since the first drugs that bind to dopamine receptors (e.g., chlorpromazine) were used in clinical medicine, there remain many unresolved mechanisms. There are three major brain dopamine pathways: the nigrostriatal (from cells in the A9 region), the mesolimbic-cortical (from cells in the A10 or ventral tegmentum), and the tuberoinfundibular (hypothalamic) system *(27)*. The first evidence for a biochemical mechanism of dopamine systems was the observation that dopamine could dose-dependently stimulate the synthesis of the second messenger cAMP *(28)* in a fashion that was antagonized by antipsychotic drugs *(29)*. Both phenothiazine and thioxanthine antipsychotics competitively inhibited the dopamine-stimulated activity of adenylate cyclase in proportion to their clinical potency, suggesting that this was the major functional mechanism of dopamine in the CNS *(29,30)*. Studies with antipsychotics in other structural families, however, showed that these other behaviorally potent antipsychotics had little potency in inhibiting dopamine-stimulated adenylate cyclase *(31)*. This discrepancy led to the hypothesis of forms of dopamine receptor. One of these was linked to stimulation of adenylate cyclase *(28)*, and bound thioxanthines and phenothiazine antipsychotics with high affinity. The other was the high affinity haloperidol-binding site *(32)*, and later was shown to inhibit adenylate cyclase. This bifurcation *(4,5)* has been codified in the use of the terms D_1-like and D_2-like receptors, respectively.

For the first decade of the molecular era of the study of dopamine receptors, the primary focus was on the discrete regional and cellular localization of the various dopamine receptors, and on how this affected the functional role of dopamine and dopaminergic drugs (*see reviews by us and many others, e.g.,9,33)*. Aside from heuristic interest, this was motivated by the fact that dopamine receptors play a role in therapy or etiology of psychiatric and neurological disorders *(34–37)*. The localization of dopamine receptors in the two principal striatal outflow pathways has been of particular interest as regards Parkinson's and other neurological diseases *(38–44)*. It should be noted that there are critical functional roles for peripheral, as well as central, dopamine receptors *(45–48)*. Thus, although the focus on classical ideas of receptor action (and interaction) was predominant for many years, this chemoarchitecture suggests that the dopamine neurons comprising these systems may possibly utilize and evoke different patterns of dopamine receptor signaling.

10.1.3 Dopamine Receptor Signaling

It is now clear that signaling through modulation of adenylate cyclases is only a fraction of how dopamine receptors modulate cellular function. The signaling mechanisms of the dopamine receptors has been extensively reviewed in recent years (e.g., an excellent one was *49)*. The isoforms within the D_1 and D_2 subclasses of dopamine receptor share some signaling properties due to similarities in structure within the class (e.g., large vs. small third cytoplasmic loop). D_1-like signaling is mediated by the heterotrimeric G proteins utilizing primarily $G\alpha_s$ and $G\alpha_{olf}$ that initially activate adenylate cyclase, and secondarily cause increases in cAMP-dependent protein kinase (PKA), and the protein phosphatase-1 inhibitor DARPP-32. D_2-like receptor signaling is mediated by $G\alpha_o$ and $G\alpha_i$ isoforms that usually inhibit, rather than stimulate, adenylate cyclase. In addition, non-G protein mechanisms are of particular relevance to dopamine receptor functional selectivity. Evidence has emerged in recent years for a non-canonical cAMP/PKA-independent signaling mechanism associated with the D_2-like receptors *(50)*.

10.1.3.1 Molecular Mechanisms of Signaling of Importance to Functional Selectivity

Recently, we provided an overview of the many molecular mechanisms that may influence functional selectivity of GPCRs *(51)*. It may be useful to note a few of the mechanisms that are very relevant to the dopamine receptors. One major contributor might be the role of homodimers, and of heterodimers with receptors of several superfamilies, e.g., D_1 with NMDA; D_5 with $GABA_A$; D_2 or D_4 with EGF (epidermal growth factor), PDGF (platelet-derived growth factor), and adenosine A_{2A} receptors to name a few. Intramembrane interactions of the D_2 receptor with neurotensin and metabotropic glutamate receptors are hypothesized to be mediated by direct receptor/receptor interactions *(52)*. The formation of heteromers can be constitutive *(53)*, or stimulated by the binding of ligands, particularly agonists *(54)*. Receptor heteromerization is thought to be a key molecular mechanism in the interaction of dopamine and adenosine or somatostatin systems *(55)*. In some cases, the formation of such heteromers may change the pharmacology of the resulting complex in significant ways *(56)*. A number of interactions between the third cytoplasmic loop of D_2-like receptors and other proteins are likely to influence D_2-like receptor signaling. D_2 and D_3 receptors (but not D_1 or D_4) bind the actin-binding protein filamin A (ABP-280), and this in turn may affect signaling *(57)*. The third cytoplasmic loop of the D_2 receptor includes a binding site for spinophilin, a scaffolding protein that also binds PP1, critical for dopamine-induced modulation of glutamate receptor activity *(58,59)*. Other proteins that can play a role include calmodulin Nek, Grb2, c-Src, and a variety of G protein regulators including various RGS (regulators of G protein signaling) proteins *(60–62)*. It is this burgeoning wealth of possible mechanisms that may provide the molecular clues to examples of functional selectivity noted below.

10.1.3.2 D_1-Like Signaling

Both D_1 and D_5 receptors couple to $G\alpha_S$, an ubiquitous G protein most commonly associated with stimulation of adenylate cyclases *(63)*, but the neostriatum (the brain region with the densest dopamine innervation and the highest expression of the D_1 receptor) has abundant expression of $G\alpha_{OLF}$, but little $G\alpha_S$ *(64)*. The nucleus accumbens and olfactory tubercle also express abundant $G\alpha_{OLF}$ and little $G\alpha_S$ *(65)*. In genetically deleted $G\alpha_{OLF}$ mice, there is no D_1-mediated stimulation of adenylate cyclase in the striatum, with concomitant diminution of D_1 agonist responses *(66)*. From such data, the conclusion might be drawn that D_1-like receptors signal through $G\alpha_S$ and $G\alpha_{OLF}$ via activation of adenylate cyclase, and subsequent cAMP effects on protein kinase A. Two decades ago, we found that D_1 binding sites in the amygdala were not associated with an appreciable stimulation of adenylate cyclase, leading us to hypothesize a novel species of D_1 receptor *(67–69)*. Molecular data over the next decade clearly demonstrate that these binding sites were actually the same D_1 receptor found in striatum, and indeed represent a nontraditional mechanism(s) of signaling *(70,71)* as shall be discussed later in this chapter.

Less is known about the $\beta\gamma$ subunits that form heterotrimers with $G\alpha_S$ or $G\alpha_{OLF}$ and their role in D_1-like signaling. In one cell line, depletion of endogenous γ_7 subunits decreases D_1-mediated stimulation of adenylate cyclase, but interestingly not D_5 effects *(72)*. In neostriatal medium spiny neurons, γ_7 is abundantly expressed in neurons that also express D_1 receptor mRNA *(72,73)*, suggesting that neostriatal D_1 receptors may signal via a G protein heterotrimer that includes $G\alpha_{OLF}$ and γ_7. Moreover, it has been hypothesized that the D_1 receptor couples to other heterotrimeric G proteins, such as $G\alpha_o$ and $G\alpha_q$ *(74–76)*, and if true, potential D_1 signaling through $G\alpha_q$ would mean that protein kinase C/Ca^{2+} mechanisms, as well as the cAMP/PKA cascade, would be critical (vide infra). It may be that D_1-like receptor activation of phospholipase C (PLC) is indirect, requiring priming from another $G\alpha_q$-coupled receptor *(77)*, and this may be region specific *(78)*. In any event, the proximal effects on stimulation of cAMP synthesis can occur via activation of several subtypes of adenylate cyclases, and this can result in effects of cAMP on PKA, subsequent effects on CREB (a cAMP response element binding protein important in synaptic plasticity) and DARPP-32 (a regulatory phosphoprotein), activation of aspects of the mitogen-activated protein (MAP) kinase cascade, and effects on many voltage and ligand-gated ion channels by various combinations of direct PKA-catalyzed phosphorylation of channel subunits and DARPP-32-mediated inhibition. D_1-like receptor signaling can be regulated by both positive feedback and feed-forward mechanisms *(79)*.

These same biochemical events can also have longer term effects on cellular function by regulating transcription. Increases in cellular cAMP causes phosphorylation of CREB at Ser^{133} in the nucleus that leads to binding of p300/CREB-binding protein (CBP), which, in turn, regulates the transcription of a great variety of genes *(80,81)*. It is known that dopaminergic stimulation *(82)*, and D_1 agonists in particular *(83)*, can affect these processes, and that they are important to a variety of responses to drugs *(84)*. In addition, less direct mechanisms can also be initiated by D_1 receptor activation, such that both extracellular signal-regulated kinase (ERK) and its downstream partner mitogen and stress-activated protein kinase 1 (MSK1) may play important roles *(85)*. Indeed, differential transcriptional changes may well be critical in improving the therapeutic profile of D_1 agonists if such ligands ever become drugs. As an example, the D_1 agonist SKF81297 caused improvement in recognition and temporal order memory performance that was associated with an increased phosphorylation of both CREB and DARPP-32 in the rat prefrontal cortex *(86,87)*. If signaling directed at this endpoint is truly the molecular mediator of working memory improvement, ligands that are functionally selective and differentially activate aspects of this complex interregulated cascade of intracellular signaling may have great therapeutic utility.

10.1.3.3 D_2-Like Signaling

A major mediator of D_2-like signaling is the $G\alpha_{i/o}$ class of G proteins that are inactivated by pertussis toxin-catalyzed ADP-ribosylation *(88,89)*. One of the complications of studying functional selectivity in the brain is the multiplicity of the D2-like isoforms. Besides the D_3 receptor, there are interesting allelic variants of the D_4 receptor, and two isoforms of the D_2 receptor due to alternative splicing yielding variants with different third cytoplasmic loops [i.e., D_{2Long} (D_{2L}) and D_{2Short} (D_{2S})]. In theory, this should influence G protein interactions and possibly result in differential G protein selection, but such differential effects are seldom found. Both receptor isoforms may be able to activate multiple $G\alpha_{i/o}$ subtypes, including $G\alpha_{i2}$, $G\alpha_{i3}$, and $G\alpha_o$ but compartmentalization and availability of different effectors, scaffolding proteins, and other regulators may be as important as the structural differences between these splice variants *(for review, see90)*. Although both D_{2S} and D_{2L} can activate the pertussis toxin-insensitive G protein $G\alpha_z$ *(91,92)*, the available data suggest that $G\alpha_o$ is likely the subtype most robustly activated by both D_{2L} and D_{2S}.

The D_3 receptor is unusual in that it tends to bind agonists with higher affinity than the other D_2-like receptors, and in a fashion that is relatively insensitive to GTP *(93)*. This may be that the D_3 receptor is constrained in a conformation with high affinity for agonists regardless of interactions with G proteins *(94)*. D_3 signaling has been hypothesized to involve $G\alpha_o$, as well as $G\alpha_z$ and $G\alpha_{q/11}$ *(95,96)*. The D_4 receptor also can activate multiple pertussis toxin-sensitive G proteins, including $G\alpha_{i2}$, $G\alpha_{i3}$, $G\alpha_o$, as well as $G\alpha_z$ and the pertussis toxin-sensitive transducin $G\alpha_{t2}$.

As with the D_1 receptor, the first signaling pathway identified for D_2-like receptors was inhibition of cAMP accumulation *(97,98)*. Both D_2 and D_4 receptors inhibit adenylate cyclase in most clonal and in situ cells *(99,100)*, whereas inhibition of adenylate cyclase by the D_3 receptor is usually undetectable without molecular manipulation. In situ, D_2-like inhibition of adenylate cyclase can lead to presynaptic/autoreceptor decreases in tyrosine hydroxylase activity and in firing of nigrostriatal dopamine neurons. Like other $G\alpha_{i/o}$-coupled receptors, D_2-like receptors modulate many signaling pathways including phospholipases, ion channels, MAP kinases, and the Na^+/H^+ exchanger, as well as adenylate cyclases *(101)*. D_2 activation has major effects on K^+ currents via dissociation of $G\beta\gamma$ subunits, and unlike the generally excitatory effect of D_1 receptor activation, D_2 stimulation decreases cell excitability by increasing K^+ currents. D_2-like receptors activate a G protein-regulated inwardly rectifying potassium channel (GIRK or Kir3), modulating several potassium currents in various tissues *(102–105)*. Dopamine release-regulating autoreceptors are coupled to potassium channels *(106)* rather than to inhibition of adenylate cyclase *(107)*, and there is robust regulation of GIRK currents by D_2 receptors in substantia nigra dopamine neurons *(108)*.

The D_2-like receptors also decrease the activity of L, N, and P/Q-type channels via pertussis toxin-sensitive G proteins *(109–113)*, and this may involve $G\beta\gamma$ actions *(114,115)*. A functional consequence of this can be regulation of neurotransmitter release via N-type Ca^{2+} channels *(116–118)*. D_2-like receptors also may regulate Na^+ channels. D_2-like receptor stimulation can either increase or decrease Na^+ currents in neostriatal neurons, perhaps depending on the subtype of D_2-like receptors expressed by a given cell *(119,120)*. In most D_2 agonist-responsive neurons, D_2-like receptor stimulation decreases Na^+ currents that may involve binding of $G\beta\gamma$ subunits to the Na^+ channels *(120)*. Agonist effects at heterologously expressed D_2, D_3, or D_4 receptors can activate the widely expressed Na^+/H^+ exchanger NHE1 *(49)*.

D_2 activation can stimulate MAP kinases, including the two isozymes of extracellular signal-regulated kinase (ERK) *(121–128)* and stress-activated protein kinase/Jun amino-terminal kinase (SAPK/JNK) *(123)*. The D_2 receptor can potentiate arachidonic acid release induced by calcium-mobilizing receptors, a response mediated by cytosolic phospholipase A2 *(129–132)*, and also directly. The D_4, but not D_3, receptor also activates this pathway *(132,133)*. Arachidonic acid and its lipoxygenase and cyclooxygenase metabolites have numerous effects on cellular function, including feedback regulation of D_2-like signaling and dopamine transport *(134–137)*. Finally, the D_2 receptor can also stimulate phospholipase D, catalyzing the hydrolysis of phosphatidylcholine to form choline and phosphatidic acid *(138,139)*.

As noted earlier, non-G protein signaling has become topically important. Most evidence relates to ß-arrestin2, a key member of the GPCR desensitization machinery, to mediate downstream signaling through PKB(AKT) and GSK3 *(140,141)*. The evidence is based on studies using a variety of genetically altered mice that were deficient in major constituents of GPCR signaling, including GPCR kinases (GRKs), ß-arrestins, the dopamine transporter (DAT), and PKB. Using these mice and a series of pharmacological compounds including atypical antipsychotics, data were evolved showing the presumably simultaneous recruitment of both protein phosphatase 2A (PP2A) and PKB by ß-arrestin2. This resulted in the dephosphorylation of PKB, and the subsequent loss of the inhibitory effect of PKB on the GSK3 signaling pathway. The latter is deemed indispensible for neuronal survival and differentiation *(140–142)*.

Mechanistically, the time course of ß-arrestin/PKB signaling at the D_2 receptor takes much longer to develop (hours vs. minutes) compared with the G protein-mediated cAMP/PKA response *(140,143,144)*. This can be interpreted in several ways. These results not only caution those using pharmacological intervention at the D_2 receptor to be mindful of the wavelets of potentially opposing responses ensuing drug administration, they also hint at likely integration of responses from other GPCR and neurotransmitter systems. It is also worth pointing out that the slow response of the ß-arrestin/PKB pathway can also be indicative of its downstream effects on gene transcription culminating in target protein transcriptional profile changes or even chromatin remodeling *(145–148)*.

10.2 Evidence for Functional Selectivity at D_2-Like Receptors

10.2.1 Hypothesized Presynaptic/Autoreceptor Selective Ligands and Functional Selectivity

On the one hand, we begin the discussion of dopamine receptor functional selectivity with examples from the D_2-like receptors, a choice that might seem out-of-order based on the order of the receptor family nomenclature. On the other hand, interest in drug discovery for what we now call D_2 receptors stretches back more than a half-century based on potential therapeutic roles in schizophrenia, Parkinson's disease, and a variety of other psychiatric and neurological conditions. Conversely, during much of this time the D_1 receptor was thought as a binding site in search of a function *(149)*.

One of the major driving forces for the D_2 drug discovery was the "dopamine hypothesis of schizophrenia" which, in its 1980s rendition, posited that excess dopamine release and/or excess sensitivity of dopaminoreceptive neurons was the cause of what are now called positive symptoms. It turns out that this hypothesis is highly relevant to the discovery and validation of the functional selectivity of dopamine receptor ligands. Although antagonism of D_2 receptors has been accepted as a way of controlling positive symptoms of schizophrenia *(32)*, a novel hypothesis was developed in the late 1970s. Dopamine autoreceptors (now known to be composed of high densities of D_2 and low densities of D_3 receptors), when activated, cause a decrease in both synthesis and release of dopamine, and a decrease in the firing of dopamine neurons. Interestingly, both dopamine and D_2 agonists tend to have higher potency at these autoreceptors vs. what are regarded as postsynaptic D_2

functions *(150)*. This would then explain why the behavioral response to a full D_2 agonist is typically biphasic with respect to dose, with inhibition seen at in low doses (the result of autoreceptor stimulation) and stimulation at higher doses (direct postsynaptic activation). One of the major mechanisms explaining this biphasic effect is greater presynaptic D_2-like receptor reserve *(151)*.

These observations led to the apparently contradictory idea of also using D_2-like agonists to cause antipsychotic effects via one of two alternative mechanisms. One approach was to use a low dose of a full agonist to achieve selective presynaptic activation. The other direction was to use a partial agonist. According to pharmacological theory, partial agonists should be far more efficacious at activating the presynaptic receptors where there is high receptor reserve *(152)*. Indeed, a plethora of in vitro and animal studies suggested that one partial agonist, (-)3-PPP (preclamol), might be an excellent candidate. Although both mechanisms have a good theoretical basis, the early clinical data were disappointing *(153–158)*. In retrospect, this was probably not unexpected because of issues such as getting the "right" presynaptic relative receptor occupancy without temporal fluctuations, or finding a partial agonist with just the "right" intrinsic activity. As discussed later, however, the "partial agonist hypothesis" was resuscitated when aripiprazole was shown to have clinical efficacy (although as we note below, we believe that the relevant mechanism is functional selectivity).

10.2.2 Early Evidence for Functional Selectivity of Dopaminergic Compounds

To our knowledge, the first clear example of dopamine receptor functional selectivity resulted from serendipitous findings with dihydrexidine, a compound designed to be a selective D_1 agonist *(1)*. After dihydrexidine was found to have D_2 affinity *(159,160)*, both it and a more D_2 selective analog (*N*-n-propyldihydrexidine; PrDHX) were characterized functionally. Both compounds competed for D_2 receptors in heterologous systems or in brain tissue with shallow, guanine-nucleotide-sensitive curves similar to typical agonists *(160–162)*. In many systems, both compounds had full intrinsic activity at inhibiting adenylate cyclase activity, effects blocked by D_2 antagonists *(161,162)*. The compounds also inhibited prolactin secretion in vivo. Yet despite properties that would clearly predict that both dihydrexidine and propylDHX were full agonists, neither ligand activated D_2-like pre/autoreceptors that mediate inhibition of dopamine neuron firing, dopamine release, or dopamine synthesis *(161,162)*. Indeed, in vivo, dihydrexidine was an antagonist of the physiological actions of apomorphine, a potent D_2 agonist *(161)*. Moreover, if dihydrexidine had both D_1 and D_2 agonist properties, it should have had behavioral effects similar to apomorphine or amphetamine, yet even at very high doses the behavioral effects seemed largely D_1-like *(163)*.

These early in vitro/in situ data suggested that dihydrexidine and propylDHX had high intrinsic activity at postsynaptic D_2-like receptors, but low intrinsic activity at presynaptic receptors *(2)*. Off-site actions were largely excluded by both receptor screening and competitive pharmacological studies *(160)*. More problematic was the possibility of off-setting actions by one or more D_2-like receptors (D_{2L}, D_{2S}, D_3, D_4). Although a D_4 contribution could be ruled out based on localization, it was possible that dihydrexidine and PrDHX were agonists at one dopamine receptor isoform (e.g., D_{2L}), but antagonists at another (e.g., D_{2S} or D_3). This possibility was ruled out because these compounds are agonists at all of the D_{2L} receptors when the canonical cAMP pathway was examined. In concert with the unexpected fact that these ligands had antagonist effects at presynaptic/autoreceptor functions (where there was higher *presynaptic* receptor reserve), it led to the hypothesis that these compounds were D_2 functionally selective *(2)*.

Evidence for this hypothesis was obtained in several model systems. The first was the pituitary lactotroph that expresses only products of the D_2 gene (DRD2), and that provided two functional endpoints that paralleled those studied in situ. As in heterologously expressed D_2 receptors, dihydrexidine was a full agonist at inhibiting adenylate cyclase, with effects blocked by D_2, but not D_1, antagonists. Yet dihydrexidine had little intrinsic activity at D_2 receptors coupled to G protein-coupled inwardly-rectifying potassium channels (GIRK), and was an effective antagonist of the functional actions of dopamine in these lactotrophs *(161)*. MN9D cells are mesencephalic-derived and the transfection of the D_{2L} receptor leads to coupling to inhibition of adenylate cyclase and inhibition of dopamine release as occurs in dopamine neurons *(164,165)*. Neither dihydrexidine nor propylDHX had effects in untransfected MN9D cells, but after D_{2L}-transfection both compounds (as well as the reference D_2 agonist quinpirole) were full agonists at inhibition of adenylate cyclase activity. These agonist effects were blocked by D_2, but not D_1, antagonists. Yet when examining the depolarization-induced release of dopamine, quinpirole caused a concentration-responsive, antagonist-reversible effect, but neither dihydrexidine nor propylDHX inhibited dopamine release. Moreover, propylDHX actually antagonized the effects of quinpirole *(162)*. In toto, these studies convinced us that "functional selectivity" was "real" *(2,161,162,166–168)*.

We were clearly aware that the prevalent theory at this time did not allow for a molecule to be both full agonist and pure antagonist at the same receptor that is not having consistent intrinsic efficacy *(169)*. It was also clear that the GPCRs could have effects via involving different signaling partners *(170)*, and indeed, on this basis the possibility of functional selectivity had been hypothesized *(171)*. Although the common theoretical conceptualizations of receptors was in the form of discrete active states *(172,173)*, our conceptualization was of a dynamical system in which ligands could induce an essentially limitless number of conformational states of a target receptor, which in turn, might sometimes cause quite different effects on the receptors signaling partners *(167,174)*. Specifically, functional selectivity has sometimes been explained as the stabilization of novel discrete active states of the receptor. Our concept was that functionally-selective ligands *induce* unique conformation states of the receptor distinct from those caused by either the endogenous ligand or by antagonists. In a sense, binding of the ligand provides the driving force needed for the receptor to assume conformations that would otherwise not be energetically favorable and not otherwise found (hence our eschewing of the term "stabilization"). At the purely pharmacological level, this may just be a difference in semantics, but at the mechanistic level it may well have impact. In a sense, one can think of xenobiotic ligands as being a combination of isosteric and allosteric elements. The issues about the molecular mechanisms mediating functional selectivity, and questions about whether this phenomenon impacts on drug action in vivo make this an exciting heuristic front.

10.2.3 Functional Selectivity In Vitro Affects Pharmacological Effects In Vivo

One of the key issues was whether functional selectivity was simply a heuristically interesting phenomena, or of relevance to drug action in vivo. We tested this idea by characterizing the behavioral effects of propylDHX. Although dihydrexidine clearly did not have the behavioral actions of a D_1/D_2 agonist *(163)*, one might argue that this was because the D_1 selectivity masked the D_2 effects. For this reason, we also characterized the actions of PrDHX, which, as noted earlier, is much more D_2/D_1 selective (similar in fact to apomorphine). Classic pharmacology would have predicted that a compound with full intrinsic activity at D_2-mediated adenylate cyclases would have behavioral effects similar to apomorphine or the prototypical D_2 agonist quinpirole. Thus, such D_2/D_3 agonists have biphasic effects, causing locomotor inhibition at low doses and locomotor stimulation at higher doses *(175)*. Yet propylDHX only caused modest locomotor inhibition across a wide range of doses, with no competing behaviors that might have interfered with locomotion *(176)*. This provided possibly the first evidence in support of the hypothesis that functionally selective compounds would have novel pharmacological actions. As is discussed below, this question is now highly topical clinically (see *Sect. 10.4.1*). Moreover, as should be obvious from the brief review in *Sect.10.1.3.3*, there is ample evidence of the D_2 receptor signaling through canonical G protein mechanisms as well as non-canonical mechanisms. The potential impact of a functionally selective ligand is illustrated in Fig. 10.1. In theory, not only should it be possible for a ligand to discriminate among canonical functions mediated by different G protein heterotrimers, but as well to affect differentially longer term drug responses.

10.3 Evidence for Functional Selectivity at D_1-Like Receptors

10.3.1 Possible Mechanisms of D_1-Receptor Signaling that Could Evoke Functional Selectivity

A clearer understanding of D_1-like signaling pathways is critical for the identification and design of functionally selective D_1 compounds. D_1-family receptors are expressed not only in the central nervous system, but peripherally in many tissues including

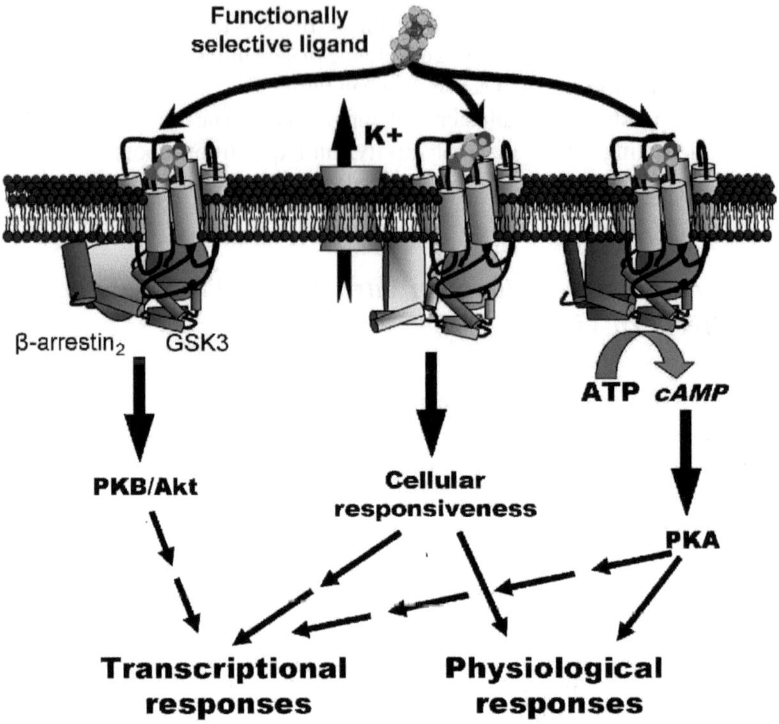

Fig. 10.1 The implications of functional selectivity at the dopamine D_2 receptor are illustrated in this cartoon. Functionally selective drugs, unlike dopamine, might differentially affect both canonical and non-canonical signaling pathways. This could result in differential acute effects illustrated by the center and right hand aspects of the figure. For example, a ligand would cause differential effects on mediators of acute action (e.g., the dopamine-induced hyperpolarization of a cell mediated via inward-rectifying potassium channels and acute actions of cAMP). As importantly, the longer term effects of drugs could also be differentially affected. Thus, changes in transcription initiated by cAMP on mechanisms like CREB could be altered by some drugs in a way different than receptor-initiated changes in transcription initiatiated by β-arrestin2/PKB/Akt

the adrenal glands, blood vessels, heart, kidney, and urinary tract *(9)*. Comparison of signaling in different tissues can be convoluted by the presence or absence of effectors (i.e., G proteins, β-arrestins, GRKs, etc.), differences in membrane dynamics (e.g., highly lipophilic neuronal cells vs. renal proximal tubular cells), or attenuation/potentiation of input signaling pathways. Here we briefly review some of the evidence for D_1-like signaling pathways, and key issues that are not fully understood.

D_1-like dopamine receptors, of the D_1 and D_5 receptor subtypes, are thought to mediate signaling through the heterotrimeric G proteins $G\alpha_s$, G_{olf}, and possibly G_q. D_1-like receptors couple prominently to the stimulation of adenylate cyclase and can stimulate additional effector pathways depending on the complement of cellular machinery within a cell *(33,49)*. Despite promising therapeutic potential for D_1

receptor agonists, few efforts have been made to investigate functional selectivity at D_1-like receptors. The most studied arm of D_1 signaling is the cAMP/PKA pathway. D_1-like receptor activation of PKA leads to a phosphorylation/dephosphorylation cycle of DARPP-32 at Thr^{34} and Thr^{35}, resulting in inhibition of protein phosphatase-1 (PP-1) *(79)*. When DARPP-32 is phosphorylated at Thr^{75} by Cdk5, it is transformed into an inhibitor of PKA, serving as a negative feedback loop. D_1-activated PKA can lead to phosphorylation of other receptors in the cell (e.g., L-Ca^{2+} channels, NMDA) and inhibition of PP-1 that then causes dephosphorylation of many substrates including the same receptors phosphorylated by PKA *(177)*. This exquisite balance provides one mechanism for tight regulation of dopaminergic signaling. One of the challenges in identifying distinct effector pathways for the study of D_1 receptor functional selectivity is the often apparent dependence of the receptor on cAMP/PKA signaling. Many D_1 receptor effectors are either conditioned, or at least modulated, by this latter pathway.

MAP kinase is an important mediator of growth and cell cycle control. Few studies have investigated its role in D_1-like receptor signaling, and such studies are complicated by the promiscuity of ERK1/2. Using a catalytically defective form of MEK1, one study reported a lack of activation by the D_1 partial agonist SKF38393; however, with wild-type MEK1, ERK1/2 was selectively activated. Interestingly, p-ERK has been found to form stable heterotrimeric complexes with the D_1 receptor and β-arrestin2 *(178)*. Nagai et al. *(179)* revealed dose-dependent D_1 activation of ERK1/2 in the mouse prefrontal cortex that was unaffected by microinjection of a D_2 antagonist and blocked by a D_1 antagonist. The mechanism(s) responsible for D_1-MAP kinase activation is unclear, but evidence suggests a dependence on the β-arrestin scaffolding protein. The D_1-MAP kinase pathway shows promise but further studies must be carried out to gain a clearer understanding of the mechanisms involved.

Another important D_1 effector pathway is that of the D_1 receptor to inhibit Na^+/H^+ (NHE) exchangers and basolateral Na^+/K^+ ATPases *(180–182)*. Binding studies in rat renal tissue have revealed the presence of D_1-like receptors in the renal cortex and vessels. Stimulation of these receptors causes vasodilatation of renal vessels, increasing blood flow to the inner cortex and medulla, leading to an increase in GFR (glomerular filtration rate) and urinary output. Inhibition of NHE via D_1-like receptors is independent of PLC activation, whereas NHE and Na^+/K^+ ATPases appear to be modulated by the PKA-cAMP pathway. Attenuation of this effect with the D_1 antagonist SCH23390 in dogs implies a specific role for D_1-like receptors *(181,183)*.

Calcium is a dynamic yet tightly regulated second-messenger pathway that is controlled differentially across non-nervous and nervous tissue. The slow IP3-mediated pathway via phospholipase C predominates in nonexcitable cells. In excitable cells, however, voltage-dependent Ca^{2+} channels (L, N, P/Q type) play a greater role, balancing Ca^{2+} intake/output from the cell. D_1-like receptors are positively coupled to L-type channels, but negatively coupled to N, P/Q-type Ca^{2+} channels *(184)*. The mechanism of D_1-stimulated inhibition of voltage-gated calcium channels is not entirely clear, but appears to be dependent on the PKA/DARPP-32 cascade *(185)*. In a study of prefrontal cortex neurons, D_1-stimulated L-type Ca^{2+}

channels lead to PKA-dependent potentiation and PKC-dependent suppression of currents *(186)*. Inhibition of T-Type Ca^{2+} currents in rat adrenal cells was found to be co-dependent on modulation by cAMP and Gβγ *(187)*.

10.3.2 Phospholipase C as a D_1 Signaling Mechanism

Phospholipase C (PLC) hydrolyzes phosphatidylinositol 4,5-bisphosphate (PIP_2) producing 1,2-diacylglycerol (DAG) and inositol 1,4,5-trisphosphate (IP_3). IP_3 binds to the IP_3 receptor, stimulating the release of Ca^{2+} from intracellular stores within the endoplasmic reticulum. DAG recruits protein kinase C (PKC) to the membrane leading to, among many other responses, NF-κB activation and actin reorganization. Phospholipase C may be activated by both $G\alpha_q$ and Gβγ. All four members of the $G\alpha_{q/11}$ family (α_q, α_{11}, α_{14}, α_{16}) have been shown to activate PLC-β isoforms but not PLC-γ, PLC-δ, or PLC-ε *(188–190)*. There is evidence from other receptor systems that agonists can induce specific receptor conformations leading to selective activation of G_s, G_i, or G_q proteins *(191–196)*. It is also clear that Gβγ plays a role. PLC-β has distinct sites for $G\alpha_q$ and Gβγ activation, and thus can be synergistically activated *(197)*. The composition of specific Gβγ subunits appears also to affect its potency for PLC-β *(198)*. This dual regulation, while a complicating factor, provides mechanisms for ligand functional selectivity. Indeed, if both cAMP/PKA and PLC signaling occur with the D_1 receptor, it provides a splendid way to exploit the principle of functional selectivity in drug discovery.

When assessing functionally selective compounds, it is crucial to identify independent signaling pathways. In LTK^- cells, D_1-linked PLC was shown to be dependent on PKA activation *(199)*. Yet the predominant PLC isoform thought to be involved, PLC-β2, is not expressed in LTK^- cells. It was found that PLC-γ was responsible for the observed activity. Therefore, the model system in question may not be predictive of all PLC-β2 expressing systems. A recent study of D_1 effects on intracellular Ca^{2+} currents supported the idea of $G\alpha_q$-mediated PLC activation, yet this mechanism was found to be co-dependent on a PKA-cAMP signal *(200)*. These studies do confound results as measurement of downstream targets is subject to positive and negative feedback inputs (i.e., PKA-mediated phosphorylation of IP_3R). In addition, measurement of intracellular Ca^{2+} currents is not directly translatable from PLC stimulation. Similar data have been found in other studies, thus there is a strong possibility of a D_1-mediated non-PLC Ca^{2+} mechanism *(201)*.

10.3.3 Implications and Complications of Purported D_1-PLC Signaling

Unlike the D_1 actions on cAMP/PKA, the importance of PLC in direct D_1 signaling is more controversial. Early evidence showed that SKF82526, a D_1 full agonist, stimulated PLC in rat renal cortical membranes *(202)*. This effect was selectively blocked by SCH23390, but unaffected by the α-adrenergic antagonists prazosin and phentolamine. Additionally, the selective D_2 agonist LY171555 did not induce IP release. Later studies have suggested concurrent $G_{q/11}$ and $G_{s/olf}$ coupling to the D_1 receptor *(74,203,204)* and a recent report indicated differential coupling by SKF83959 *(205)*. Panchalingam and Undie *(204)* reported D_1 receptor-selective G protein activation in striatal membranes using three phenylbenzazepines they hypothesized were pathway selective [SKF38393 (G_q/G_s); SKF85174 (G_s); SKF83959 (G_q)]. The addition of deoxycholate affected G_q/PLC vs. G_s/adenylate cyclase signaling, and this was interpreted as being due to a change in membrane dynamics and thus receptor conformation *(206)*.

Non-cyclase D_1-mediated signaling clearly occurs *(67,70,71,207)*. Further evidence for a cAMP/PKA-independent signaling pathway was shown in studies of adenylate cyclase V-deficient mice *(208,209)*. While 85–90% of cyclase activity was abrogated, locomotion was enhanced. Such findings are direct evidence for the possibility of discovering D_1 functionally selective compounds. Our group originally (and mistakenly) hypothesized that these non-adenylate cyclase linked effects might represent a D_1-like receptor from a different superfamily but with similar pharmacology *(67)*. We rapidly realized that this was more likely the same gene product utilizing different signaling machinery *(69–71,207)*. Interestingly, another group subsequently resurrected the hypothesis of a novel "D_1-like" receptor *(210,211)*, and those data are very relevant to the issue of D_1 functional selectivity.

The data in support of the hypothesis for a novel D_1-like receptor came from D_1 ligand activation of PLC. As noted earlier, the obvious mechanism of such effects on PLC would be the direct activation of $G\alpha_q$ with subsequent activation of PLC *(212,213)*. There are, however, recurrent issues with the design of the studies that have led to this hypothesis as it relates to the role of D_1 receptor. For reasons that are unclear, the concentrations of the drugs used in the published experiments have always been supra-pharmacological (10 μM or greater, orders of magnitude higher than the affinities of the compounds). Moreover, although there is a good correlation of binding affinity of these ligands and their potency at activating adenylate cyclase, this was not found in correlation of binding and PLC activation. Of course, one of the expressions of functional selectivity can be large changes in potency as well as intrinsic activity *(51,168,174,214,215)*. In this case, however, the combination of high required drug concentrations and SAR discordance may be a red flag.

In this regard, such studies were repeated in D_1 knockout mice, and identical PLC activation was found. As expected *(216)*, D_1-mediated adenylate cyclase activity was abolished in these mice *(210)*. The conclusion from this study was that this was evidence for a novel D_1 receptor *(210)*, yet the alternate hypothesis (effects through a non-dopamine receptor or receptors) would seem to merit equal consideration. In support of the latter, none of the known or orphan GPCRs has been matched to this proposed new D_1-like receptor. Moreover, in D_1 knockout mice, there is a profound decrease in phenylbenzazepine binding sites *(217)*, making it highly unlikely that the pharmacological properties would be unaffected if this was a D_1 effect. Indeed, this purported D_1-like $G\alpha_q$-coupled receptor does not react with a D_1 receptor antibody *(75)*. On the one hand, together, the evidence may favor the hypothesis that these PLC effects are mediated by a non-dopamine receptor binding site that can be activated with low potency by one class of dopamine ligands *(71)*. On the other hand, there are some recent data that suggest that phenylbenzazepine-based D_1 ligands actually may signal through this pathway, and indeed, in a functionally-selective fashion. A postulated functional D_1/D_2 dimer has been reported to be activated by one member of this chemical class (SKF83959), but not another (SKF83822). Interestingly, this correlates with the suggestion that SKF83959 selectively activates D_1-linked $G\alpha_q$ whereas SKF83822 selectively affects $G\alpha_s$. *(205,218)*. These types of data support the importance of PLC/$G\alpha_q$ signaling that is mediated by D_1 receptor heterodimers (e.g., with D_2) *(219)*. This arena clearly deserves great attention regardless of some of the skepticism about the overall importance of direct $G\alpha_q$ activation by D_1 receptors. If this pathway is indeed important, it may well be that certain phenylbenzazepines (and possibly D_1 ligands of other chemical classes) are highly functionally selective *(220,221)*, and may provide a route to novel new drugs.

10.3.4 Direct Evidence for Functional Selectivity at the D_1-Like Receptors

Functional selectivity has been difficult to demonstrate for D_1-like dopamine receptors because of the lack of clear effectors coupled to the receptor. The clearest evidence of D_1 functional selectivity was shown in two recent studies comparing the functional endpoints of adenylate cyclase activation and receptor internalization. The first study explored the relationship between agonist structure, receptor affinity, and efficacy of adenylate cyclase activation and receptor internalization in response to 13 agonists from three different structural classes *(220)*. This study identified several D_1 agonists that activate adenylate cyclase with great efficacy but fail to cause receptor internalization. Interestingly, internalization efficacy was found to be independent of agonist structural class and agonist affinity. This study revealed interesting disparities in the ability of synthetic D_1 agonists to regulate receptor trafficking, and suggested that, at least for the D_1 receptor, functional selectivity is not predictable by simple structural examination.

A subsequent study compared the ability of dopamine and two structurally dissimilar agonists, A77636 (chemically an isochroman) and dinapsoline (DNS, chemically an isoquinoline) in regulating receptor internalization and trafficking with that of dopamine *(221)*. These compounds are full agonists at activating adenylate cyclase, and reach steady-state internalization by 30 min. DNS exhibited efficacy similar to dopamine in causing internalization 1 h after agonist treatment, while A77636 caused significantly greater internalization. Investigation of post-endocytic agonist effects on receptor trafficking revealed significant differences in agonist regulation of receptor trafficking. Dopamine caused the D_1 receptor to recycle back to the cell surface within 1 h, whereas the D_1 receptor persisted intra-cellularly up to 48 h after removal of A77636. Surprisingly, DNS caused the receptor to recycle back to the membrane after 48 h. Pulse chase experiments, and use of actinomycin D to inhibit new protein biosynthesis, demonstrated that cell surface recovery was not due to synthesis of new proteins. Together, these data indicate that these agonists target the D_1 receptor to different intracellular trafficking pathways. Experiments revealed a slow dissociation rate of A77636 from the D_1 receptor, suggesting that ligand-receptor interactions distal to the binding pocket may dictate the ability of an agonist to cause receptor internalization and regulate long-term receptor trafficking. A77636 is notorious for producing an extremely rapid motor tolerance *(222)*. Although the mechanisms of this tolerance are unknown, it is interesting to note that A77636 elicits profound in vivo tolerance within 24 h *(223)*, whereas DNS does not induce such tolerance in a rat model of Parkinson's disease *(224)*.

10.4 Dopamine Receptor Drug Discovery and Functional Selectivity

One of the important questions inherent in this chapter and those of the others in this volume is whether or not there are practical implications for functional selectivity, or whether it is simply an interesting laboratory epiphenomenon. Clearly, we believe the former is true, and this viewpoint seems to be almost universally embraced by other scientists who have studied functional selectivity. We feel there is now ample evidence that functional selectivity may impact on the discovery of new drugs, and as well as the understanding of some of the differences in older molecules. The dopamine receptors, as well as adrenergic, serotonergic, opioid, and cannabinoid receptors, provide some very interesting data relevant to this point. The following sections will review this as regards a few prominent dopaminergically-affected disorders.

10.4.1 D_2 Functionally Selective Drugs and Schizophrenia

The serendipitous observations that led to the discovery of chlorpromazine *(225)* led to the finding that the antipsychotic effects were due to antidopaminergic actions *(226)*. This was later shown to result from blockade of D_2, not D_1, dopamine receptors *(32)*. As of this date, there is no effective antipsychotic drug that does not have some degree of D_2 antagonism as part of its pharmacological profile (we note, of course, that many promising and novel non-dopaminergic drugs are in various stages of clinical testing). As discussed earlier, one of the novel hypotheses of interest to the field was that drugs with dopamine agonist properties might decrease dopamine neurotransmission, and thus have a dopamine antagonist-like effect *(227)*. This idea was revived with the approval of aripiprazole, an antipsychotic drug proffered by its developers as the first high affinity, low intrinsic activity partial D_2 agonist. Although the compound has effects on several other receptors, many of the leading figures in schizophrenia biology have taken to calling aripiprazole the first "dopamine stabilizer" based on its D_2 partial agonist properties *(228–230)*.

According to this view, in situations of high extracellular dopamine concentrations (e.g., in mesolimbic areas involved in positive symptoms), the partial agonist properties of aripiprazole compete with dopamine and cause partial antagonism offering clinical benefit. Conversely, in situations where extracellular dopamine concentrations are low (e.g., in dopamine circuits involved in working memory), the drug can occupy additional receptors and cause partial activation. On its face, this is a cogent argument, combining classic pharmacological logic about mechanisms of partial agonism with recent information about the biology of schizophrenia.

On the one hand, there is no doubt that in many assay systems, aripiprazole looks like a low-to-moderate intrinsic activity partial agonist *(231–233)* as required by this prevalent hypothesis. On the other hand, much of the available data are difficult to reconcile with this partial agonist hypothesis. First, the intrinsic activity and potency of aripiprazole for the D_2-mediated inhibition of cAMP accumulation is very dependent on the cell line being studied. The drug demonstrates weak partial agonist activity in the CHO-D_{2L} cell line, but strong partial agonist activity in HEK-D_{2L} cells *(231–233)*. Moreover, aripiprazole has been shown to have markedly different potencies at two D_{2L}-mediated functions within the same cell line *(215)*. Similarly, in some systems, aripiprazole completely antagonizes both D_2 agonist-mediated GTPγS binding and GIRK channel activity *(233)*, whereas it is a full agonist in situ for D_2-mediated inhibition of tyrosine hydroxylase *(234)*.

These data indicate that aripiprazole may elicit D_2-mediated functional effects that encompass the whole range of classic pharmacological traits. It is such large variations in intrinsic activity and/or potency, not explicable by other mechanisms, that suggest the drug is "functionally selective" *(51)* and not a simple partial agonist. These in vitro findings can also be placed in context of the actions of aripiprazole in vivo. Here there are some data that are irreconcilable with the partial agonist hypothesis. One of the clearest examples is the effect of aripiprazole in the unilaterally lesioned 6-hydroxydopamine (6-OHDA) treated rat *(235)*. Both full and partial dopamine agonists cause the test animal to turn with high frequency in a tight contralateral fashion (i.e., leftward turning if the lesion was on the right-side nigrostriatal pathway). The robust rotation in this model is a result of relative receptor/cellular hypersensitivity of the target receptors on the lesioned, dopamine-depleted, side. As a partial agonist, aripiprazole also should cause such contralateral rotation, but it does not *(234)*. Moreover, it is a pure antagonist of the turning caused by known dopamine D_2 agonists *(234)*. Here then is a situation in which aripiprazole is a pure antagonist in a system with low dopamine tone, findings directly contradictory to the partial agonist hypothesis, but completely explicable by functional selectivity. Another less direct example relates to Parkinson's disease (PD). Many PD patients develop psychotic side effects as a result of their use of levodopa and/or dopamine agonists (the latter largely working via D_2 receptors). By similar reasoning as espoused for schizophrenia, aripiprazole (as a partial agonist) should be very useful in treating these psychotic symptoms when added to the dopaminergic regimen of the PD patient. In fact, aripiprazole not only lacks effectiveness in treating the psychoses, but also it tends to worsen motor function *(236)*.

In our opinion, rather than being a "simple" partial agonist, we conceptualize aripiprazole as a functionally selective D_2 ligand. The drug, of course, is not selective for a single target. It has high affinity for several other receptors (e.g., 5-HT_{1A}) and modest affinity for several more *(233)*. Nonetheless, the interest in this compound revolves around its D_2 action. Because its intrinsic activity varies markedly depending on the signaling environment of the D_2 receptor *(51,231,233)*, we believe the parsimonious explanation to integrate the in vitro and in vivo data is functional selectivity at the D_2 receptor. If this hypothesis is true, a corollary is that compounds with similar D_2 partial agonist effects at adenylate cyclase might have very different clinical effects in treating psychoses. This latter hypothesis may be testable as other "partial agonists" (e.g., bifeprunox) are brought in clinical trials.

10.4.2 Potential Utility of D_1 Functionally Selective Drugs

The D_1-like receptors have been implicated as therapeutic targets for numerous central nervous system disorders such as Parkinson's disease *(237,238)*, schizophrenia, dysfunction of learning and memory *(239,240)*, and attention deficit hyperactivity disorder (ADHD) *(241)*. One of the more exciting prospects of functional selectivity is the design of D_1 receptor drugs that interact with specific signaling pathways to enhance beneficial effects while avoiding unwanted side effects. D_1 receptor agonists have proven to be effective in the treatment of several disorders, but their clinical utility may be limited by side effects such as hypotension and tachycardia *(9)*.

Excessive stimulation of peripheral D_1 receptors can result in hypotension and tachycardia, thereby precluding the use of high doses of D_1 agonists to treat disorders such as Parkinson's disease. Dopamine, via D_1-like receptors, can modulate blood pressure by regulating renal sodium excretion and controlling the resistance of arteries *(242,243)*. At low concentrations, dopamine can act through D_1-like receptors to relax smooth muscle cells via protein kinase A *(243–247)* leading to decreases in blood pressure *(248)*. Although the D_1-linked signaling cascade(s) regulating sodium transport is ambiguous, evidence suggests that D_1 receptors modulate sodium resorption primarily via the Na^+/H^+-exchanger and Na^+-K^+-ATPase *(249)*. Design of a D_1 agonist that is less efficacious at these (or other as yet unidentified) transduction pathway(s) could permit the use of high doses of drug in patients.

The clinical promise of functional selective compounds lies in their postulated ability to selectively activate specific signaling pathways leading to decreased side effects while retaining therapeutic efficacy. SKF83822, a high affinity D_1 agonist, has been reported to activate adenylate cyclase selectively, but not PLC *(211)*, albeit, vide supra. Unlike typical D_1 agonists, SKF83822 induces neither intense grooming in rats (O'Sullivan 2004) nor oral dyskinesia in nonhuman primates *(250)*. More intensive study must be performed to confirm the functionally selective actions of SKF83822 in other systems, but these findings imply that differential behavioral effects can be induced by selective activation of signaling pathways. Further work must be done to elucidate the transduction mechanisms underlying D_1-agonism induced side effects; however, the potential clinical utility of a functionally selective D_1 agonist is clear. Design of such compounds will require increased knowledge of D_1 signaling pathways as well as a greater understanding of the structural characteristics that confer functional selective properties.

These issues are of particular relevance to Parkinson's disease. In the early 1980s, the involvement of D_1 receptors in motor control, and their interaction with D_2 receptors was first demonstrated *(251)*. Although a possible role for D_1 agonists in the treatment of PD studies had been suggested, the two D_1 agonists then available (SKF38393 and CY208-243) failed to demonstrate dramatic anti-parkinsonian actions in the MPTP primate model or humans (*see review in 237)*. Both SKF38393 and CY208-243 were partial D_1 agonists, and dihydrexidine, the first full D_1 agonist *(1,159,160)* did cause dramatic antiparkinson effects *(238)*. Later reports with other full D_1 agonists have confirmed these profound antiparkinson effects in nonhuman primates *(252–257)*. Most importantly, dramatic antiparkinson effects of D_1 also have been seen in humans. ABT-431, a drug chemically and pharmacologically similar to dihydrexidine *(257,258)*, caused antiparkinson effects comparable to the maximum improvement from levodopa *(259,260)*. The consistent and extensive nonhuman primate literature, coupled with the limited human studies outlined earlier, suggest that the D_1 receptor may be a useful Parkinson's target.

Full D_1 agonists have not, however, yet been approved for clinical use. Although one major limiting factor is chemical (the need to have a catechol moiety for all known full agonists), there also have been side effect issues that may relate to D_1 functional selectivity. The two most important issues may be the induction or elicitation of dyskinesias and acute hypotension. Dyskinesias induced by levodopa remain as one of the troubling consequences of long-term therapy, and the mechanistic literature is often conflicting. Although it may be that a D_1 agonist of proper pharmacokinetic properties will prevent the induction of dyskinesias, it also may be that a functionally selective drug might be more useful. As noted earlier, it is even more likely that a functionally selective D_1 ligand might decrease the hypotension associated with dopamine agonists.

Finally, D_1-ligands are believed to have great potential as an enhancer of working memory and other cognitive functions. Activation of D_1 receptors in the prefrontal cortex can improve working memory process *(239,240,261–266)*. Although there is overwhelming data supporting the beneficial effects of D_1 stimulation in cognition and memory, there is also a clear dose-dependency for these effects, and, paradoxically, higher doses have been demonstrated actually to impair memory performance in aged monkeys *(266)*. The involved mechanisms by which D_1 receptors affect memory and cognition are not fully understood, but it may be that a functionally selective D_1 ligand might avoid the biphasic dose-response relationships that clearly could be troublesome if a D_1 ligand was used for cognitive enhancement. Such examples illustrate how elucidation of mechanisms of functional selectivity of dopamine receptor ligands may have clinical relevance as well as intrinsic heuristic value.

Acknowledgments This work was supported, in part, by Public Health Service research grants MH082441, MH40537, NS39036, ES007126, and GM007040.

Declaration of Conflict of Interest: Richard Mailman has a potential conflict of interest based on intellectual property assigned to the University of North Carolina.

References

1. Lovenberg TW, Brewster WK, Mottola DM et al. Dihydrexidine, a novel selective high potency full dopamine D-1 receptor agonist. Eur J Pharmacol 1989; 166:111–113.
2. Mottola DM, Cook LL, Jones SR, Booth RG, Nichols DE, Mailman RB. Dihydrexidine, a selective dopamine receptor agonist that may discriminate postsynaptic D_2 receptors. Soc Neurosci Abstr 1991;818.
3. Lawler CP, Watts VJ, Booth RG, Southerland SB, Mailman RB. Discrete functional selectivity of drugs: OPC-14597 a selective antagonist for post-synaptic dopamine D2 receptors. Soc Neurosci Abstr 1994; 20:525.
4. Garau L, Govoni S, Stefanini E, Trabucchi M, Spano PF. Dopamine receptors: pharmacological and anatomical evidences indicate that two distinct dopamine receptor populations are present in rat striatum. Life Sci 1978; 23:1745–1750.
5. Kebabian JW, Calne DB. Multiple receptors for dopamine. Nature 1979; 277:93–96.
6 Neve KA, Neve RL. The dopamine receptors. Totowa, N.J.: Humana Press, 1997.
7. Jenner P, Demirdemar R. Dopamine receptor sub-types: from basic sciences to clinical applications. Burke, VA: IOS Press, 1997.
8. Sealfon SC, Olanow CW. Dopamine receptors: from structure to behavior. Trends Neurosci 2000; 23:S34–S40.
9. Huang X, Lawler CP, Lewis MM, Nichols DE, Mailman RB. D1 dopamine receptors. Int Rev Neurobiol 2001; 48:65–139.
10. Dearry A, Gingrich JA, Falardeau P, Fremeau RT, Jr., Bates MD, Caron MG. Molecular cloning and expression of the gene for a human D1 dopamine receptor. Nature 1990; 347:72–76.
11. Monsma FJ, Jr., Mahan LC, McVittie LD, Gerfen CR, Sibley DR. Molecular cloning and expression of a D1 dopamine receptor linked to adenylyl cyclase activation. Proc Natl Acad Sci USA 1990; 87:6723–6727.
12. Zhou QY, Grandy DK, Thambi L et al. Cloning and expression of human and rat D_1 dopamine receptors. Nature 1990; 347:76–80.
13. Sunahara RK, Guan HC, O'Dowd BF et al. Cloning of the gene for a human dopamine D_5 receptor with higher affinity for dopamine than D_1. Nature 1991; 350:614–619.
14. Dal Toso R, Sommer B, Ewert M et al. The dopamine D_2 receptor: two molecular forms generated by alternative splicing. EMBO J 1989; 8:4025–4034.
15. Giros B, Sokoloff P, Martres MP, Riou JF, Emorine LJ, Schwartz JC. Alternative splicing directs the expression of two D_2 dopamine receptor isoforms. Nature 1989; 342:923–926.
16. Monsma FJJ, McVittie LD, Gerfen CR, Mahan LC, Sibley DR. Multiple D_2 dopamine receptors produced by alternative RNA splicing. Nature 1989; 342:926–929.
17. Chio CL, Hess GF, Graham RS, Huff RM. A second molecular form of D_2 dopamine receptor in rat and bovine caudate nucleus. Nature 1990; 343:266–269.
18. Sokoloff P, Giros B, Martres MP, Bouthenet ML, Schwartz JC. Molecular cloning and characterization of a novel dopamine receptor (D_3) as a target for neuroleptics. Nature 1990; 347:146–151.
19. van Tol HH, Bunzow JR, Guan HC et al. Cloning of the gene for a human dopamine D_4 receptor with high affinity for the antipsychotic clozapine. Nature 1991; 350:610–614.
20. O'Malley KL, Harmon S, Tang L, Todd RD. The rat dopamine D4 receptor: sequence, gene structure, and demonstration of expression in the cardiovascular system. New Biol 1992; 4:137–146.

21. Mill J, Curran S, Kent L et al. Attention deficit hyperactivity disorder (ADHD) and the dopamine D4 receptor gene: evidence of association but no linkage in a UK sample. Mol Psychiatry 2001; 6:440–444.
22. Schmidt LA, Fox NA, Perez-Edgar K, Hu S, Hamer DH. Association of DRD4 with attention problems in normal childhood development. Psychiatr Genet 2001; 11:25–29.
23. Swanson JM, Sunohara GA, Kennedy JL et al. Association of the dopamine receptor D4 (DRD4) gene with a refined phenotype of attention deficit hyperactivity disorder (ADHD): a family-based approach. Mol Psychiatry 1998; 3:38–41.
24. Jonsson EG, Ivo R, Gustavsson JP et al. No association between dopamine D4 receptor gene variants and novelty seeking. Mol Psychiatry 2002; 7:18–20.
25. Jonsson EG, Ivo R, Forslund K et al. No association between a promoter dopamine D(4) receptor gene variant and schizophrenia. Am J Med Genet 2001; 105:525–528.
26. Kotler M, Manor I, Sever Y et al. Failure to replicate an excess of the long dopamine D4 exon III repeat polymorphism in ADHD in a family-based study. Am J Med Genet 2000; 96:278–281.
27. Ungerstedt U. Stereotaxic mapping of the monoamine pathways in the rat brain. Acta Physiol Scand Suppl 1971; 367:1–48.
28. Kebabian JW, Petzold GL, Greengard P. Dopamine-sensitive adenylate cyclase in caudate nucleus of rat brain, and its similarity to the "dopamine receptor". Proc Natl Acad Sci USA 1972; 69:2145–2149.
29. Clement-Cormier YC, Kebabian JW, Petzold GL, Greengard P. Dopamine-sensitive adenylate cyclase in mammalian brain: a possible site of action of antipsychotic drugs. Proc Natl Acad Sci USA 1974; 71:1113–1117.
30. Iversen LL. Dopamine receptors in the brain. Science 1975; 188:1084–1089.
31. Trabucchi M, Longoni R, Fresia P, Spano PF. Sulpiride: a study of the effects on dopamine receptors in rat neostriatum and limbic forebrain. Life Sci 1975; 17:1551–1556.
32. Creese I, Burt DR, Snyder SH. Dopamine receptor binding predicts clinical and pharmacological potencies of antischizophrenic drugs. Science 1976; 192:481–483.
33. Mailman RB, Huang X. Dopamine receptor pharmacology. In: Koller W, Melamed E, editors. Handbook of Clinical Neurology (3rd Series) Parkinson's Disease and Related Disorders. Amsterdam: Elsevier, 2007: 77–105.
34. Schwartz JC, Diaz J, Pilon C, Sokoloff P. Possible implications of the dopamine D(3) receptor in schizophrenia and in antipsychotic drug actions. Brain Res Brain Res Rev 2000; 31:277–287.
35. Tarazi FI, Baldessarini RJ. Dopamine D4 receptors: significance for molecular psychiatry at the millennium. Mol Psychiatry 1999; 4:529–538.
36. Wilson JM, Sanyal S, van Tol HH. Dopamine D2 and D4 receptor ligands: relation to antipsychotic action. Eur J Pharmacol 1998; 351:273–286.
37. Levant B. The D3 dopamine receptor: neurobiology and potential clinical relevance. Pharmacol Rev 1997; 49:231–252.
38. Robertson GS, Vincent SR, Fibiger HC. D1 and D2 dopamine receptors differentially regulate c-fos expression in striatonigral and striatopallidal neurons. Neuroscience 1992; 49:285–296.
39. Starr MS. Glutamate/dopamine D1/D2 balance in the basal ganglia and its relevance to Parkinson's disease. Synapse 1995; 19:264–293.
40. Gerfen CR. Dopamine-mediated gene regulation in models of Parkinson's disease. Ann Neurol 2000; 47:S42–S50.
41. Gerfen CR. Molecular effects of dopamine on striatal-projection pathways. Trends Neurosci 2000; 23:S64–S70.
42. Albin RL, Young AB, Penney JB. The functional anatomy of basal ganglia disorders. Trends Neurosci 1989; 12:366–375.
43. DeLong MR. Primate models of movement disorders of basal ganglia origin. Trends Neurosci 1990; 13:281–285.
44. Graybiel AM. Neurotransmitters and neuromodulators in the basal ganglia. Trends Neurosci 1990; 13:244–254.
45. Hussain T, Lokhandwala MF. Renal dopamine receptors and hypertension. Exp Biol Med (Maywood) 2003; 228:134–142.

46. Jose PA, Eisner GM, Felder RA. Dopamine and the kidney: a role in hypertension? Curr Opin Nephrol Hypertens 2003; 12:189–194.
47. Amenta F, Ricci A, Rossodivita I, Avola R, Tayebati SK. The dopaminergic system in hypertension. Clin Exp Hypertens 2001; 23:15–24.
48. Velasco M, Contreras F, Cabezas GA, Bolivar A, Fouillioux C, Hernandez R. Dopaminergic receptors: a new antihypertensive mechanism. J Hypertens Suppl 2002; 20:S55–S58.
49. Neve KA, Seamans JK, Trantham-Davidson H. Dopamine receptor signaling. J Recept Signal Transduct Res 2004; 24:165–205.
50. Beaulieu JM, Gainetdinov RR, Caron MG. The Akt-GSK-3 signaling cascade in the actions of dopamine. Trends Pharmacol Sci 2007; 28:166–172.
51. Urban JD, Clarke WP, von Zastrow M et al. Functional selectivity and classical concepts of quantitative pharmacology. J Pharmacol Exp Ther 2007; 320:1–13.
52. Rimondini R, Fuxe K, Ferre S. Multiple intramembrane receptor-receptor interactions in the regulation of striatal dopamine D2 receptors. NeuroReport 1999; 10:2051–2054.
53. Canals M, Marcellino D, Fanelli F et al. Adenosine A2A-dopamine D2 receptor-receptor heteromerization: qualitative and quantitative assessment by fluorescence and bioluminescence energy transfer. J Biol Chem 2003; 278:46741–46749.
54. Rocheville M, Lange DC, Kumar U, Patel SC, Patel RC, Patel YC. Receptors for dopamine and somatostatin: Formation of hetero-oligomers with enhanced functional activity. Science 2000; 288:154–157.
55. Agnati LF, Ferre S, Lluis C, Franco R, Fuxe K. Molecular mechanisms and therapeutical implications of intramembrane receptor/receptor interactions among heptahelical receptors with examples from the striatopallidal GABA neurons. Pharmacol Rev 2003; 55:509–550.
56. Leonard SK, Mailman RB. Multi-modal binding of D_1-like dopamine antagonists to D_1/D_5 receptors and to D_2-A_{2A} receptor heteromers. Neuropsychopharmacology 2005 (in press).
57. Li M, Bermak JC, Wang ZW, Zhou QY. Modulation of dopamine D_2 receptor signaling by actin-binding protein (ABP-280). Mol Pharmacol 2000; 57:446–452.
58. Smith FD, Oxford GS, Milgram SL. Association of the D2 dopamine receptor third cytoplasmic loop with spinophilin, a protein phosphatase-1-interacting protein. J Biol Chem 1999; 274:19894–19900.
59. Yan Z, Hsieh-Wilson L, Feng J et al. Protein phosphatase 1 modulation of neostriatal AMPA channels: regulation by DARPP-32 and spinophilin. Nat Neurosci 1999; 2:13–17.
60. Zhong H, Neubig RR. Regulator of G protein signaling proteins: novel multifunctional drug targets. J Pharmacol Exp Ther 2001; 297:837–845.
61. Rahman Z, Schwarz J, Gold SJ et al. RGS9 modulates dopamine signaling in the basal ganglia. Neuron 2003; 38:941–952.
62. Boutet-Robinet EA, Finana F, Wurch T, Pauwels PJ, De Vries L. Endogenous RGS proteins facilitate dopamine D_{2S} receptor coupling to G_{alpha0} proteins and Ca^{2+} responses in CHO-K1 cells. FEBS Lett 2003; 533:67–71.
63. Sidhu A. Coupling of D1 and D5 dopamine receptors to multiple G proteins: implications for understanding the diversity in receptor-G protein coupling. Mol Neurobiol 1998; 16:125–134.
64. Zhuang X, Belluscio L, Hen R. G(olf)alpha mediates dopamine D1 receptor signaling. J Neurosci 2000; 20:RC91.
65. Herve D, Le Moine C, Corvol JC et al. Galpha(olf) levels are regulated by receptor usage and control dopamine and adenosine action in the striatum. J Neurosci 2001; 21:4390–4399.
66. Corvol JC, Studler JM, Schonn JS, Girault JA, Herve D. Galpha(olf) is necessary for coupling D1 and A2a receptors to adenylyl cyclase in the striatum. J Neurochem 2001; 76:1585–1588.
67. Mailman RB, Schulz DW, Kilts CD, Lewis MH, Rollema H, Wyrick S. Multiple forms of the D_1 dopamine receptor: its linkage to adenylate cyclase and psychopharmacological effects. Psychopharmacol Bull 1986; 22:593–598.
68. Mailman RB, Schulz DW, Kilts CD, Lewis MH, Rollema H, Wyrick S. The multiplicity of the D1 dopamine receptor. Adv Exp Med Biol 1986; 204:53–72.
69. Kilts CD, Anderson CM, Ely TD, Mailman RB. The biochemistry and pharmacology of mesoamygdaloid dopamine neurons. Ann N Y Acad Sci 1988; 537:173–187.

70. Leonard SK, Petitto JM, Anderson CM et al. D_1 dopamine receptors in the amygdala exhibit unique properties. Ann NY Acad Sci 2003; 985:536–539.
71. Leonard SK, Anderson CM, Lachowicz JE, Schulz DW, Kilts CD, Mailman RB. Amygdaloid D_1 receptors are not linked to stimulation of adenylate cyclase. Synapse 2003; 50:320–333.
72. Wang Q, Jolly JP, Surmeier JD et al. Differential dependence of the D1 and D5 dopamine receptors on the G protein gamma 7 subunit for activation of adenylylcyclase. J Biol Chem 2001; 276:39386–39393.
73. Watson JB, Coulter PM, Margulies JE et al. G-protein gamma 7 subunit is selectively expressed in medium-sized neurons and dendrites of the rat neostriatum. J Neurosci Res 1994; 39:108–116.
74. Wang HY, Undie AS, Friedman E. Evidence for the coupling of Gq protein to D1-like dopamine sites in rat striatum: possible role in dopamine-mediated inositol phosphate formation. Mol Pharmacol 1995; 48:988–994.
75. Jin LQ, Wang HY, Friedman E. Stimulated D_1 dopamine receptors couple to multiple Galpha proteins in different brain regions. J Neurochem 2001; 78:981–990.
76. Kimura K, White BH, Sidhu A. Coupling of human D-1 dopamine receptors to different guanine nucleotide binding proteins. Evidence that D-1 dopamine receptors can couple to both Gs and G(o). J Biol Chem 1995; 270:14672–14678.
77. Lezcano N, Mrzljak L, Eubanks S, Levenson R, Goldman-Rakic P, Bergson C. Dual signaling regulated by calcyon, a D1 dopamine receptor interacting protein. Science 2000; 287: 1660–1664.
78. Lezcano N, Bergson C. D1/D5 dopamine receptors stimulate intracellular calcium release in primary cultures of neocortical and hippocampal neurons. J Neurophysiol 2002; 87:2167–2175.
79. Greengard P, Allen PB, Nairn AC. Beyond the dopamine receptor: the DARPP-32/protein phosphatase-1 cascade. Neuron 1999; 23:435–447.
80. Andrisani OM. CREB-mediated transcriptional control. Crit Rev Eukaryot Gene Expr 1999; 9:19–32.
81. Sands WA, Palmer TM. Regulating gene transcription in response to cyclic AMP elevation. Cell Signal 2008; 20:460–466.
82. Konradi C, Cole RL, Heckers S, Hyman SE. Amphetamine regulates gene expression in rat striatum via transcription factor CREB. J Neurosci 1994; 14:5623–5634.
83. Hyman SE, Cole RL, Konradi C, Kosofsky BE. Dopamine regulation of transcription factor-target interactions in rat striatum. Chem Senses 1995; 20:257–260.
84. Minowa MT, Lee SH, Mouradian MM. Autoregulation of the human D1A dopamine receptor gene by cAMP. DNA Cell Biol 1996; 15:759–767.
85. Brami-Cherrier K, Valjent E, Garcia M, Pages C, Hipskind RA, Caboche J. Dopamine induces a PI3-kinase-independent activation of Akt in striatal neurons: a new route to cAMP response element-binding protein phosphorylation. J Neurosci 2002; 22:8911–8921.
86. Hotte M, Thuault S, Lachaise F et al. D1 receptor modulation of memory retrieval performance is associated with changes in pCREB and pDARPP-32 in rat prefrontal cortex. Behav Brain Res 2006; 171:127–133.
87. Bateup HS, Svenningsson P, Kuroiwa M et al. Cell type-specific regulation of DARPP-32 phosphorylation by psychostimulant and antipsychotic drugs. Nat Neurosci 2008; 11:932–939.
88. Kurose H, Katada T, Amano T, Ui M. Specific uncoupling by islet-activating protein, pertussis toxin, of negative signal transduction via alpha-adrenergic, cholinergic, and opiate receptors in neuroblastoma x glioma hybrid cells. J Biol Chem 1983; 258:4870–4875.
89. Bokoch GM, Katada T, Northup JK, Hewlett EL, Gilman AG. Identification of the predominant substrate for ADP-ribosylation by islet activating protein. J Biol Chem 1983; 258:2072–2075.
90. Neve KA, DuRand CJ, Teeter MM. Structural analysis of the mammalian D_2, D_3, and D_4 dopamine receptors. In: Sidhu A, Laruelle M, Vernier P, editors. Dopamine Receptors and Transporters: Function, Imaging, and Clinical Implication. New York: Marcel Dekker Inc., 2003: 77–144.
91. Wong YH, Conklin BR, Bourne HR. Gz-mediated hormonal inhibition of cyclic AMP accumulation. Science 1992; 255:339–342.

92. Obadiah J, Avidor-Reiss T, Fishburn CS et al. Adenylyl cyclase interaction with the D2 dopamine receptor family; differential coupling to Gi, Gz, and Gs. Cell Mol Neurobiol 1999; 19:653–664.
93. Sokoloff P, Andrieux M, Besancon R et al. Pharmacology of human dopamine D_3 receptor expressed in a mammalian cell line: comparison with D2 receptor. Eur J Pharmacol 1992; 225:331–337.
94. Vanhauwe JF, Josson K, Luyten WH, Driessen AJ, Leysen JE. G-protein sensitivity of ligand binding to human dopamine D_2 and D_3 receptors expressed in Escherichia coli: clues for a constrained D(3) receptor structure. J Pharmacol Exp Ther 2000; 295:274–283.
95. Newman-Tancredi A, Cussac D, Audinot V, Pasteau V, Gavaudan S, Millan MJ. G protein activation by human dopamine D3 receptors in high-expressing Chinese hamster ovary cells: A guanosine-5′-O-(3-[35S]thio)- triphosphate binding and antibody study. Mol Pharmacol 1999; 55:564–574.
96. Zaworski PG, Alberts GL, Pregenzer JF, Bin Im W, Slightom JL, Gill GS. Efficient functional coupling of the human D3 dopamine receptor to G(o) subtype of G proteins in SH-SY5Y cells. Br J Pharmacol 1999; 128:1181–1188.
97. De Camilli P, Macconi D, Spada A. Dopamine inhibits adenylate cyclase in human prolactin-secreting pituitary adenomas. Nature 1979; 278:252–254.
98. Stoof JC, Kebabian JW. Opposing roles for D-1 and D-2 dopamine receptors in efflux of cyclic AMP from rat neostriatum. Nature 1981; 294:366–368.
99. Huff RM. Signaling pathways modulated by dopamine receptors. In: Neve KA, Neve RL, editors. The Dopamine Receptors. Totowa, NJ: Humana Press, 1997: 167–192.
100. Oak JN, Oldenhof J, van Tol HH. The dopamine D_4 receptor: one decade of research. Eur J Pharmacol 2000; 405:303–327.
101. Huff RM, Chio CL, Lajiness ME, Goodman LV. Signal transduction pathways modulated by D_2-like dopamine receptors. Adv Pharmacol (New York) 1998; 42:454–457.
102. Oxford GS, Wagoner PK. The inactivating K + current in GH3 pituitary cells and its modification by chemical reagents. J Physiol (Lond) 1989; 410:587–612.
103. Lacey MG, Mercuri NB, North RA. Dopamine acts on D_2 receptors to increase potassium conductance in neurones of the rat substantia nigra zona compacta. J Physiol (Lond) 1987; 392:397–416.
104. Liu L, Shen RY, Kapatos G, Chiodo LA. Dopamine neuron membrane physiology: characterization of the transient outward current (IA) and demonstration of a common signal transduction pathway for IA and IK. Synapse 1994; 17:230–240.
105. Greif GJ, Lin YJ, Liu JC, Freedman JE. Dopamine-modulated potassium channels on rat striatal neurons: specific activation and cellular expression. J Neurosci 1995; 15: 4533–4544.
106. Cass WA, Zahniser NR. Potassium channel blockers inhibit D2 dopamine, but not A1 adenosine, receptor-mediated inhibition of striatal dopamine release. J Neurochem 1991; 57:147–152.
107. Memo M, Missale C, Carruba MO, Spano PF. D2 dopamine receptors associated with inhibition of dopamine release from rat neostriatum are independent of cyclic AMP. Neurosci Lett 1986; 71:192–196.
108. Davila V, Yan Z, Craciun LC, Logothetis D, Sulzer D. D3 dopamine autoreceptors do not activate G-protein-gated inwardly rectifying potassium channel currents in substantia nigra dopamine neurons. J Neurosci 2003; 23:5693–5697.
109. Lledo PM, Homburger V, Bockaert J, Vincent JD. Differential G protein-mediated coupling of D_2 dopamine receptors to K^+ and Ca^{2+} currents in rat anterior pituitary cells. Neuron 1992; 8:455–463.
110. Seabrook GR, Knowles M, Brown N et al. Pharmacology of high-threshold calcium currents in GH4C1 pituitary cells and their regulation by activation of human D2 and D4 dopamine receptors. Br J Pharmacol 1994; 112:728–734.
111. Seabrook GR, Mcallister G, Knowles MR et al. Depression of high-threshold calcium currents by activation of human D2 (short) dopamine receptors expressed in differentiated NG108-15 cells. Br J Pharmacol 1994; 111:1061–1066.

112. Kuzhikandathil EV, Oxford GS. Activation of human D3 dopamine receptor inhibits P/Q-type calcium channels and secretory activity in AtT-20 cells. J Neurosci 1999; 19:1698–1707.
113. Okada Y, Miyamoto T, Toda K. Dopamine modulates a voltage-gated calcium channel in rat olfactory receptor neurons. Brain Res 2003; 968:248–255.
114. Yan Z, Song WJ, Surmeier J. D2 dopamine receptors reduce N-type Ca2 + currents in rat neostriatal cholinergic interneurons through a membrane-delimited, protein-kinase-C- insensitive pathway. J Neurophysiol 1997; 77:1003–1015.
115. Zamponi GW, Snutch TP. Decay of prepulse facilitation of N type calcium channels during G protein inhibition is consistent with binding of a single Gbeta subunit. Proc Natl Acad Sci USA 1998; 95:4035–4039.
116. Dunlap K, Luebke JI, Turner TJ. Exocytotic Ca2 + channels in mammalian central neurons. Trends Neurosci 1995; 18:89–98.
117. Koga E, Momiyama T. Presynaptic dopamine D2-like receptors inhibit excitatory transmission onto rat ventral tegmental dopaminergic neurones. J Physiol 2000; 523 Pt 1:163–173.
118. Svensson E, Wikstrom MA, Hill RH, Grillner S. Endogenous and exogenous dopamine presynaptically inhibits glutamatergic reticulospinal transmission via an action of D2-receptors on N-type Ca2 + channels. Eur J Neurosci 2003; 17:447–454.
119. Surmeier DJ, Eberwine J, Wilson CJ, Cao Y, Stefani A, Kitai ST. Dopamine receptor subtypes colocalize in rat striatonigral neurons. Proc Natl Acad Sci USA 1992; 89:10178–10182.
120. Surmeier DJ, Kitai ST. D1 and D2 dopamine receptor modulation of sodium and potassium currents in rat neostriatal neurons. Prog Brain Res 1993; 99:309–324.
121. Voyno-Yasenetskaya T, Conklin BR, Gilbert RL, Hooley R, Bourne HR, Barber DL. G alpha 13 stimulates Na-H exchange. J Biol Chem 1994; 269:4721–4724.
122. Huff RM. Signal transduction pathways modulated by the D2 subfamily of dopamine receptors. Cell Signal 1996; 8:453–459.
123. Luo Y, Kokkonen GC, Wang X, Neve KA, Roth GS. D_2 dopamine receptors stimulate mitogenesis through pertussis toxin-sensitive G proteins and Ras-involved ERK and SAP/JNK pathways in rat C6-D_{2L} glioma cells. J Neurochem 1998; 71:980–990.
124. Welsh GI, Hall DA, Warnes A, Strange PG, Proud CG. Activation of microtubule-associated protein kinase (Erk) and p70 S6 kinase by D2 dopamine receptors. J Neurochem 1998; 70:2139–2146.
125. Choi EY, Jeong D, Won K, Park Baik JH. G protein-mediated mitogen-activated protein kinase activation by two dopamine D2 receptors. Biochem Biophys Res Commun 1999; 256:33–40.
126. Ghahremani MH, Forget C, Albert PR. Distinct roles for $Galpha_{i2}$ and Gbetagamma in signaling to DNA synthesis and Galpha(i)3 in cellular transformation by dopamine D_{2S} receptor activation in BALB/c 3T3 cells. Mol Cell Biol 2000; 20:1497–1506.
127. Oak JN, Lavine N, van Tol HH. Dopamine D_4 and D_{2L} receptor stimulation of the mitogen-activated protein kinase pathway is dependent on trans-activation of the platelet-derived growth factor receptor. Mol Pharmacol 2001; 60:92–103.
128. Kim SJ, Kim MY, Lee EJ, Ahn YS, Baik JH. Distinct regulation of internalization and mitogen-activated protein kinase activation by two isoforms of the dopamine D2 receptor. Mol Endocrinol 2004; 18:640–652.
129. Kanterman RY, Mahan LC, Briley EM et al. Transfected D_2 dopamine receptors mediate the potentiation of arachidonic acid release in Chinese hamster ovary cells. Mol Pharmacol 1991; 39:364–369.
130. Piomelli D, Pilon C, Giros B, Sokoloff P, Martres MP, Schwartz JC. Dopamine activation of the arachidonic acid cascade as a basis for D_1/D_2 receptor synergism. Nature 1991; 353:164–167.
131. Chio CL, Drong RF, Riley DT, Gill GS, Slightom JL, Huff RM. D_4 dopamine receptor-mediated signaling events determined in transfected Chinese hamster ovary cells. J Biol Chem 1994; 269:11813–11819.
132. Vial D, Piomelli D. Dopamine D_2 receptors potentiate arachidonate release via activation of cytosolic, arachidonate-specific phospholipase A2. J Neurochem 1995; 64:2765–2772.

133. Nilsson CL, Hellstrand M, Ekman A, Eriksson E. Both dopamine and the putative dopamine D3 receptor antagonist PNU-99194A induce a biphasic inhibition of phorbol ester-stimulated arachidonic acid release from CHO cells transfected with the dopamine D3 receptor. Life Sci 1999; 64:939–951.
134. Piomelli D, Greengard P. Lipoxygenase metabolites of arachidonic acid in neuronal transmembrane signalling. Trends Pharmacol Sci 1990; 11:367–373.
135. DiMarzo V, Piomelli D. Participation of prostaglandin E2 in dopamine D_2 receptor-dependent potentiation of arachidonic acid release. J Neurochem 1992; 59:379–382.
136. L'hirondel M, Cheramy A, Godeheu G, Glowinski J. Effects of arachidonic acid on dopamine synthesis, spontaneous release, and uptake in striatal synaptosomes from the rat. J Neurochem 1995; 64:1406–1409.
137. Zhang L, Reith ME. Regulation of the functional activity of the human dopamine transporter by the arachidonic acid pathway. Eur J Pharmacol 1996; 315:345–354.
138. Mitchell R, McCulloch D, Lutz E et al. Rhodopsin-family receptors associate with small G proteins to activate phospholipase D. Nature 1998; 392:411–414.
139. Senogles SE. The D2s dopamine receptor stimulates phospholipase D activity: a novel signaling pathway for dopamine. Mol Pharmacol 2000; 58:455–462.
140. Beaulieu JM, Tirotta E, Sotnikova TD et al. Regulation of Akt signaling by D2 and D3 dopamine receptors in vivo. J Neurosci 2007; 27:881–885.
141. Beaulieu JM, Marion S, Rodriguiz RM et al. A beta-arrestin2 signaling complex mediates lithium action on behavior. Cell 2008; 132:125–136.
142. Frebel K, Wiese S. Signalling molecules essential for neuronal survival and differentiation. Biochem Soc Trans 2006; 34:1287–1290.
143. Beaulieu JM, Sotnikova TD, Yao WD et al. Lithium antagonizes dopamine-dependent behaviors mediated by an AKT/glycogen synthase kinase 3 signaling cascade. Proc Natl Acad Sci USA 2004; 101:5099–5104.
144. Beaulieu JM, Sotnikova TD, Marion S, Lefkowitz RJ, Gainetdinov RR, Caron MG. An Akt/beta-arrestin2/PP2A signaling complex mediates dopaminergic neurotransmission and behavior. Cell 2005; 122:261–273.
145. Alimohamad H, Rajakumar N, Seah YH, Rushlow W. Antipsychotics alter the protein expression levels of beta-catenin and GSK-3 in the rat medial prefrontal cortex and striatum. Biol Psychiatry 2005; 57:533–542.
146. Li X, Rosborough KM, Friedman AB, Zhu W, Roth KA. Regulation of mouse brain glycogen synthase kinase-3 by atypical antipsychotics. Int J Neuropsychopharmacol 2007; 10:7–19.
147. Beaulieu JM, Caron MG. Beta-arrestin goes nuclear. Cell 2005; 123:755–757.
148. Kang J, Shi Y, Xiang B et al. A nuclear function of beta-arrestin1 in GPCR signaling: regulation of histone acetylation and gene transcription. Cell 2005; 123:833–847.
149. Laduron P. Commentary: Dopamine-sensitive adenylate cyclase as a receptor site. In: Kaiser C, Kebabian JW, editors. Dopamine Receptors. Washington DC: American Chemical Society, 1983: 22.
150. Feenstra MG, Sumners C, Goedemoed JH, de Vries JB, Rollema H, Horn AS. A comparison of the potencies of various dopamine receptor agonists in models for pre- and postsynaptic receptor activity. Naunyn Schmiedebergs Arch Pharmacol 1983; 324:108–115.
151. Meller E, Bohmaker K, Namba Y, Friedhoff AJ, Goldstein M. Relationship between receptor occupancy and response at striatal dopamine autoreceptors. Mol Pharmacol 1987; 31:592–598.
152. Kenakin TP. Pharmacologic analysis of drug-receptor interaction. 3rd ed. Philadelphia: Lippincott-Raven Publishers, 1997.
153. Tamminga CA, Schaffer MH, Smith RC, Davis JM. Schizophrenic symptoms improve with apomorphine. Science 1978; 200:567–568.
154. Corsini GU, Pitzalis GF, Bernardi F, Bocchetta A, Del ZM. The use of dopamine agonists in the treatment of schizophrenia. Neuropharmacology 1981; 20:1309–1313.
155. Smith RC, Tamminga C, Davis JM. Effect of apomorphine on schizophrenic symptoms. J Neural Transm 1977; 40:171–176.

156. Tamminga CA, Gotts MD, Thaker GK, Alphs LD, Foster NL. Dopamine agonist treatment of schizophrenia with N-propylnorapomorphine 3. Arch Gen Psychiatry 1986; 43:398–402.
157. Lahti AC, Weiler MA, Corey PK, Lahti RA, Carlsson A, Tamminga CA. Antipsychotic properties of the partial dopamine agonist (-)-3-(3-hydroxyphenyl)-*N*-n-propylpiperidine(preclamol) in schizophrenia. Biol Psychiatry 1998; 43:2–11.
158. Tamminga CA, Cascella NG, Lahti RA, Lindberg M, Carlsson A. Pharmacologic properties of (-)-3PPP (preclamol) in man. J Neural Transm Gen Sect 1992; 88:165–175.
159. Brewster WK, Nichols DE, Riggs RM et al. trans-10,11-dihydroxy-5,6,6a,7,8,12b-hexahydrobenzo[a]phenanthridine: a highly potent selective dopamine D1 full agonist. J Med Chem 1990; 33:1756–1764.
160. Mottola DM, Brewster WK, Cook LL, Nichols DE, Mailman RB. Dihydrexidine, a novel full efficacy D_1 dopamine receptor agonist. J Pharmacol Exp Ther 1992; 262:383–393.
161. Mottola DM, Kilts JD, Lewis MM et al. Functional selectivity of dopamine receptor agonists. I. Selective activation of postsynaptic dopamine D2 receptors linked to adenylate cyclase. J Pharmacol Exp Ther 2002; 301:1166–1178.
162. Kilts JD, Connery HS, Arrington EG et al. Functional selectivity of dopamine receptor agonists. II. Actions of dihydrexidine in D2L receptor-transfected MN9D cells and pituitary lactotrophs. J Pharmacol Exp Ther 2002; 301:1179–1189.
163. Darney KJ, Jr., Lewis MH, Brewster WK, Nichols DE, Mailman RB. Behavioral effects in the rat of dihydrexidine, a high-potency, full-efficacy D1 dopamine receptor agonist. Neuropsychopharmacology 1991; 5:187–195.
164. O'Hara CM, Uhland-Smith A, O'Malley KL, Todd RD. Inhibition of dopamine synthesis by dopamine D2 and D3 but not D4 receptors. J Pharmacol Exp Ther 1996; 277:186–192.
165. Tang L, Todd RD, O'Malley KL. Dopamine D_2 and D_3 receptors inhibit dopamine release. J Pharmacol Exp Ther 1994; 270:475–479.
166. Mailman RB, Nichols DE, Lewis MM, Blake BL, Lawler CP. Functional effects of novel dopamine ligands: dihydrexidine and Parkinson's disease as a first step. In: Jenner P, Demirdemar R, editors. Dopamine Receptor Subtypes: From Basic Science to Clinical Application. IOS Stockton Press, 1998: 64–83.
167. Mailman RB, Nichols DE, Tropsha A. Molecular drug design and dopamine receptors. In: Neve KA, Neve RL, editors. The Dopamine Receptors. Totowa, NJ: Humana Press, 1997: 105–133.
168. Gay EA, Urban JD, Nichols DE, Oxford GS, Mailman RB. Functional selectivity of D2 receptor ligands in a Chinese hamster ovary hD2L cell line: evidence for induction of ligand-specific receptor states. Mol Pharmacol 2004; 66:97–105.
169. Stephenson RP. A modification of receptor theory. Br J Pharmacol 1956; 11:379–393.
170. Kenakin T. Drugs and receptors. An overview of the current state of knowledge. Drugs 1990; 40:666–687.
171. Roth BL, Chuang DM. Multiple mechanisms of serotonergic signal transduction. Life Sci 1987; 41:1051–1064.
172. De Lean A, Stadel JM, Lefkowitz RJ. A ternary complex model explains the agonist-specific binding properties of the adenylate cyclase-coupled beta-adrenergic receptor. J Biol Chem 1980; 255:7108–7117.
173. Leff P, Scaramellini C, Law C, McKechnie K. A three-state receptor model of agonist action. Trends Pharmacol Sci 1997; 18:355–362.
174. Mailman RB, Gay EA. Novel mechanisms of drug action: functional selectivity at D_2 dopamine receptors (a lesson for drug discovery). Med Chem Res 2004; 13:115–126.
175. Eden RJ, Costall B, Domeney AM et al. Preclinical pharmacology of ropinirole (SK&F 101468-A) a novel dopamine D2 agonist. Pharmacol Biochem Behav 1991; 38:147–154.
176. Smith HP, Nichols DE, Mailman RB, Lawler CP. Locomotor inhibition, yawning and vacuous chewing induced by a novel dopamine D2 post-synaptic receptor agonist. Eur J Pharmacol 1997; 323:27–36.
177. Snyder GL, Fienberg AA, Huganir RL, Greengard P. A dopamine/D1 receptor/protein kinase A/dopamine- and cAMP-regulated phosphoprotein (Mr 32 kDa)/protein phosphatase-1 pathway regulates dephosphorylation of the NMDA receptor. J Neurosci 1998; 18:10297–10303.

178. Chen J, Rusnak M, Luedtke RR, Sidhu A. D1 dopamine receptor mediates dopamine-induced cytotoxicity via the ERK signal cascade. J Biol Chem 2004; 279:39317–39330.
179. Nagai T, Takuma K, Kamei H et al. Dopamine D1 receptors regulate protein synthesis-dependent long-term recognition memory via extracellular signal-regulated kinase 1/2 in the prefrontal cortex. Learn Mem 2007; 14:117–125.
180. Ogimoto G, Yudowski GA, Barker CJ et al. G protein-coupled receptors regulate Na+, K + -ATPase activity and endocytosis by modulating the recruitment of adaptor protein 2 and clathrin. Proc Natl Acad Sci USA 2000; 97:3242–3247.
181. Pinto dOP, Chibalin AV, Katz AI, Soares-da-Silva P, Bertorello AM. Short-term vs. sustained inhibition of proximal tubule Na,K-ATPase activity by dopamine: cellular mechanisms. Clin Exp Hypertens 1997; 19:73–86.
182. Xu J, Li XX, Albrecht FE, Hopfer U, Carey RM, Jose PA. Dopamine(1) receptor, G(salpha), and Na(+)-H(+) exchanger interactions in the kidney in hypertension. Hypertension 2000; 36:395–399.
183. Frederickson ED, Bradley T, Goldberg LI. Blockade of renal effects of dopamine in the dog by the DA1 antagonist SCH 23390. Am J Physiol 1985; 249:F236–F240.
184. Surmeier DJ, Bargas J, Hemmings HC, Jr., Nairn AC, Greengard P. Modulation of calcium currents by a D1 dopaminergic protein kinase/phosphatase cascade in rat neostriatal neurons. Neuron 1995; 14:385–397.
185. Fienberg AA, Hiroi N, Mermelstein PG et al. DARPP-32: regulator of the efficacy of dopaminergic neurotransmission. Science 1998; 281:838–842.
186. Young CE, Yang CR. Dopamine D1/D5 receptor modulates state-dependent switching of soma-dendritic Ca2 + potentials via differential protein kinase A and C activation in rat prefrontal cortical neurons. J Neurosci 2004; 24:8–23.
187. Drolet P, Bilodeau L, Chorvatova A, Laflamme L, Gallo-Payet N, Payet MD. Inhibition of the T-type Ca2 + current by the dopamine D1 receptor in rat adrenal glomerulosa cells: requirement of the combined action of the G betagamma protein subunit and cyclic adenosine 3′,5′-monophosphate. Mol Endocrinol 1997; 11:503–514.
188. Kozasa T, Hepler JR, Smrcka AV et al. Purification and characterization of recombinant G16 alpha from Sf9 cells: activation of purified phospholipase C isozymes by G-protein alpha subunits. Proc Natl Acad Sci USA 1993; 90:9176–9180.
189. Hepler JR, Kozasa T, Smrcka AV et al. Purification from Sf9 cells and characterization of recombinant Gq alpha and G11 alpha. Activation of purified phospholipase C isozymes by G alpha subunits. J Biol Chem 1993; 268:14367–14375.
190. Lee CH, Park D, Wu D, Rhee SG, Simon MI. Members of the Gq alpha subunit gene family activate phospholipase C beta isozymes. J Biol Chem 1992; 267:16044–16047.
191. Cordeaux Y, Ijzerman AP, Hill SJ. Coupling of the human A1 adenosine receptor to different heterotrimeric G proteins: evidence for agonist-specific G protein activation. Br J Pharmacol 2004; 143:705–714.
192. Harikumar KG, Chattopadhyay A. Differential discrimination of G-protein coupling of serotonin(1A) receptors from bovine hippocampus by an agonist and an antagonist. FEBS Lett 1999; 457:389–392.
193. Gazi L, Nickolls SA, Strange PG. Functional coupling of the human dopamine D(2) receptor with Galpha(i1), Galpha(i2), Galpha(i3) and Galpha(o) G proteins: evidence for agonist regulation of G protein selectivity. Br J Pharmacol 2003; 138:775–786.
194. Berg KA, Maayani S, Goldfarb J, Scaramellini C, Leff P, Clarke WP. Effector pathway-dependent relative efficacy at serotonin type 2A and 2C receptors: evidence for agonist-directed trafficking of receptor stimulus. Mol Pharmacol 1998; 54:94–104.
195. Brink CB, Wade SM, Neubig RR. Agonist-directed trafficking of porcine $alpha_{2A}$-adrenergic receptor signaling in Chinese hamster ovary cells: l-isoproterenol selectively activates G(s). J Pharmacol Exp Ther 2000; 294:539–547.
196. Bonhaus DW, Chang LK, Kwan J, Martin GR. Dual activation and inhibition of adenylyl cyclase by cannabinoid receptor agonists: evidence for agonist-specific trafficking of intracellular responses. J Pharmacol Exp Ther 1998; 287:884–888.

197. Runnels LW, Scarlata SF. Determination of the affinities between heterotrimeric G protein subunits and their phospholipase C-beta effectors. Biochemistry 1999; 38:1488–1496.
198. Boyer JL, Graber SG, Waldo GL, Harden TK, Garrison JC. Selective activation of phospholipase C by recombinant G-protein alpha- and beta gamma-subunits. J Biol Chem 1994; 269:2814–2819.
199. Yu PY, Eisner GM, Yamaguchi I, Mouradian MM, Felder RA, Jose PA. Dopamine D1A receptor regulation of phospholipase C isoform. J Biol Chem 1996; 271:19503–19508.
200. Dai R, Ali MK, Lezcano N, Bergson C. A crucial role for cAMP and protein kinase A in D1 dopamine receptor regulated intracellular calcium transients. Neurosignals 2008; 16:112–123.
201. Lin CW, Miller TR, Witte DG et al. Characterization of cloned human dopamine D_1 receptor-mediated calcium release in 293 cells. Mol Pharmacol 1995; 47:131–139.
202. Felder CC, Jose PA, Axelrod J. The dopamine-1 agonist, SKF 82526, stimulates phospholipase-C activity independent of adenylate cyclase. J Pharmacol Exp Ther 1989; 248:171–175.
203. Mannoury la CC, Vidal S, Pasteau V, Cussac D, Millan MJ. Dopamine D1 receptor coupling to Gs/olf and Gq in rat striatum and cortex: a scintillation proximity assay (SPA)/antibody-capture characterization of benzazepine agonists. Neuropharmacology 2007; 52:1003–1014.
204. Panchalingam S, Undie AS. Optimized binding of [^{35}S]GTPgammaS to Gq-like proteins stimulated with dopamine D_1-like receptor agonists. Neurochem Res 2000; 25:759–767.
205. Rashid AJ, So CH, Kong MM et al. D1-D2 dopamine receptor heterooligomers with unique pharmacology are coupled to rapid activation of Gq/11 in the striatum. Proc Natl Acad Sci USA 2007; 104:654–659.
206. Panchalingam S, Undie AS. Physicochemical modulation of agonist-induced [35s] GTPgammaS binding: implications for coexistence of multiple functional conformations of dopamine D1-like receptors. J Recept Signal Transduct Res 2005; 25:125–146.
207. Mailman RB, Ferris RM, Tang FL et al. Erythrosine (Red No. 3) and its nonspecific biochemical actions: what relation to behavioral changes? Science 1980; 207:535–537.
208. Iwamoto T, Okumura S, Iwatsubo K et al. Motor dysfunction in type 5 adenylyl cyclase-null mice. J Biol Chem 2003; 278:16936–16940.
209. Lee FJS, Xue S, Pei L, Vukusic B, Chéry N, Wang Y et al. Dual Regulation of NMDA Receptor functions by direct protein-protein interactions with the dopamine D_1 receptor. Cell 2002; 111:219–230.
210. Friedman E, Jin LQ, Cai GP et al. D1-like dopaminergic activation of phosphoinositide hydrolysis is independent of D1A dopamine receptors: evidence from D1A knockout mice. Mol Pharmacol 1997; 51:6–11.
211. Undie AS, Weinstock J, Sarau HM, Friedman E. Evidence for a distinct D_1-like dopamine receptor that couples to activation of phosphoinositide metabolism in brain. J Neurochem 1994; 62:2045–2048.
212. Undie AS, Friedman E. Stimulation of a dopamine D_1 receptor enhances inositol phosphates formation in rat brain. J Pharmacol Exp Ther 1990; 253:987–992.
213. Pacheco MA, Jope RS. Comparison of [3H]phosphatidylinositol and [3H]phosphatidylinositol 4,5-bisphosphate hydrolysis in postmortem human brain membranes and characterization of stimulation by dopamine D1 receptors. J Neurochem 1997; 69:639–644.
214. Mailman RB. GPCR functional selectivity has therapeutic impact. Trends Pharmacol Sci 2007; 28:390–396.
215. Urban JD, Vargas GA, von Zastrow M, Mailman RB. Aripiprazole has functionally selective actions at dopamine D(2) receptor-mediated signaling pathways. Neuropsychopharmacology 2007; 32:67–77.
216. Drago J, Gerfen CR, Lachowicz JE et al. Altered striatal function in a mutant mouse lacking D1A dopamine receptors. Proc Natl Acad Sci USA 1994; 91:12564–12568.
217. Montague DM, Striplin CD, Overcash JS, Drago F, Lawler CP, Mailman RB. Quantification of D_{1B} (D_5) receptors in dopamine D_{1A} receptor-deficient mice. Synapse 2001; 39: 319–322.
218. Lee SP, So CH, Rashid AJ et al. Dopamine D1 and D2 receptor co-activation generates a novel phospholipase C-mediated calcium signal. J Biol Chem 2004; 279:35671–35678.

219. Rashid AJ, O'Dowd BF, Verma V, George SR. Neuronal Gq/11-coupled dopamine receptors: an uncharted role for dopamine. Trends Pharmacol Sci 2007; 28:551–555.
220. Ryman-Rasmussen JP, Nichols DE, Mailman RB. Differential activation of adenylate cyclase and receptor internalization by novel dopamine D1 receptor agonists. Mol Pharmacol 2005; 68:1039–1048.
221. Ryman-Rasmussen JP, Griffith A, Oloff S et al. Functional selectivity of dopamine D(1) receptor agonists in regulating the fate of internalized receptors. Neuropharmacology 2007; 52:562–575.
222. Asin KE, Bednarz L, Nikkel A, Perner R. Rotation and striatal c-fos expression after repeated, daily treatment with selective dopamine receptor agonists and levodopa. J Pharmacol Exp Ther 1995; 273:1483–1490.
223. Lin CW, Bianchi BR, Miller TR et al. Persistent activation of the dopamine D_1 receptor contributes to prolonged receptor desensitization: studies with A-77636. J Pharmacol Exp Ther 1996; 276:1022–1029.
224. Gulwadi AG, Korpinen CD, Mailman RB, Nichols DE, Sit SY, Taber MT. Dinapsoline: characterization of a D_1 dopamine receptor agonist in a rat model of Parkinson's disease. J Pharmacol Exp Ther 2001; 296:338–344.
225. Delay J, Deniker P, Harl JM. Therapeutic method derived from hiberno-therapy in excitation and agitation states. Ann Med Psychol (Paris) 1952; 110:267–273.
226. Carlsson A, Lindqvist M. Effect of chlorpromazine and haloperidol on formation of 3-methoxytyramine and normetanephrine in mouse brain. Acta Pharmacol Toxicol (Copenh) 1963; 20:140–144.
227. Tamminga CA, Carlsson A. Partial dopamine agonists and dopaminergic stabilizers, in the treatment of psychosis. Curr Drug Targets CNS Neurol Disord 2002; 1:141–147.
228. Tamminga CA. Partial dopamine agonists in the treatment of psychosis. J Neural Transm 2002; 109:411–420.
229. Stahl SM. Dopamine system stabilizers, aripiprazole, and the next generation of antipsychotics, part 1, "Goldilocks" actions at dopamine receptors. J Clin Psychiatry 2001; 62:841–842.
230. Lieberman JA. Dopamine partial agonists: a new class of antipsychotic. CNS Drugs 2004; 18:251–267.
231. Lawler CP, Prioleau C, Lewis MM et al. Interactions of the novel antipsychotic aripiprazole (OPC-14597) with dopamine and serotonin receptor subtypes. Neuropsychopharmacology 1999; 20:612–627.
232. Burris KD, Molski TF, Xu C et al. Aripiprazole, a novel antipsychotic, is a high-affinity partial agonist at human dopamine D2 receptors. J Pharmacol Exp Ther 2002; 302:381–389.
233. Shapiro DA, Renock S, Arrington E et al. Aripiprazole, a novel atypical antipsychotic drug with a unique and robust pharmacology. Neuropsychopharmacology 2003; 28:1400–1411.
234. Kikuchi T, Tottori K, Uwahodo Y et al. 7-(4-[4-(2,3-Dichlorophenyl)-1-piperazinyl] butyloxy)-3,4-dihydro-2(1H)-qui nolinone (OPC-14597), a new putative antipsychotic drug with both presynaptic dopamine autoreceptor agonistic activity and postsynaptic D2 receptor antagonistic activity. J Pharmacol Exp Ther 1995; 274:329–336.
235. Ungerstedt U, Arbuthnott GW. Quantitative recording of rotational behavior in rats after 6-hydroxy- dopamine lesions of the nigrostriatal dopamine system. Brain Res 1970; 24: 485–493.
236. Friedman JH, Berman RM, Goetz CG et al. Open-label flexible-dose pilot study to evaluate the safety and tolerability of aripiprazole in patients with psychosis associated with Parkinson's disease. Mov Disord 2006; 21:2078–2081.
237. Mailman R, Huang X, Nichols DE. Parkinson's disease and D1 dopamine receptors. Curr Opin Investig Drugs 2001; 2:1582–1591.
238. Taylor JR, Lawrence MS, Redmond DE, Jr. et al. Dihydrexidine, a full dopamine D1 agonist, reduces MPTP-induced parkinsonism in monkeys. Eur J Pharmacol 1991; 199:389–391.
239. Arnsten AF, Cai JX, Murphy BL, Goldman-Rakic PS. Dopamine D1 receptor mechanisms in the cognitive performance of young adult and aged monkeys. Psychopharmacology (Berl) 1994; 116:143–151.

240. Goldman-Rakic PS, Castner SA, Svensson TH, Siever LJ, Williams GV. Targeting the dopamine D_1 receptor in schizophrenia: insights for cognitive dysfunction. Psychopharmacology (Berl) 2004; 174:3–16.
241. Heijtz RD, Kolb B, Forssberg H. Motor inhibitory role of dopamine D1 receptors: implications for ADHD. Physiol Behav 2007; 92:155–160.
242. Chatziantoniou C, Ruan X, Arendshorst WJ. Defective G protein activation of the cAMP pathway in rat kidney during genetic hypertension. Proc Natl Acad Sci USA 1995; 92:2924–2928.
243. Zeng C, Wang D, Asico LD et al. Aberrant D1 and D3 dopamine receptor transregulation in hypertension. Hypertension 2004; 43:654–660.
244. Edwards RM. Response of isolated renal arterioles to acetylcholine, dopamine, and bradykinin. Am J Physiol 1985; 248:F183–F189.
245. Hall AS, Bryson SE, Vaughan PF, Ball SG, Balmforth AJ. Pharmacological characterization of the dopamine receptor coupled to cyclic AMP formation expressed by rat mesenteric artery vascular smooth muscle cells in culture. Br J Pharmacol 1993; 110:681–686.
246. Tamaki T, Hura CE, Kunau RT, Jr. Dopamine stimulates cAMP production in canine afferent arterioles via DA1 receptors. Am J Physiol 1989; 256:H626–H629.
247. Zeng C, Wang D, Yang Z et al. Dopamine D1 receptor augmentation of D3 receptor action in rat aortic or mesenteric vascular smooth muscles. Hypertension 2004; 43:673–679.
248. Ventura HO, Messerli FH, Frohlich ED et al. Immediate hemodynamic effects of a dopamine-receptor agonist (fenoldopam) in patients with essential hypertension. Circulation 1984; 69:1142–1145.
249. Hussain T, Lokhandwala MF. Renal dopamine receptor function in hypertension. Hypertension 1998; 32:187–197.
250. Peacock L, Gerlach J. Aberrant behavioral effects of a dopamine D1 receptor antagonist and agonist in monkeys: evidence of uncharted dopamine D1 receptor actions. Biol Psychiatry 2001; 50:501–509.
251. Mailman RB, Schulz DW, Lewis MH, Staples L, Rollema H, DeHaven DL. SCH-23390: a selective D1 dopamine antagonist with potent D2 behavioral actions. Eur J Pharmacol 1984; 101:159–160.
252. Grondin R, Bedard PJ, Britton DR, Shiosaki K. Potential therapeutic use of the selective dopamine D_1 receptor agonist, A-86929: an acute study in parkinsonian levodopa-primed monkeys. Neurology 1997; 49:421–426.
253. Asin KE, Domino EF, Nikkel A, Shiosaki K. The selective dopamine D1 receptor agonist A-86929 maintains efficacy with repeated treatment in rodent and primate models of Parkinson's disease. J Pharmacol Exp Ther 1997; 281:454–459.
254. Kebabian JW, Britton DR, DeNinno MP et al. A-77636: a potent and selective dopamine D1 receptor agonist with antiparkinsonian activity in marmosets. Eur J Pharmacol 1992; 229:203–209.
255. Goulet M, Madras BK. D(1) dopamine receptor agonists are more effective in alleviating advanced than mild parkinsonism in 1-methyl-4-phenyl-1,2,3, 6-tetrahydropyridine-treated monkeys. J Pharmacol Exp Ther 2000; 292:714–724.
256. Gnanalingham KK, Hunter AJ, Jenner P, Marsden CD. Selective dopamine antagonist pretreatment on the antiparkinsonian effects of benzazepine D1 dopamine agonists in rodent and primate models of Parkinson's disease--the differential effects of D1 dopamine antagonists in the primate. Psychopharmacology (Berl) 1995; 117:403–412.
257. Shiosaki K, Jenner P, Asin KE et al. ABT-431: the diacetyl prodrug of A-86929, a potent and selective dopamine D_1 receptor agonist: in vitro characterization and effects in animal models of Parkinson's disease. J Pharmacol Exp Ther 1996; 276:150–160.
258. Michaelides MR, Hong Y, DiDomenico SJ et al. (5aR,11bS)-4,5,5a,6,7,11b-hexahydro-2-propyl-3-thia-5-azacyclopent-1- ena[c]-phenanthrene-9,10-diol (A-86929): a potent and selective dopamine D1 agonist that maintains behavioral efficacy following repeated administration and characterization of its diacetyl prodrug (ABT-431). J Med Chem 1995; 38:3445–3447.

259. Rascol O, Blin O, Thalamas C et al. ABT-431, a D1 receptor agonist prodrug, has efficacy in Parkinson's disease. Ann Neurol 1999; 45:736–741.
260. Rascol O, Nutt JG, Blin O et al. Induction by dopamine D_1 receptor agonist ABT-431 of dyskinesia similar to levodopa in patients with Parkinson's disease. Arch Neurol 2001; 58:249–254.
261. Lidow MS, Goldman-Rakic PS, Gallager DW, Rakic P. Distribution of dopaminergic receptors in the primate cerebral cortex: quantitative autoradiographic analysis using [3H]raclopride, [3H]spiperone and [3H]SCH23390. Neuroscience 1991; 40:657–671.
262. Steele TD, Hodges DB, Jr., Levesque TR, Locke KW. D1 agonist dihydrexidine releases acetylcholine and improves cognitive performance in rats. Pharmacol Biochem Behav 1997; 58:477–483.
263. Hersi AI, Rowe W, Gaudreau P, Quirion R. Dopamine D_1 receptor ligands modulate cognitive performance and hippocampal acetylcholine release in memory-impaired aged rats. Neuroscience 1995; 69:1067–1074.
264. Cai JX, Arnsten AF. Dose-dependent effects of the dopamine D1 receptor agonists A77636 or SKF81297 on spatial working memory in aged monkeys. J Pharmacol Exp Ther 1997; 283:183–189.
265. Schneider JS, Sun ZQ, Roeltgen DP. Effects of dihydrexidine, a full dopamine D-1 receptor agonist, on delayed response performance in chronic low dose MPTP-treated monkeys. Brain Res 1994; 663:140–144.
266. Castner SA, Williams GV, Goldman-Rakic PS. Reversal of antipsychotic-induced working memory deficits by short-term dopamine D1 receptor stimulation. Science 2000; 287: 2020–2022.

Chapter 11
Functional Selectivity at Receptors for Cannabinoids and Other Lipids

Allyn C. Howlett

Abstract CB_1 and CB_2 cannabinoid receptors are associated with Gi/o proteins to activate signal transduction pathways that include inhibition of adenylyl cyclase, activation of mitogen activated protein kinase (MAPK), and regulation of ion channels (CB_1 only). Agonists for these receptors include structurally diverse cannabinoid, aminoalkylindole, and eicosanoid ligands. Arylpyrazole ligands generally behave as competitive antagonists and inverse agonists to block constitutive activity of the cannabinoid receptors. Allosteric regulators of the CB_1 receptor have been identified. One mechanism for functional selectivity of these ligands to direct signal transduction is the ability of certain ligands to behave as agonists for some Gαi subtypes and as inverse agonists for other Gαi subtypes. Biochemical and modeling studies suggest that selectivity can be attributed to conformational changes initiated by interactions within the 7-transmembrane helical bundle that ultimately modulate either the juxtamembrane C-terminal domain to activate Gαi3 or Gαo, and intracellular loop 3 domains to activate Gαi1, Gαi2, or Gαs. Additional selectivity in signaling pathways is conferred by accessory proteins including CRIP1a, FAN, β-arrestin, and GASP1. Close physical association of the CB_1 receptor with other pharmacologically distinct GPCRs, including D2 dopamine, opioid, orexin 1, and $GABA_B$ receptors, allows ligands acting on one partner to alter the efficacy of ligands for the partner receptors. These mechanisms of ligand-directed functional selectivity can be utilized in the design of pharmacological agents that initiate signal transduction pathways of therapeutic relevance.

Keywords Allosteric regulation, Anandamide (arachidonylethanolamide), 2-arachidonoylglycerol (2-AG), CP55940, Endocannabinoids, G proteins, G protein-coupled receptors (GPCRs), Heterodimeric receptors, Rimonabant (SR141716), Steric trigger mechanism of activation, Δ^9-tetrahydrocannabinol (THC), WIN55212-2,

---

A.C. Howlett
Department of Physiology and Pharmacology, Wake Forest University Health Sciences, Winston-Salem, NC 27101
e-mail: ahowlett@wfubmc.edu

K. Neve (ed.), *Functional Selectivity of G Protein-Coupled Receptor Ligands*,
DOI: 10.1007/978-1-60327-335-0_11,

11.1 Cannabinoid and Endocannabinoid Ligands and Receptors

CB_1 and CB_2 cannabinoid receptors were initially identified in brain and immune systems, respectively, and were named based on their ability to respond to Δ^9-tetrahydrocannabinol (THC), the psychoactive and antinociceptive component in *Cannabis sativa* extracts *(1)*. Subsequent academic and pharmaceutical drug development resulted in structurally similar classical cannabinoid agonists (HU210), structurally modified nonclassical cannabinoid agonists (CP55940), and structurally unrelated aminoalkylindole agonists (WIN55212-2) *(1,2)* (Fig. 11.1). Endogenous agonists for the cannabinoid receptors, including anandamide (arachidonylethanolamide) and 2-arachidonoylglycerol (2-AG), were later characterized *(3)*. These agonists stimulated pertussis toxin-sensitive Gi/o proteins to promote signal transduction (including inhibition of adenylyl cyclase and activation of p42/p44 mitogen activated protein kinase (MAPK)), with limited selectivity for CB_1 vs. CB_2 receptors *(4)*. CB_1 but not CB_2 receptors signal via inhibition of voltage-gated Ca^{2+} channels *(4)*. These signal transduction responses can be competitively antagonized by aryl pyrazole ligands including rimonabant (SR141716), AM251, and taranabant at the CB_1 receptor, and SR144528 and AM630 at the CB_2 receptor *(2,5–7)*. CB_1 and CB_2 receptors appear to exert effects on signal transduction pathways in the absence of exogenous agonist stimulation, which can be reversed by antagonists that behave as inverse agonists *(8)*. Neutral antagonists, VCHSR and AM4113, have been reported, which fail to alter the Ca^{2+} channel regulation and cyclic AMP accumulation, respectively *(9,10)*. Thus, a wide spectrum of ligands can influence cannabinoid receptor activity, ranging from agonists that are highly efficacious at stimulating Gi/o-mediated responses to those inverse agonists able to preclude constitutive activation of G proteins.

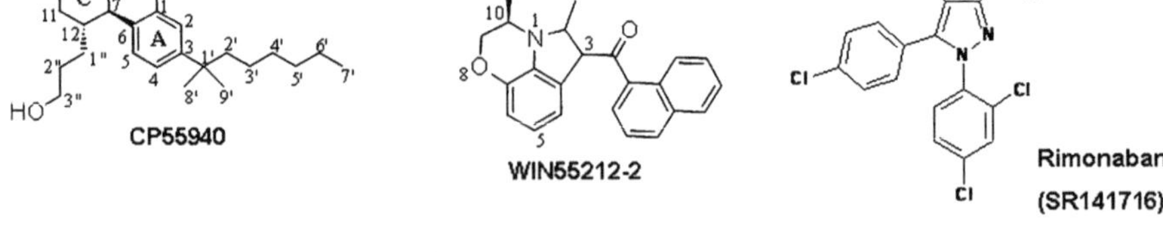

Fig. 11.1 Cannabinoid receptor ligands

11.2 Signal Transduction Pathways Utilized by CB_1 and CB_2 Receptors

The range of signal transduction mechanisms initiated by activation of the CB_1 and CB_2 cannabinoid receptors has been reviewed *(4,11)*. The predominant class of G proteins with which the CB_1 receptor interacts is the Gi/o class. Inhibition of adenylyl cyclase and the concomitant reduction in cyclic AMP-stimulated protein kinase A (PKA) activity appear to be responsible for a number of subsequent pathways. For example, a net decrease in dephosphorylation can increase voltage-dependent current flow at A-type K^+ channels *(12)*, which can impact neurotransmitter release in response to depolarizing stimuli. In contrast, cyclic AMP and PKA activity were not involved in cannabinoid agonist-activated inwardly rectifying K^+ currents (K_{ir}) in AtT-20 pituitary tumor cells and Xenopus oocytes that exogenously expressed CB_1 receptors *(13–15)*. CB_1-mediated inhibition of PKA activity resulted in increased tyrosine phosphorylation of focal adhesion kinase (pp125FAK) *(16)* and FAK-related non-kinase (FRNK) *(17)*, which are critical determinants of cytoskeletal neurite remodeling regulated by integrins *(16,18)*. These kinases also serve as scaffolding proteins, coupling to p130-Cas and Fyn, which subsequently activate the mitogen-activated protein kinase (MAPK) pathways that regulate gene expression *(19,20)*.

In contrast to the generally observed inhibition of adenylyl cyclase, CB_1 receptor agonists are able to increase cyclic AMP accumulation in globus pallidus slice preparations and in cell model systems *(21,22)* (evaluated previously *(4,11)*. The CB_1 receptor-mediated cyclic AMP production can result from stimulation of adenylyl cyclase isoforms 2/4/7, in which Gβγ augments the Gαs response *(23)*. This response is blocked by pertussis toxin, since Gβγ are derived from the dissociation of Gi/o proteins. In neurons or CHO cells expressing recombinant CB_1 receptors, cannabinoid-induced cyclic AMP production required pertussis toxin treatment to observe the response *(22,24,25)*, suggesting that in the absence of Gi/o proteins, Gs can interact with the CB_1 receptor.

Sequelae to CB_1 agonist-stimulated cyclic AMP levels include isotopic Ca^{2+} influx into neuroblastoma cells via a mechanism involving PKA activation *(26)*. In dorsal striatum, nucleus accumbens, and striatal projection neurons, CB_1 receptor stimulation by agonists led to phosphorylation of DARPP-32 at the PKA site *(27,28)*. Another report comparing WIN55212-2-modulation of K-type K^+ channels in cultured hippocampal cells to modulation by PKA and PKC suggested a role for stimulation of PKA in the cannabinoid response *(29)*.

Evidence has been reported that supports and refutes CB_1 receptor regulation of phospholipase C and subsequent Ca^{2+} mobilization *(30–33)*. In cerebellar granule cells, CB_1 receptor stimulation in low Mg^{2+} media mobilized Ca^{2+} via a caffeine and IP_3 receptor-sensitive mechanism, which augmented the response to depolarization by glutamate receptors or high K^+ *(34)*. In contrast, cannabinoid agonists inhibited neurotransmitter-stimulated inositol phospholipid production in hippocampal preparations *(35)*. CB_1-mediated attenuation of PKA and the resulting net decrease in ryanodine channel phosphorylation led to a sustained suppression of NMDA-excitotoxic Ca^{2+}-mediated Ca^{2+} release and cell death in cultured hippocampal neurons *(36)*.

Ion channels can be regulated by the CB_1 receptor either directly via Gi/o protein regulation (e.g., Ca^{2+} currents) or indirectly via phosphorylation/dephosphorylation mechanisms. CB_1 agonists inhibited N-type voltage-gated Ca^{2+} channels in differentiated neuroblastoma cells in a pertussis toxin-sensitive manner *(37–41)*. P/Q-type Ca^{2+} currents were inhibited in rat cortical and cerebellar preparations and in AtT-20 pituitary cells expressing recombinant CB_1 receptors *(13,42)*. L-type Ca^{2+} currents in cat brain arterial smooth muscle cells were inhibited by CB_1 agonists, which correlated with vascular relaxation in cat cerebral arterial rings *(43)*. The mechanism of inhibition of Ca^{2+} channels can be attributed to G proteins and the interaction of Gβγ with the ion channel proteins *(40,41,44–46)*.

MAPK activation by CB_1 agonists has been observed in a wide variety of cell types and various host cells expressing recombinant CB_1 receptors *(47–50)*. In CHO cells expressing recombinant CB_1 receptors, p42/p44 MAPK activation was dependent on CB_1 receptor-stimulated Gi/o dissociation, which could supply the Gβγ dimer as a scaffold for proteins in the MAPK activation complex *(47)*. In U373 MG astrocytoma cells and CHO cells expressing recombinant CB_1 receptors, activation of phosphatidylinositol-3-kinase (PI3K) and phosphorylation of membrane inositol phospholipids recruited protein kinase B (PKB, also known as Akt) to initiate the raf-1, MAP-ERK Kinase (MEK), p42/p44 MAPK cascade *(49,51,52)*. In N1E-115 neuroblastoma cells *(53)* and in hippocampal slices *(20)*, CB_1-mediated attenuation of PKA activity and net dephosphorylation of c-raf permitted raf kinase to serve as an activator of MEK in the p42/p44 MAPK activation module. p38 MAPK was activated by cannabinoid receptor agonists in CHO cells expressing recombinant CB_1 receptors *(54)*, in human vein endothelial cells *(55)*, and mouse hippocampal slices *(56)*. Jun N-terminal kinases (JNK1 and JNK2) were activated by CB_1 agonists via Gi/o, PI3K, and ras in CHO cells expressing recombinant CB_1 receptors *(54)*.

CB_1 receptor interaction with tyrosine kinase receptors is suggested by reports of transactivation mechanisms initiated by cannabinoid agonists. In CHO cells expressing recombinant CB_1 or CB_2 receptors, SR141716 and SR144528, respectively, blocked the pertussis toxin-sensitive MAPK activation by insulin receptors or insulin-like growth factor receptors via a mechanism believed to involve sequestration of Gi/o proteins *(57,58)*. In CB_1 receptor-expressing CHO cells, cannabinoid agonist transactivation of platelet-derived growth factor (PDGF) receptors was implicated in the JNK activation mechanism *(54)*. In glioblastoma or lung carcinoma cells, cannabinoid transactivation of epidermal growth factor (EGF) receptors led to MAPK and PKB/Akt activation *(59)*. 2-AG and anandamide evoked TrkB receptor tyrosine phosphorylation in cultured cortical interneurons, and coimmunoprecipitation of TrkB receptors coexpressed with CB_1 receptors in PC12 cells indicated that the two proteins formed a complex as a result of the cannabinoid stimulation *(60)*.

It should be noted, however, that anandamide inhibited the TrkA receptor-induced Rap1/B-Raf/ERK activation in the nerve growth factor-stimulated PC12 and neuronal stem cell differentiation process *(61)*. CB_1 receptor-Gi/o activation in neuroblastoma or PC12 cells led to vascular endothelial growth factor (VEGF) receptor or non-receptor Src or Fyn tyrosine phosphorylation, respectively, which led to PKC and MAPK activation, resulting in Ca^{2+} influx *(62–64)*. In a neuroblastoma cell model, stimulation of the CB_1 receptor/Gi/o activated Rap1, Ral, and Rac, and phosphorylated tyrosine kinases Src and JNK, which activate Stat3, a transcription factor modulating gene expression and neurite extension *(65–67)*.

CB_1 receptor-mediated production and release of nitric oxide (NO) via neuronal NO synthase (nNOS) or endothelial (eNOS) has been demonstrated for a variety of tissues including saphenous vein segments *(68)*, endothelial cells *(69–71)*, monocytes *(72)*, brain slice preparations *(73)*, and neuroblastoma cells *(74)*. NO and peroxynitrite increased cellular uptake of anandamide and 2-AG in human endothelial cells, HEK cells, and C6 glioma cells *(71,75,76)*, which would decrease extracellular anandamide where it can stimulate CB_1 receptors, but increase intracellular anandamide where it can regulate TRPV1 channels *(76)*. Stimulation of NO-sensitive guanylyl cyclase and cyclic GMP production in N18TG2 neuroblastoma cells was detected in response to cannabinoid agonists *(77,78)*. The reports of increased NO in response to cannabinoid agonists contrast with the experimental findings that cannabinoid agonists were neuroprotective against depolarization-induced Ca^{2+} influx and subsequent NOS activation in cerebellar granule cells *(79)*. In those studies, the response was attributed to the CB_1 receptor-mediated inhibition of voltage-gated Ca^{2+} channels *(79)*. Similarly, in mouse cortical neurons treated with high concentrations of NMDA for 6 h, concurrent treatment with WIN55212-2 precluded NO generation and excitotoxic damage *(80)*. The effects of WIN55212-2 could be reversed by dibutyryl cyclic AMP *(80)*, suggesting that prolonged attenuation of PKA by the CB_1 agonist could alter the pattern of phosphorylation of nNOS from that of an excitotoxic reaction *(81,82)*.

Cannabinoid agonists repress inducible NOS (iNOS) expression and NO production in response to an inflammatory stimulus in saphenous vein endothelial cells *(68)*, RAW264.7 macrophages *(83)*, microglial cells *(84)*, and astrocytes *(85–87)*. The mechanism appears to involve either the CB_1 or CB_2 receptor, and a reduction in cellular cyclic AMP *(68,83,86,87)* and/or the release of the cytokine interleukin-1 receptor antagonist (IL-1ra), which itself suppresses iNOS expression *(88)*.

Considering the multiple complex interactions of cellular signal transduction pathways and networks that can be regulated following stimulation of cannabinoid receptors, it is risky to assign "functional selectivity" as a property of the ligand's modulation of a biological response that is relatively far removed from the original CB_1 receptor stimulus. It is important to consider concurrent cellular events, such as depolarization, increases in intracellular Ca^{2+} levels, and regulation of multiple kinases and phosphatases as independent but coexistent modulators of a cellular endpoint. Our continued commentary on functional selectivity will concentrate on the impact that the ligand exerts on the receptor and its associated proteins.

11.3 Functional Selectivity for G Proteins

11.3.1 Orthosteric Ligands Selectively Regulate Coupling to Gi/o

A primary mechanism by which the CB_1 receptor initiates signal transduction selectivity is in agonist-directed coupling to different Gα subtypes. Glass and Northup reported differences in agonist efficacy to promote the GDP exchange for [^{35}S]GTPγS on purified Gαi (all subtypes) and Gαo proteins reconstituted with recombinant CB_1 receptors expressed in Sf9 cell membranes *(89)*. There were no differences in the [^{35}S] GTPγS-binding properties promoted by the most efficacious (HU210) or the least efficacious (Δ^9-THC) agonists, both of which stimulated Gαi or Gαo similarly. However, WIN55212 and anandamide near-maximally stimulated Gαi, but only partially (70%) stimulated Gαo. Prather and colleagues *(90)* reported that in rat cerebellum membranes, the ED_{50} for WIN55212-2 to stimulate [^{32}P]azidoanilido-GTP binding to Gαi1 and Gαo3 was approximately 40-fold less than for binding to Gαo2, confirming the notion of agonist-directed selectivity for G protein activation.

Howlett's laboratory tested the hypothesis that structurally distinct ligands would exhibit differential ability to regulate CB_1 receptor–Gα protein interactions, with dissociation of the agonist (A)-R-$G\alpha_{GDP}\beta\gamma$ ternary complex in detergent solution as the endpoint *(91)*. CB_1 receptor–Gαi/o complexes exist in detergent solution without the requirement for agonists to stabilize a ternary complex *(92–95)*. The aminoalkylindole WIN55212-2, the cannabinoid desacetyllevonantradol, and the eicosanoid (R)-methanandamide promoted an equilibrium mixture of CB_1 receptors bound to G proteins (ARGαi or RGαi) plus free CB_1 receptors (AR or R) *(91)* (Fig. 11.2). Agonists in the presence of GTPγS promoted a transient AR*$G\alpha_{GTP}\gamma_S$ ternary complex, which dissociated to A + R + $G\alpha_{GTP}\gamma_S$ + βγ. This could be observed as a decrease in the ratio of Gα to CB_1 receptor bands in the coimmunoprecipitation and Western blot experiment. Theoretically, in the presence of a competitive antagonist, the dissociation of the AR*$G\alpha_{GTP}\beta\gamma$ complex would be interrupted in a competitive manner. If that antagonist behaved as an inverse agonist (I), binding to the RG complexes would stabilize an IR°G association even in the presence of GTPγS.

Consistent with the notion of functional selectivity, the three agonists promoted dissociation of the complex at different rates and extents, depending upon the Gαi subtype. The highly efficacious agonist WIN55212-2 promoted dissociation of the CB_1 receptor from all three Gαi subtypes, indicating that this ligand was a functional agonist for all three Gαi subtypes *(91)*. As predicted, GTPγS shifted the equilibrium to a state in which the G protein was dissociated from CB_1 receptors in the presence of WIN55212-2. In contrast, desacetyllevonantradol promoted dissociation of receptors only from Gαi1 and Gαi2, indicating that the cannabinoid ligand served as a functional agonist for those receptors coupled to the Gαi1/2, but was not able to elicit a response from RGαi3 complexes *(91)*. In the presence of GTPγS, CB_1 receptors remained bound to Gαi3 proteins, and, as had been observed in radioligand binding studies, these receptors remained in a high-affinity state for desacetyllevonantradol *(92)*. (R)-methanandamide promoted dissociation of receptors only from Gαi3, limiting its

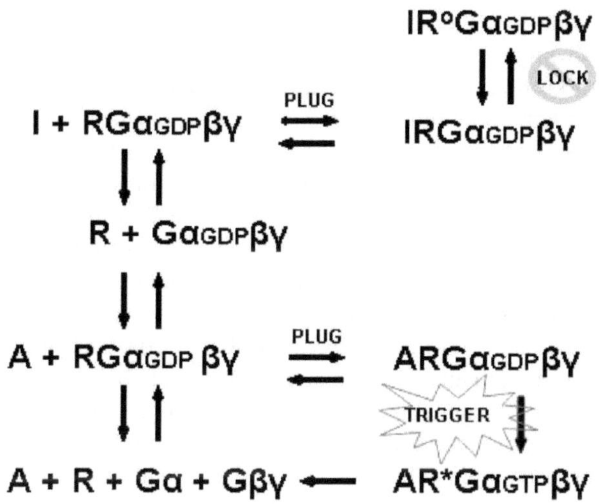

Fig. 11.2 Equilibrium mixture of receptors and G proteins in the presence of an agonist or an inverse agonist

agonist efficacy to RGαi3 complexes but not the RGαi1/2 complexes. Extrapolating to cellular signal transduction, these findings would suggest that CB_1 receptors in an environment with a preponderance of Gαi1 or Gαi2 would respond to WIN55212-2 or a cannabinoid ligand. CB_1 receptors in an environment with a preponderance of Gαi3 would respond to WIN55212-2 or an anandamide analog.

Constitutive activity has been reported as the ability of the CB_1 receptor to trigger Gαi/o activation in the absence of stimulation by exogenous agonists (for review, see *(8)*). Those ligands that can reverse the constitutive activity, such as rimonabant, are referred to as antagonist/inverse agonists *(44,45,57)*. In the initial studies describing constitutive activity for CB_1 receptors, basal [^{35}S]GTPγS binding was increased, MAPK activation was increased, and adenylyl cyclase activation was decreased in CHO cells expressing recombinant CB_1 receptors compared with nontransfected cells *(57)*. Rimonabant in the 30 nM concentration range reversed the basal [^{35}S]GTPγS binding imposed by exogenous CB_1 receptors *(57,96–99)*. Higher concentrations of rimonabant were required for the inverse agonist effect on [^{35}S]GTPγS binding in rat brain membranes *(100)* and in cell models expressing native CB_1 receptors *(44,100,101)*. G protein sequestration was also observed in investigations of the inhibition of N-type Ca^{2+} channels by CB_1 receptors, in which exogenous CB_1 receptors commandeered a pool of Gαi/o proteins that were required for the exchange of Gβγ subunits to regulate Ca^{2+} channel activity by other receptors *(44,45)*. Bouaboula and colleagues proposed a mechanism for the inverse agonism in which the CB_1 receptor sequesters Gα1 proteins in an "inactive" conformational state, R°, which could be selected or induced by the inverse agonist (I) in a ternary complex with inactive G proteins (IR°G_{GDP}), thereby reducing the fraction of RG_{GDP} complex that could spontaneously convert to R*G. *(57)*.

In detergent solution, the competitive antagonist rimonabant did not facilitate dissociation of G proteins from CB_1 receptors, as would be expected of a ligand with limited or no efficacy *(91)*. Furthermore, rimonabant maintained a significant fraction of CB_1 receptors bound to G proteins in the presence of GTPγS, consistent with its serving as an "inverse agonist" by conversion of the RG_{GDP} complex to a sustainable $IR^{o}G_{GDP}$ complex. A similar scenario was observed for desacetyllevonantradol to preclude the ability of GTPγS to drive forward the dissociation of Gαi3, and (R)-methanandamide to drive forward the dissociation of Gαi1 and Gαi2 *(91)*. Judging by the ability to disrupt the dissociation of $G\alpha i_{GTP}\gamma_S$, the inverse agonist efficacy to promote the isomerization to $IR^{o}G_{GDP}$ was greater for the desacetyllevonantradol-CB_1-Gαi3 and (R)-methanandamide-CB_1-Gαi1/2 complexes than it was for complexes of rimonabant-CB_1 with any of the Gαi subtypes *(91)*. These results provide evidence for the cannabinoid and endocannabinoid ligands to differentially behave as agonists or as inverse agonists depending upon the Gαi subtype to which the CB_1 receptor is coupled. The physiological implications are that signal transduction events promoted predominantly by certain subtypes of Gαi/o proteins could be excluded from stimulation in response to the ligand.

The notion that the CB_1 receptor ligands are able to behave as either agonists or inverse agonists depending on the subtype of the G protein is supported by observations of relative efficacy in the research literature. WIN55212-2 generally behaves as a full agonist in signal transduction assays (see *(11)* for review), consistent with its ability to dissociate all three RGαi subtypes. In contrast, anandamide behaves as a weak partial agonist (see *(1,3)* for review and original references), consistent with the failure of (R)-methanandamide to dissociate CB_1 receptors associated with Gαi1 and Gαi2. This observation could explain some of the differences observed between stimulation of endogenous anandamide release at a synapse, as opposed to stimulation of the CB_1 receptors by exogenous cannabinoid agonists. If anandamide's agonist action is limited to signal transduction pathways directed by Gαi3, then its response would be very different from a cannabinoid drug activating signal transduction pathways directed by Gαi1/2. WIN55212-2 would be predicted to initiate a much broader range of signal transduction pathways directed by all three Gαi subtypes, and hence, is considered to be a "full" agonist.

11.3.2 Allosteric Ligands Regulate CB_1 Receptor-Mediated Responses

The notion that a ligand can modify CB_1 receptor coupling to G proteins by binding to loci that are distinct from the agonist binding pocket has gained support from identification of allosteric regulators of the CB_1 signal transduction response (Org27569, Org27759, and Org29657) *(102)* and PSNCBAM-1 *(103)* (see reviews by Pertwee and Ross *(8,104)*). These regulators could antagonize the CP55940 or WIN55212-2-stimulated [^{35}S]GTPγS binding and other responses in a noncompetitive manner *(102,103)*.

Radioligand-binding studies demonstrated that the allosteric inhibitors could increase the binding of [^{3}H]-CP55940 by decreasing the rate of agonist dissociation *(102,103)*. The increase in receptor affinity for the agonist is consistent with a shift toward receptors in the "high affinity state" for agonists, which would be those in the R or RG_{GDP} state (see discussion of efficacy *(11)*). Thus, it can be expected that an allosteric inhibitor of agonist-stimulated responses would facilitate agonist binding with high affinity but would preclude the ability of the agonist to trigger the receptor release of G proteins as well as the agonist ligand. The allosteric ligands could displace [^{3}H]rimonabant in the same concentration range as observed for the increased binding of [^{3}H]-CP55940 *(102,103)*. It should be noted that the competition did not achieve maximal displacement of [^{3}H]-rimonabant for either class of allosteric inhibitors *(102,103)*. One explanation for these results is that these allosteric inhibitors occupy a site distinct from those of the orthosteric agonists and antagonists. Further studies need to be done to determine whether the binding site for rimonabant is shared with the allosteric regulators, which would be consistent with both ligands using a similar mechanism to maintain the CB_1 receptor in a state that is unable to trigger G protein activation. It is interesting to speculate that rimonabant could occupy two regions within the CB_1 receptor: one responsible for competitive antagonism of the agonist, and the other in common with the allosteric antagonists. Although the Org compounds appeared to exhibit no inverse agonist effect on signal transduction in the reported signaling assays *(102)*, PSNCBAM-1 exerted a distinct decrease in [^{35}S]GTPγS binding in HEK293 cells expressing exogenous CB_1 receptors, typical of an "inverse agonist" response. These properties resemble those of CP272871, which behaved as a competitive antagonist and an inverse agonist to augment forskolin-stimulated cyclic AMP production and inhibit [^{35}S]GTPγS binding in N18TG2 cells endogenously expressing CB_1 receptors *(101)*. CP272871 also exhibited a complex, noncompetitive antagonism of the maximum response that could not be explained by either competitive antagonism or inverse agonism. Development of CB_1-binding ligands that could differentially affect the activity of distinct agonists in a selective manner would provide an excellent means to address therapeutic selectivity. Functional selectivity within the allosteric ligand classes might also be possible if the ligands could be shown to block activation of specific G proteins.

11.4 Structural Basis of Functional Selectivity for G Proteins

Domains of the CB_1 receptor that selectively interact with Gαi/o proteins have been identified using coimmunoprecipitation procedures *(105)* (Fig. 11.3). CB_1 receptors immunoprecipitated from CHAPS-solubilized rat brain or N18TG2 neuroblastoma membranes remained coupled to each of the Gαi/o subtypes in the absence of exogenous agonists *(92,95,106)*. The CB_1 receptor-Gα association was disrupted by pertussis toxin treatment, supporting the assumption that these complexes represent a functional equilibrium of free receptor with the receptor-G protein complex *(94,95)*. This tendency toward RG "precoupling" could form the basis for CB_1 receptor sequestration of G proteins observed in intact cell studies.

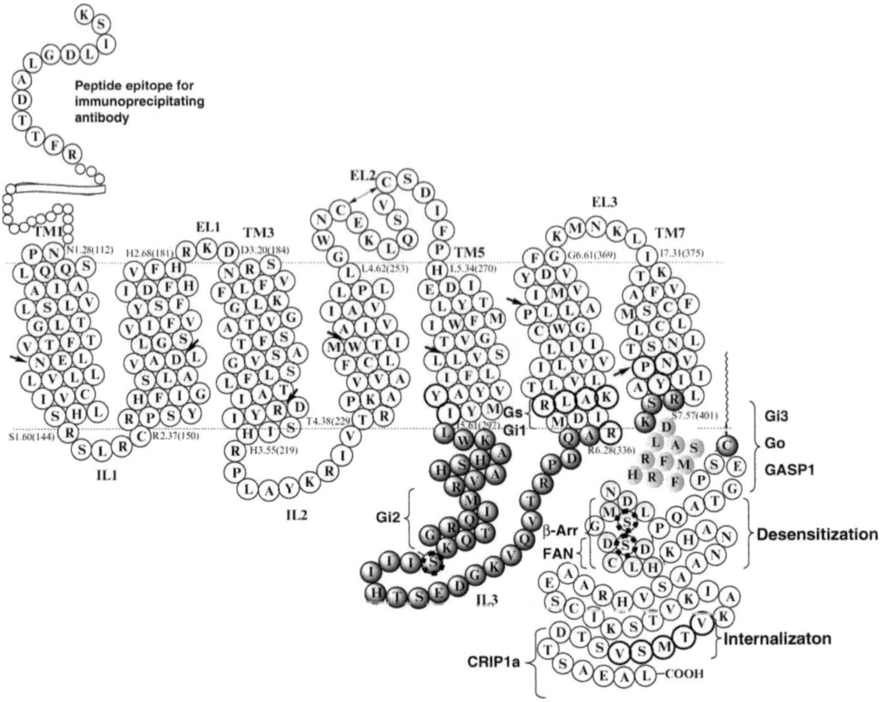

Fig. 11.3 Helical net configuration of the human CB_1 receptor. Amino acid residues are denoted by single letter abbreviations, and regions of interest described in the text are highlighted in *grey* or by *bold circles*. This figure is a modification of Fig. 11.1 in Mukhopadhyay et al., 2002 *(74)*, from which details regarding the numbering can be found

11.4.1 CB_1 Helix 8 (H8) Interactions with Gαo or Gαi3

One region of the CB_1 receptor that is of particular importance in conferring activity to G proteins is the juxtamembrane C-terminal domain, which is demarcated from the extension of the TM7 to the C415 palmitoylation site that anchors this region into the cell membrane, and encompasses a helical structure (H8). A juxtamembrane C-terminal peptide (R401 to E417) was able to stimulate [^{35}S]GTPγS binding and inhibit adenylyl cyclase, demonstrating that this domain in isolation is able to trigger G protein activation *(94,105,107)*.

The juxtamembrane C-terminal peptide competed effectively for the CB_1 receptor interaction with Gαo or Gαi3 in immunoprecipitates from solubilized rat brain or N18TG2 membrane preparations; however, the peptide failed to disrupt the CB_1 receptor interaction with Gαi1 or Gαi2 *(95,106,108)*. These studies support the hypothesis that the C-terminal juxtamembrane domain serves as the primary locus of CB_1 receptor interaction with Gαo or Gαi3. Evidence to support this hypothesis comes from studies of CB_1 receptor mutants truncated at the membrane surface of the TM7 were devoid of Ca^{2+} channel inhibition by CB_1 receptors *(109)*, as would be expected if Gαi3 or Gαo mediated this response.

The juxtamembrane C-terminal domain of the CB_1 receptor is predicted to form an intracellular helical segment *(110)*, and biophysical studies have indicated that depending upon the environment, the helix can exist either as an α-helix or a 3_{10}-helix. Circular dichroism (CD) spectroscopy indicated that the juxtamembrane C-terminal peptide exhibited amorphous character in sodium phosphate buffer or methanol, but helical character in Na dodecylsulfate (SDS) and dipalmitoylphosphatidylglycerol (DPPG) micelles *(107)*. The peptide also exhibited 65% helical character in dodecylphosphocholine (DPC), a zwitterionic membrane mimetic *(111)*. Nuclear magnetic resonance (NMR) studies performed on peptide fragment rCB_1 I397-G418 in DPC micelles showed α-helical structure between residues I397-L400 (the TM7 domain), a bend at R401, and a juxtamembrane α-helix at S402 to F413 *(111,112)*. The clean separation between TM7 and H8 is denoted by a bend at R401, inasmuch as an R401A modification in the peptide resulted in the appearance of a single helix that continued through F413 *(111)*. Nuclear Overhauser enhancement (NOE) analyses suggested that H8 is aligned parallel to the micelle surface *(111)*, with a salt bridge between D404 and H407 *(112)*. When the C-terminal juxtamembrane peptide was examined in SDS, a negatively-charged membrane mimetic, helical character was observed *(113)*. However, the lack of NOEs between the α proton of residue i and amide proton of i + 4 indicated that in the negatively-charged environment, the peptide existed primarily as a 3_{10}-helix between L405 and F413 *(113)*.

The H8 amphipathic helix comprises a multibasic sequence formed by three Arg residues and a Lys residue. These cationic charges are important for activity because deletion of R401 or acylation of K403 resulted in decreased ligand efficacy *(107)*. NOE data that showed close through-space interactions between A399 and L405 indicate that these cationic residues would be positioned toward the cellular space, facing the G protein *(112)*. Studies of ^{31}P-^{1}H NMR NOE showed close through-space interactions between phosphate headgroups and R406 and R410 in DPC micelles *(111)*, consistent with the amphipathic helix being embedded deeply into the membrane bilayer to maintain an interaction with the lipid headgroups and yet allow for the cationic residues to point toward the cytosol (or the G protein) *(114)*.

The facile conversion between α-helix and 3_{10}-helix induced by the negative charges comprising the SDS micelles could reflect a conformational change that might occur upon interaction with a patch of negative residues on the G protein surface. On the basis of these observations, it was proposed that a close interaction with the negatively-charged G protein surface would optimize conditions for a helical structure of CB_1 H8 that could interconvert as the receptor-G protein complex is activated and dissociates. A conformational change in the TM7 domain that repositions the C-terminal end at the membrane interface could initiate a remodeling of the helix H-bonding pattern beginning at a nucleation site at the TM7-H8 elbow (see *(113)* for discussion). The two helical structures form sterically and electrostatically distinct charge clusters *(113)*. The α-helix formed a compact, closely packed cluster of positively-charged residues (K403, R406, H407, R410) at the N-terminal end *(113)*. In contrast, the 3_{10}-helix was extended in length, and the positively-charged residues were distributed over a larger surface area. The reorientation of the electrostatic potential induced by a conformational change in this domain could either behave as an on/off switch for G protein activation or as a modulator that transmits a distinct signal that could be recognized by different G proteins depending upon the charge distribution profile.

It has been hypothesized that rigid body movement that repositions the TM7 helix terminus laterally at the cytoplasmic surface is associated with the activation transition of a G protein (see *(113)* for discussion and references). As proof of this principle, mutations that would alter the elbow region to the CB_1 H8 were constructed and tested for activity. A conserved motif in the class A GPCRs that impacts the TM7-H8 angle is the $NPXXY(X)_{5,6}F$ motif *(115)*. Phe is replaced by Leu in the CB_1 receptor, and by Ile in the CB_2 receptor, leading Abood, Reggio, and colleagues to hypothesize that this motif could confer unique signaling properties to the CB_1 receptor *(116)*. To test this, mutations L(404)F and L(404)I were compared with the wild-type CB_1 receptor *(116)*.

Computational modeling studies placed H8 within the phosphate/glycerol region of the bilayer parallel to the membrane surface, with hydrophobic residues facing toward the helical bundle and polar residues extending toward the cytosol *(116)*. Both L404I and L404F mutant H8 domains appeared to be closer to the cytosolic surface than the wild-type H8, and the L404F mutant was positioned closer to the TM1 helix in the model. There was greater opportunity for H-bonding between the H8 residues and the elbow region for the L404F and L404I mutants than for the wild-type receptor, suggesting that the mutation restricts flexibility in the H8 region *(116)*.

The biological impact of L404F and L404I mutations in the CB_1 receptor were determined in HEK293 cells expressing comparable receptor expression levels and similar ligand-binding affinities for CP55940 and rimonabant *(116)*. The two mutant receptors showed a significant reduction in $[^{35}S]GTP\gamma S$ binding to G proteins in response to HU210 (L404I only), CP55940, or WIN5212-2, consistent with a restriction in maximal activity such as might be observed if certain populations of G proteins were excluded from stimulation by the receptor *(116)*. Both mutant receptors failed to coimmunoprecipitate Gαi3 in CHAPS detergent, although associations with Gαi1 and Gαi2 remained intact *(116)*. The mutation in this region identified the importance of H8 for CB_1 receptor internalization kinetics, but not for G protein sequestration, supporting the premise that H8 structure is critical to functional selectivity to direct some signal transduction processes but not others *(116)*.

It was further hypothesized that mutation of H8 would significantly impact cellular signal transduction pathways expected to be directed by Gαo1 or Gαi3. To investigate agonist responses contributed by these Gαi/o subtypes, superior cervical ganglial neurons expressing N-type Ca^{2+} channels were injected with pertussis toxin-resistant Gαo1 or Gαi3 proteins, such that pertussis toxin treatment would inactivate other Gαi/o proteins, allowing only the resistant subtypes to remain functional. In cells expressing their native population of Gαi/o proteins, L404F and L404I receptors exhibited significantly reduced magnitude but a more rapid time course of the Ca^{2+} current inhibition in response to WIN55212-2 *(116)*. Interestingly, the reduced magnitude was overcome in the Gαo1-replete cells, suggesting that providing an abundance of the preferred Gαo1 could overcome the defective receptor-G protein coupling. Gαi3-replete cells also inhibited Ca^{2+} currents, but the magnitude was less than for the native Gαi/o proteins or the Gαo1-replete cells, and was even less when stimulated by the L404I mutant *(116)*. However, the Gαi3-replete cells exhibited a much greater degree of constitutive activity for the wild-type CB_1 receptors to suppress Ca^{2+} current, which could be exaggerated in those cells that also expressed H8 mutants. This finding, combined with the observation that the H8 mutants failed to coimmunoprecipitate Gαi3, suggests that perturbation of the H8 region results in altered kinetics of the receptor-Gαi3 activation reaction in the absence of an agonist. These studies suggest that Gαi3 is not the preferred Gαi/o subtype for WIN55212-2-directed Gβγ release to Ca^{2+} channels. However, it is the CB_1 receptor–Gαi3 complex that appears to release Gβγ in the absence of stimulation by an agonist.

11.4.2 CB_1 IL3 Interactions with Gαi1, Gαi2, and Gαs

The third intracellular loop (IL3) of most class A GPCRs is important for G protein coupling. The CB_1 receptor IL3, comprising 38 amino acids (L298 to A335), was investigated by determining function of three peptides comprising the N-terminal side, center, and C-terminal side. Combinations of these IL3 peptides were able to disrupt the CB_1 receptor association with Gαi1 or Gαi2, but not Gαo or Gαi3, in CHAPS detergent extracts *(95,106)*, demonstrating selectivity for Gαi/o subtypes. Although IL3 peptides exhibited limited activity compared with the juxtamembrane C-terminal peptide, the IL3 C-terminal domain was particularly involved in G protein activation *(105)*. A 9-mer peptide derived from the IL3 C-terminal domain stimulated GTPase activity of Gαi1 *(108)*.

The structure of the CB_1 receptor IL3 region has been characterized. A YXXIXXL/A motif at the C-terminal end of TM5 is found in the CB_1 receptor (**Y**MY**I**LW**KA** (294–301)), and this sequence was demonstrated by NMR to form an amphipathic α-helix in SDS micelles *(117)*. A peptide comprising the entire IL3 domain was shown to be helical at the N-terminal side distal to TM5 (i.e., W299 to R307 and I309 to S316), with limited structure except for a Gly turn within the middle third, and helical structure within the C-terminal third approximately two turns proximal to TM6 *(117)*. The presence of a cationic patch, **R**MDI**R**LA**K** (R336 to K343), embedded at the membrane-cytosol interface suggests that a functional interaction with negatively-charged regions of certain G proteins might be possible *(118)*. It was predicted that the N-terminal amphipathic α-helix would interact with the G protein *(117)*. NMR analysis of this C-terminal peptide in the presence of Gαi1 indicated that the peptide was coupled as a helix *(108)*.

CB_1 receptor interaction with Gαs was indicated by studies in which pertussis toxin-mediated ADP-ribosylation precluded the interaction of Gαi/o with CB_1 receptors *(22,24,25)*. Under those conditions, cyclic AMP was increased in response to cannabinoid agonists in primary neuronal cultures and CHO cells expressing recombinant CB_1 receptors. The order of potency for agonist stimulation was the same as for inhibition of cyclic AMP production; however, the efficacies of cannabinoid receptor agonists for regulation of Gαs were not as great as for regulation of Gαi *(22)*.

Kendall and colleagues *(119)* studied a double mutation L341A/A342L, which converted the wild-type sequence of the CB_1 receptor to the signature motif for Gαs coupling in the β2-adrenergic receptor (Ala-Leu-Lys-Thr) *(120)*. This change caused the mutated receptor to exhibit selectively for Gαs rather than Gαi, but also conferred constitutive activity to stimulate Gαs. The CB_1 C-terminal IL3 domain in complex with Gαi1 showed a helical structure by NMR, whereas the peptide with the Ala-Leu motif formed a single turn structure *(108)*. This finding implies that Gαs can interact with the wild-type CB_1 receptor via a modified IL3 conformation resembling that of the Gαs-sensitive mutant.

11.5 Receptor Conformational Induction: The Steric Trigger Mechanism

The finding that structurally distinct ligands could differentially regulate RG complexes suggests that an isomerization of ARG to AR*G can be induced differentially by these ligands, qualifying them as agonists for these G protein subtypes. The many studies showing competition in radioligand-binding studies predict that the ligands would occupy mutually exclusive space within the receptor-binding pocket. The amino acid residues that are influenced by occupation of the ligand within its binding site would not be expected to be the same for each class of ligand, or even individual compounds within a given structural class of CB_1 receptor ligands. It can be envisioned that adjustments in TM3–TM6 would selectively modify the IL3 juxtamembrane region, which would involve Gαi1 and Gαi2. Adjustments in the position of TM7 and H8 would be expected to involve Gαi3 and Gαo.

Activation of a GPCR–G protein complex requires a sequence of transitions that must overcome a series of energy barriers to achieve GDP-GTP exchange and release of G proteins from the receptor. CB_1 cannabinoid receptor homology models based on the ground-state structure of rhodopsin have been useful to describe docking of ligands into the inactive state. Modifications that predict the interaction of ligands with the receptor upon activation have been developed based on biophysical studies of other GPCRs *(116,121–124)*. Comparable to the steric trigger mechanism by which the retinal conformational change is transmitted to the binding pocket of rhodopsin to promote a protein conformational change, it was predicted that cannabinoid agonist conformational changes leading to lower free energy states of the ligand would concomitantly exert steric clash with amino acid residues in the receptor. It was assumed that the agonist docking conformation constrained for optimal interaction with the binding pocket of the inactive ground-state CB_1 receptor would not be the lowest free energy conformation that the unconstrained ligand could achieve *(125)*. In the process of achieving its lowest free energy state, the ligand would exert steric clash with residues within the TM bundle, which would release inter-helical bonds and trigger multiple, serial conformational changes within the CB_1 receptor.

Chemically distinct ligands would be expected to interact with different residues within the binding space, and thereby initiate different routes of transition to unique receptor conformations that selectively promote interaction with one G protein as opposed to another. Helical translocations in TM3 and TM6 have been predicted from biophysical studies of rhodopsin and the β-adrenergic receptor, in which it is believed that the conserved E(or D)-R-Y motif at the intracellular side of TM3 maintains the GPCR in the ground state by an ionic lock with a TM6 E(or D) residue (see*(125)* and *(126)* for original references).

Reggio's laboratory modeled the CB_1 receptor structure to conform to either ground state R or an agonist-activated state R*. In the ground state, the TM6 was kinked at the G-P in the conserved GPCR CWXP motif to maintain the TM3 R3.50(214) to TM6 D6.30(338) ionic bridge that locks these helices in their inactive position *(126)*. The Reggio agonist-activated R* state exhibited a straightening of the TM6 helix G-P kink, accompanied by counter-clockwise turns in TM3 and TM6 and movement of these two helices away from each other at the juxtamembrane interface *(96,126)*. The impetus to increase the probability of straightening the TM6 helix within the C-W-G-P sequence includes a "rotamer toggle switch" transition at W6.48(356) that is coordinated with a rotameric transition in F3.36(200) in TM3 *(121,126)*.

11.5.1 Cannabinoid Agonist Activation Mechanism

Structural selectivity is observed for the interaction between the cannabinoid phenolic hydroxyl or the anandamide carboxamide oxygen with TM3 K3.28(192), an interaction that is not possible for the aminoalkylindole WIN55212-2 *(127)*. TM7 S7.39 is a binding site for classical and nonclassical cannabinoids, but not WIN55212-2 or rimonabant *(123)*. This latter site can induce a bend in TM7 that may be important for triggering a response. In a TM2 D2.63N mutation that neutralizes the negatively-charged residue, signal transduction was severely reduced for classical and nonclassical cannabinoid agonists, but minimally affected for WIN55212-2 *(123)*.

Shim and Howlett *(122,125)* described a mechanism by which the highly potent, stereoselective, nonclassical cannabinoid CP55244 could bind in the CB_1 receptor ground state as the "plug" by an interaction between the phenolic hydroxyl with TM3 K3.28(192), the D-ring hydroxyl secured by other H-bonding interactions, the ACD tricyclic ring moiety secured by hydrophobic residues on TMs2-3-7, and the alkyl side chain interacting with F3.36(200) and W6.48(356). They identified all CP55244 torsion angles for which the rotational energy barriers would be possible to overcome on the pathway toward achieving a lowest energy conformation. Molecular dynamics simulations of CP55244 demonstrated an initial shift to an intermediate state, from which two equally probably low energy states could exist for the molecule. Those two states required movement along the torsion angles associated with the alkyl side chain, which served as the "steric trigger" *(125)*. As the two states were achieved, steric hindrance with amino acid residues within the binding pocket would require the receptor to shift in conformation from its ground state. For CP55244, those regions of the binding pocket that were perturbed included residues on TM3 and TM6 *(125)*.

11.5.2 Endocannabinoid Agonist Activation Mechanism

The Reggio laboratory developed a model for selective CB_1 receptor interactions with the eicosanoid anandamide locked in a U/J-shaped conformation by an intramolecular H-bond locking the carboxamide oxygen and the ethanolamide hydroxyl *(126)*. In this conformation, several methyl groups of anandamide interact with V3.32(196), and the carboxamide oxygen H-bonds with K3.28(192). This would allow the acyl chain to participate in hydrophobic and C-H pi interactions with residues within the TMs2-3-7 region *(126)*.

Howlett's and Thomas' laboratories limited the number of probable anandamide conformations by constraining this flexible eicosanoid to the cannabinoid shape of 11-OH-hexahydrocannabinol *(128)* or ^{9}-tetrahydrocannabinol *(129)*. On the basis of the predicted structure of a twisted or bent configuration, a series of constrained analogs in which the homoallyl double bond series was replaced by an unsaturated ring were developed, and tested for biological activity in CB_1 signal transduction assays *(130)*. By allowing the carboxamide oxygen of these constrained analogs to interact with the CB_1 K3.28(192), the most potent and efficacious compound could establish interactions with the same residues in TMs3-4-5-6 that CP55244 utilized, whereas less efficacious analogs formed fewer or less energetically favorable interactions *(131)*. Thus, as least one mechanism by which anandamide and other eicosanoid ligands activate the CB_1 receptor would be similar to that of the non-classical cannabinoid ligands.

Reggio's laboratory reasoned that the hydrophobic nature of the endocannabinoids is suited to a mechanism by which these ligands orient within the lipid bilayer to establish the initial contact with the CB_1 receptor. They identified the V6.43/I6.46 pair, which exists as a β-X-X-β motif (β is a beta branching amino acid) just proximal to the C-W-G-P kink in TM6 on the hydrophobic surface of the inactive CB_1 receptor *(126)*. This motif can form a groove into which the alkyl tail of anandamide can establish a foot-hold, so that the anandamide carboxamide NH can H-bond with the P6.50 backbone carbonyl oxygen. This mechanism was possible for anandamide, but not for inactive analogs that do not possess adequate flexibility to establish both of these interactions with the protein. It was envisioned that as the alkyl tail interacts with the V-X-X-I groove, the TM6 would straighten and rotate to allow anandamide to slip into the TM bundle. This mechanism of structural selectivity may have general applicability to other GPCRs for lipid ligands (see discussion by *(132)*).

11.5.3 Aminoalkylindole Agonist Activation Mechanism

The aminoalkylindole WIN55212-2 was able to bind at multiple loci within the CB_1 receptor binding pocket *(96,133,134)*, explaining why this ligand could initiate an agonist signal to all subtypes of Gαi/o as well as to Gαs. WIN55212-2 does not require K3.28(192) to bind to the CB_1 receptor *(127)*. In contrast to the cannabinoid and eicosanoid ligands, WIN55212-2 required extensive aromatic stacking of the naphthyl ring with a cluster of Phe and Trp residues in TMs3-5-6, which were also required for interactions with rimonabant *(96,133)*. The WIN55212-2 indole ring interacted with a TM5 T5.39(275) *(96)*. The Shim model proposed that a moiety of WIN55212-2 interacts with the same hydrophobic binding pocket as the cannabinoid alkyl side chain, which serves as the steric trigger. This binding requirement allowed for two alternative binding modes, each of which would be associated with a different pathway to a lower energy state for the ligand utilizing the aroyl moiety as the steric trigger. By one configuration, WIN55212-2 could interact with an aromatic cluster composed of F3.36(200), W6.48(356), F5.42(278), and Y5.39(275) in TMs3, 5, and 6 *(134)*, residues shown by the earlier mutation studies to be required for WIN55212 interactions *(96)*. This would result in a steric trigger mechanism involving Y5.39(275). The resulting destabilization of TM5 could potentially impact interactions between IL3 and Gαi1/Gαi2. By the second configuration, WIN55212-2 could interact with a cluster of Phe and His residues in TMs 2 and 7 *(134)*. This would result in a steric trigger mechanism involving F2.61(174). The TM2 disruption could potentially impact the proposed H-bonding network that maintains the TM2 and TM7 helices in an inactive state *(135)*, thereby modulating the H8 conformation and its interactions with Gαi3 and Gαo. Thus, WIN55212-2 could behave as a "full agonist" by virtue of its capability of initiating diverse conformational induction mechanisms that can accommodate a broader range of G proteins.

11.5.4 Diarylpyrazole Antagonist/Inverse Agonist Mechanism

To elucidate a mechanism by which the receptor can respond to rimonabant in maintaining the inactive state, the Reggio laboratory and collaborators applied both receptor mutation and computational modeling studies using both an inactive (R) CB_1 homology model based on rhodopsin's ground state structure, and an "active" (R*) CB_1 model in which modifications have been made based on biophysical studies of activation of rhodopsin, β-adrenergic receptors, and other GPCRs *(124)*. These researchers suggested that the binding site for rimonabant lies within an aromatic microdomain involving TMs3-4-5-6 as well as direct aromatic stacking interactions with F3.36(200), Y5.39(275), and W5.43(279) *(96)*.

The preferred binding of rimonabant to the inactive R state of the CB_1 receptor was characterized by ligand aromatic stacking with F3.36 and W6.48, which stabilized TM6 in its locked position. This position allowed the carboxamide oxygen of the C3 substituent of rimonabant to H-bond with K3.28(192), an interaction that was forbidden in the R* active conformation *(9)*. After prohibiting this interaction with a K3.28A mutation, rimonabant continued to serve as a competitive antagonist, but was no longer able to perform as an inverse agonist to block the tonic suppression of the Ca^{2+} current in superior cervical ganglion cells. Analogs of rimonabant, such as 5-(4-chlorophenyl)-3-[(E)-2-cyclohexylethenyl]-1-(2,4-dichlorophenyl)-4-me thyl-1H-pyrazole (VCHSR), which are not able to participate in the H-bonding interaction with K3.28(192), behaved as competitive neutral antagonists, but were not able to function as inverse agonists *(9,124)*. These studies of mechanism provide valuable insight for the development of novel ligands that can selectively behave as competitive antagonists but not inverse agonists. Development of ligands that can selectively antagonize interactions with specific G protein subtypes could be therapeutically useful in curtailing cellular functions with greater selectivity.

11.6 CB_1 Receptor Accessory Proteins

Modulation of the output of CB_1 receptor conformational changes can occur via binding partners other than the G proteins that transduce the signaling to the effector enzymes or channels. Although not currently well characterized, it is likely that, in combination with other scaffolding and interacting proteins, these accessory proteins may direct divergent signaling pathways in cells.

11.6.1 CRIP1a: Cannabinoid Receptor Interacting Protein1a

Nie and Lewis found that deletion of the C-terminal distal to the H8 domain slowed the time to peak Ca^{2+} current inhibition, augmented the tonic inhibition of Ca^{2+} currents, and promoted the ability of the CB_1 receptor to sequester G proteins *(109,136)*. They hypothesized that the C-terminal tail serves an auto-inhibitory function, perhaps by binding to accessory proteins. Using the CB_1 receptor distal C-terminal as bait, they identified a pair of splice variant proteins, CRIP1a and CRIP1b *(137)*. CRIP1a possesses a palmitoylation site that could potentially localize it to the membrane, and a PDZ class I ligand at its C-terminal that could allow interaction with PDZ-containing proteins. CRIP1a could bind to a glutathione S-transferase (GST)-fusion protein attached to the CB_1 C-terminal tail, and could be coimmunoprecipitated with the CB_1 receptor *(137)*. Although neither CRIP1a nor CRIP1b altered agonist-stimulated Ca^{2+} channel inhibition, CRIP1a (but not CRIP1b) was able to attenuate the rimonabant-reversible tonic inhibitory action of CB_1 on Ca^{2+} channels *(137)*. The finding that Gαi3 was responsible for the tonic inhibition of Ca^{2+} channels by CB_1 receptors in the absence of agonists *(116)* suggests that CRIP1a may exert some selectivity in the type of G protein with which it can compete.

The function of CB_1 receptors to regulate neurotransmitter release by suppression of Ca^{2+} currents makes the CB_1–CRIP1a complex relevant to a vital neuroregulatory function. CRIP has the potential to modulate other effectors and signaling pathways associated with the CB_1–G protein complex (e.g., adenylyl cyclase, MAPK), as well as other neuromodulator receptor systems that have not yet been identified.

11.6.2 FAN: Factor Associated with Neutral Sphingomyelinase

Factor associated with neutral sphingomyelinase (FAN) is a protein that directs the agonist-stimulated CB_1 receptor signal transduction activity to sphingomyelin hydrolysis to generate the second messenger ceramide in astrocytes *(138)*. Gαi/o proteins were not required for this response, as it was not blocked by pertussis toxin treatment. After treatment of cells with a CB_1 agonist, FAN coimmunoprecipitated in a complex with CB_1 receptors, and rimonabant could antagonize this agonist-stimulated association *(138)*. A binding site motif on the C-terminal of the CB_1 receptor bears some homology with the neutral sphingomyelinase activation domain of other FAN-associating proteins *(138)*.

11.6.3 β-Arrestin

Prolonged agonist stimulation of CB_1 receptors led to desensitization of signal transduction through inwardly rectifying K^+ channels by a mechanism that can be supported by G protein receptor kinase 3 (GRK3) and β-arrestin2 *(139)*. A CB_1 receptor C-terminal tail domain distal to H8 was required for the desensitization of the response, and this domain included two Ser residues that would be potential substrates for GRK phosphorylation. Phosphorylation of these Ser residues was important for desensitization of the G protein-dependent K^+-channel response as well as the MAPK activation, inasmuch as the desensitization could be precluded by S425A/S539A mutations *(139,140)*. However, phosphorylation of these residues was not necessary for β-arrestin2 recruitment and the internalization process *(139,140)*. The divergence of these responses could be a source of functional selectivity by agonists. A diphosphopeptide derived from this domain could bind to β-arrestin2 and assume a conformation with a very flexible Gly residue separating two helices *(141)*, suggesting that structure can direct function in this region of the receptor.

11.6.4 GASP1: GPCR-Associated Sorting Protein 1

GPCR-associated sorting protein 1 (GASP1) is a protein involved in post-endocytic targeting of the CB_1 receptor to the lysosome for degradation *(142,143)*. This protein mediated down-regulation of internalized CB_1 receptors upon prolonged exposure to agonists. As a consequence, receptors that had been internalized from neuronal cell bodies were not redistributed to axons where they could subserve signal transduction functions at their presynaptic loci *(144)*. GASP1 was coimmunoprecipitated with the CB_1 receptor, and could bind to a GST fusion protein linked to a CB_1 receptor C-terminal peptide *(142)*. Of particular importance for functional selectivity, GASP1 could interact with conserved residues in H8, thereby competing with G proteins that interact with the CB_1 receptor at this locus *(145)*.

11.7 Functional Selectivity via CB_1 Receptor Association and Heterodimerization

Recent studies have provided evidence that GPCR association or heterodimerization can influence efficacy for ligands that stimulate one member of the pair. Although discussion of this phenomenon is not comprehensive, a few examples will be described in which signal transduction has been functionally modulated by the interaction between partners.

11.7.1 CB_1-D_2 Receptors

When CB_1 receptors were transiently expressed in HEK293 cells stably expressing D_2 receptors, CP55940 stimulated cyclic AMP production, although this stimulation required greater concentrations of agonist than would have been required for inhibition of cyclic AMP by CB_1 receptors expressed alone *(146)*. Evidence for shared pools of Gαi proteins comes from data indicating that the D_2 dopamine receptors could prevent coupling of Gi/o to CB_1 receptors, such that CP55940 could stimulate an increase in cyclic AMP production via Gαs. Increasing free Gαi, both by overexpression of Gαi1 and by desensitization of the D_2 dopamine receptors with quinpirole, were able to reverse the response *(146)*.

In a similar HEK293 cell model expressing both CB_1 and D_2 receptors, inhibition of forskolin-activated cyclic AMP production could be reversed by increasing concentrations of the combined agonists *(147)*. Simultaneous stimulation by submaximal concentrations of agonists for both receptors evoked production of cyclic AMP as well as activation of MAPK *(147)*. Pertussis toxin attenuated the inhibition but not the stimulation of cyclic AMP production, consistent with Gαs-mediation of the stimulation component. The MAPK activation was also pertussis toxin-insensitive, indicating that Gαi/o proteins were not mediating this response *(147)*. Overexpression of Gαi1 attenuated the cyclic AMP production, suggesting that the D_2 dopamine receptors could sequester Gαi proteins and thereby allow CB_1 receptor interaction with alternative G proteins, particularly Gαs *(146)*. In the presence of the combined agonists (but not either one alone), the D_2 and CB_1 receptors formed coimmunoprecipitable complexes *(147)*, indicative of a stable protein association of these two GPCRs in their agonist-bound states. Close association of the receptors was also observed in fluorescence resonance energy transfer (FRET) studies *(148)*.

The shift in G protein activation from Gαi to Gαs by heterodimerization can explain observations in neurons and brain preparations. In cultured striatal cells, costimulation of dopaminergic and CB_1 receptors converted the response from a Gαi-mediated inhibition of cyclic AMP production to a Gαs-mediated increase in cyclic AMP production *(24)*. WIN55212-2 could increase cyclic AMP accumulation in globus pallidus slices *(21)*. Reports of cannabinoid-stimulated PKA signal transduction suggest that dual cannabinoid and dopaminergic costimulation can promote a fundamentally different signal transduction output than would have been observed by either receptor alone *(27,28)*.

11.7.2 CB_1-Opioid Receptors (OR)

Opioid and cannabinoid receptors exogenously expressed in COS-7 cells could compete for a common pool of G proteins in [^{35}S]GTPγS-binding assays *(149,150)*. However, of note, in N18TG2 cells, which endogenously express both receptors, sharing of G proteins was not evident. The inducible expression of CB_1 receptors in stably μ-OR-expressing HEK293 cells increased basal constitutive [^{35}S]GTPγS binding activity in membrane preparations *(151)*. Because the basal levels in the CB_1-expressing cells were high, further stimulation above basal by the μ-OR agonist DAMGO was attenuated. The increased basal [^{35}S]GTPγS binding was reversed by the CB_1 antagonist LY320135, so that the apparently attenuated DAMGO-stimulated level was readjusted to pre-CB_1 expression levels *(151)*. Similar results were observed in a permeabilized-cell assay for [^{35}S]GTPγS binding in CB_1-μ-OR-HEK293 cells, in which WIN55212-2 attenuated the morphine-stimulated G protein activation when expressed as a percent of basal *(152)*. Mutual suppression of agonist-stimulated [^{35}S]GTPγS binding by CB_1 agonist WIN55212-2 and μ-OR agonist DAMGO was observed in SK-N-SH cells and striatal membranes, endogenously expressing systems *(152)*.

The results of the CB_1 receptor effects on the ability of μ-OR to stimulate G protein is variably translated to successive signal transduction events. Coexpression of CB_1 receptors with stably expressed μ-OR in HEK293 cells had no influence on the inhibition by DAMGO of forskolin-activated adenylyl cyclase *(151)*. In CB_1-μ-OR-HEK293 cells, combined CB_1 and μ-OR agonists suppressed the activation of MAPK below the level stimulated by either agonist alone *(152)*. In the absence of agonist stimulation, the expression of CB_1 receptors was sufficient to attenuate the μ-OR-stimulated MAPK activation in response to morphine or DAMGO, and this attenuation could be reversed by rimonabant but not by the neutral antagonist O-2050 *(151)*. In Neuro-2A cells that endogenously express CB_1 receptors and stably express exogenous μ-OR, combined CB_1 and μ-OR agonists suppressed Src and STAT3 phosphorylation and neurite outgrowth below levels that were stimulated by either agonist alone.

Some evidence supports the notion that the influence of CB_1 receptors is due to receptor dimerization. CB_1 and μ-, κ-, or δ-opioid receptors (ORs) coexpressed in HEK293 cells increased the bioluminescence resonance energy transfer (BRET) signal, indicating that these receptors were in close association *(152)*. However, populations of receptors that are compartmentalized in different cellular locations suggest that receptor association may be a dynamic phenomenon *(151)*. It has been demonstrated that agonist-induced MAPK phosphorylation outcomes are under multiple and complex regulatory influences in different cell types *(63)*, suggesting that implications based on the assumptions of heterodimerization should be made with caution.

11.7.3 CB_1-Orexin-1 (OX1) Receptors

The orexin-1 (OX1) and the CB_1 receptors have exhibited a plasma membrane colocalization when coexpressed in CHO cells *(153)*. When expressed individually in CHO cells, both exhibited a short time-course to peak MAPK activation. Although coexpression of both receptors had little or no effect on the response to CP55940, the presence of the CB_1 receptor increased the OX1 receptor sensitivity to orexin A by shifting the ED_{50} to the left by two orders of magnitude *(153)*. Both CB_1 antagonist rimonabant and pertussis toxin treatment blocked the MAPK response to CP55940, and reversed the increase in potency of orexin A to that observed in the absence of CB_1 receptor expression, indicating that a functional CB_1–$G\alpha i/o$ complex was required *(153)*. Signal transduction selectivity was observed, inasmuch as expression of CB_1 receptors did not affect the ability of orexin A to stimulate inositol triphosphate production via $G\alpha q$, nor did it confer the ability to inhibit adenylyl cyclase via $G\alpha i$ *(153)*.

Studies that support close association of the receptors include observations of clustering in immunoelectron microscopy of the two exogenously expressed receptors in CHO cells *(153)*. FRET studies demonstrated close association of the two exogenously expressed receptors in HEK293 cells *(154)*. The presence of the CB_1 receptor influenced cellular localization of the OX1 receptor by increasing the population present within intracellular vesicles. Sustained treatment with the CB_1 ligand rimonabant or the OX1 ligand SB674042 redistributed both receptors to the cell surface *(154)*. Sustained exposure to these receptor-selective antagonists also cross-inhibited agonist-stimulated MAPK phosphorylation in a reciprocal manner with a similar time course *(154)*, suggesting that cellular translocation may influence signal transduction activity.

11.7.4 CB_1-$GABA_B$ Receptors

In cerebellar granule cells, endogenously expressed CB_1 receptors and γ-aminobutyric acid ($GABA_B$) receptors shared a common pool of adenylyl cyclase rather than a common pool of G proteins *(155,156)*. Evidence for the functional association of CB_1 and $GABA_B$ receptors is based on the cross-inhibition of $[^{35}S]GTP\gamma S$ binding in hippocampal membranes *(157)*. Low nM concentrations of the $GABA_B$ antagonist phaclofen (which would be below the concentrations required for $GABA_B$ antagonism) were able to noncompetitively attenuate the response to WIN55212-2. Conversely, the CB_1 antagonist AM251 competitively inhibited the response to the $GABA_B$ agonist SKF97541. These responses were not observed in cortex or spinal cord membranes, indicating that the responding cells within the hippocampus are especially conducive to CB_1-$GABA_B$ receptor interactions *(157)*.

11.8 Conclusions and Caveats

The studies summarized in this review have examined multiple ligands regulating the CB_1 receptor–Gαi/o protein complex to direct multiple signal transduction pathways, and the impact that associated proteins may have on ligand-directed signal selectivity. Studies of the CB_1 receptor signal transduction have surpassed those of the CB_2 receptor, largely because of the high density of receptors found in the brain, and the multiple physiological effects that can be investigated in the brain and other tissues. Both cannabinoid receptors direct interactions with Gαi/o proteins, and traditionally have been stimulated by the same agonist ligands. It is expected that developing pharmacological selectivity in CB_1 or CB_2 ligands will stimulate more intensive investigations of CB_2 signal transduction in the future. It is likely that other putative cannabinoid receptors will be characterized, and although they may be stimulated by cannabinoid or endocannabinoid ligands, their signal transduction mechanisms are likely to differ from those of the CB_1 and CB_2 receptors. A confound in the interpretation of cannabinoid receptor signal transduction studies is that promiscuity in endocannabinoid signaling can appear to be functional selectivity in response to stimulation of CB_1 or CB_2 receptors. Anandamide and related lipid mediators have the potential to serve as modulators of many enzymes and channels that trigger cellular signal transduction events that can masquerade as functional selectivity in the cellular response. For example, anandamide can activate TRPV1 channels *(158,159)*; 2-AG and *N*-arachidonoyldopamine can inhibit voltage-gated Na^+ channels *(160)*; methanandamide can allosterically inhibit nicotinic receptor subunits *(161)*; THC and anandamide can potentiate the glycine receptor channel *(162)*. It remains for future studies to sort out the caveats and confounds, as future efforts at therapeutic ligand development build on our current understanding of agonist-directed functional selectivity in the cannabinoid receptors.

References

1. Howlett AC, Barth F, Bonner TI, et al. International Union of Pharmacology. XXVII. Classification of Cannabinoid Receptors. Pharmacol Rev 2002;54:161–202.
2. Pertwee RG. Pharmacological actions of cannabinoids. Handb Exp Pharmacol 2005;1-51.
3. Di Marzo V, De Petrocellis L, Bisogno T. The biosynthesis, fate and pharmacological properties of endocannabinoids. Handb Exp Pharmacol 2005;147-185.
4. Howlett AC. Cannabinoid receptor signaling. Handb Exp Pharmacol 2005;53-79.
5. Barth F, Rinaldi-Carmona M. The development of cannabinoid antagonists. Curr Med Chem 1999;6:745–755.
6. Thomas BF, Zhang Y, Brackeen M, Page KM, Mascarella SW, Seltzman HH. Conformational characteristics of the interaction of SR141716A with the CB_1 cannabinoid receptor as determined through the use of conformationally constrained analogs. *AAPS J* 2006;8:E665–E671.
7. Hagmann WK. The discovery of taranabant, a selective cannabinoid-1 receptor inverse agonist for the treatment of obesity. Arch Pharm (Weinheim) 2008;341:405–411.

8. Pertwee RG. Inverse agonism and neutral antagonism at cannabinoid CB_1 receptors. Life Sci 2005;76:1307–1324.
9. Hurst DP, Lynch DL, Barnett-Norris J et al. N-(piperidin-1-yl)-5-(4-chlorophenyl)-1-(2,4-dichlorophenyl)-4-methyl-1H-p yrazole-3-carboxamide (SR141716A) interaction with LYS 3.28(192) is crucial for its inverse agonism at the cannabinoid CB_1 receptor. Mol Pharmacol 2002;62:1274–1287.
10. Chambers AP, Vemuri VK, Peng Y et al. A neutral CB_1 receptor antagonist reduces weight gain in rat. Am J Physiol Regul Integr Comp Physiol 2007;293:R2185–R2193.
11. Howlett AC. Efficacy in CB_1 receptor-mediated signal transduction. Br J Pharmacol 2004;142:1209–1218.
12. Mu J, Zhuang SY, Hampson RE, Deadwyler SA. Protein kinase-dependent phosphorylation and cannabinoid receptor modulation of potassium A current (I_A) in cultured rat hippocampal neurons. Pflugers Arch 2000;439:541–546.
13. Mackie K, Lai Y, Westenbroek R, Mitchell R. Cannabinoids activate an inwardly rectifying potassium conductance and inhibit Q-type calcium currents in AtT20 cells transfected with rat brain cannabinoid receptor. J Neurosci 1995;15:6552–6561.
14. Henry DJ, Chavkin C. Activation of inwardly rectifying potassium channels (GIRK1) by co-expressed rat brain cannabinoid receptors in Xenopus oocytes. Neurosci Lett 1995;186:91–94.
15. McAllister SD, Griffin G, Satin LS, Abood ME. Cannabinoid receptors can activate and inhibit G protein-coupled inwardly rectifying potassium channels in a Xenopus oocyte expression system. J Pharmacol Exp Ther 1999;291:618–626.
16. Derkinderen P, Toutant M, Burgaya F et al. Regulation of a neuronal form of focal adhesion kinase by anandamide. Science 1996;273:1719–1722.
17. Zhou D, Song ZH. CB_1 cannabinoid receptor-mediated tyrosine phosphorylation of focal adhesion kinase-related non-kinase. FEBS Lett 2002;525:164–168.
18. Zhou D, Song ZH. CB_1 cannabinoid receptor-mediated neurite remodeling in mouse neuroblastoma N1E-115 cells. J Neurosci Res 2001;65:346–353.
19. Derkinderen P, Toutant M, Kadare G, Ledent C, Parmentier M, Girault JA. Dual role of Fyn in the regulation of FAK + 6,7 by cannabinoids in hippocampus. J Biol Chem 2001;276:38289–38296.
20. Derkinderen P, Valjent E, Toutant M et al. Regulation of extracellular signal-regulated kinase by cannabinoids in hippocampus. J Neurosci 2003;23:2371–2382.
21. Maneuf YP, Brotchie JM. Paradoxical action of the cannabinoid WIN 55,212-2 in stimulated and basal cyclic AMP accumulation in rat globus pallidus slices. Br J Pharmacol 1997; 120:1397–1398.
22. Bonhaus DW, Chang LK, Kwan J, Martin GR. Dual activation and inhibition of adenylyl cyclase by cannabinoid receptor agonists: evidence for agonist-specific trafficking of intracellular responses. J Pharmacol Exp Ther 1998;287:884–888.
23. Rhee MH, Bayewitch M, Avidor-Reiss T, Levy R, Vogel Z. Cannabinoid receptor activation differentially regulates the various adenylyl cyclase isozymes. J Neurochem 1998;71:1525–1534.
24. Glass M, Felder CC. Concurrent stimulation of cannabinoid CB_1 and dopamine D_2 receptors augments cAMP accumulation in striatal neurons: evidence for a Gs linkage to the CB_1 receptor. J Neurosci 1997;17:5327–5333.
25. Felder CC, Joyce KE, Briley EM et al. LY320135, a novel cannabinoid CB_1 receptor antagonist, unmasks coupling of the CB_1 receptor to stimulation of cAMP accumulation. J Pharmacol Exp Ther 1998;284:291–297.
26. Bash R, Rubovitch V, Gafni M, Sarne Y. The stimulatory effect of cannabinoids on calcium uptake is mediated by Gs GTP-binding proteins and cAMP formation. Neurosignals 2003;12:39–44.
27. Andersson M, Usiello A, Borgkvist A et al. Cannabinoid action depends on phosphorylation of dopamine- and cAMP-regulated phosphoprotein of 32 kDa at the protein kinase A site in striatal projection neurons. J Neurosci 2005;25:8432–8438.
28. Borgkvist A, Marcellino D, Fuxe K, Greengard P, Fisone G. Regulation of DARPP-32 phosphorylation by Delta9-tetrahydrocannabinol. Neuropharmacology 2008;54:31–35.

29. Hampson RE, Mu J, Deadwyler SA. Cannabinoid and kappa opioid receptors reduce potassium K current via activation of G(s) proteins in cultured hippocampal neurons. J Neurophysiol 2000;84:2356–2364.
30. Felder CC, Veluz JS, Williams HL, Briley EM, Matsuda LA. Cannabinoid agonists stimulate both receptor- and non-receptor-mediated signal transduction pathways in cells transfected with and expressing cannabinoid receptor clones. Mol Pharmacol 1992;42:838–845.
31. Felder CC, Joyce KE, Briley EM et al. Comparison of the pharmacology and signal transduction of the human cannabinoid CB_1 and CB_2 receptors. Mol Pharmacol 1995;48:443–450.
32. Sugiura T, Kodaka T, Kondo S et al. 2-Arachidonoylglycerol, a putative endogenous cannabinoid receptor ligand, induces rapid, transient elevation of intracellular free Ca^{2+} in neuroblastoma x glioma hybrid NG108-15 cells. Biochem Biophys Res Commun 1996;229:58–64.
33. Sugiura T, Kodaka T, Kondo S et al. Is the cannabinoid CB_1 receptor a 2-arachidonoylglycerol receptor? Structural requirements for triggering a Ca^{2+} transient in NG108-15 cells. J Biochem (Tokyo) 1997;122:890–895.
34. Netzeband JG, Conroy SM, Parsons KL, Gruol DL. Cannabinoids enhance NMDA-elicited Ca^{2+} signals in cerebellar granule neurons in culture. J Neurosci 1999;19:8765–8777.
35. Nah SY, Saya D, Vogel Z. Cannabinoids inhibit agonist-stimulated formation of inositol phosphates in rat hippocampal cultures. Eur J Pharmacol 1993;246:19–24.
36. Zhuang SY, Bridges D, Grigorenko E et al. Cannabinoids produce neuroprotection by reducing intracellular calcium release from ryanodine-sensitive stores. Neuropharmacology 2005;48:1086–1096.
37. Caulfield MP, Brown DA. Cannabinoid receptor agonists inhibit Ca current in NG108-15 neuroblastoma cells via a pertussis toxin-sensitive mechanism. Br J Pharmacol 1992;106:231–232.
38. Mackie K, Hille B. Cannabinoids inhibit N-type calcium channels in neuroblastoma-glioma cells. Proc Natl Acad Sci USA 1992;89:3825–3829.
39. Mackie K, Devane WA, Hille B. Anandamide, an endogenous cannabinoid, inhibits calcium currents as a partial agonist in N18 neuroblastoma cells. Mol Pharmacol 1993;44:498–503.
40. Priller J, Briley EM, Mansouri J, Devane WA, Mackie K, Felder CC. Mead ethanolamide, a novel eicosanoid, is an agonist for the central (CB_1) and peripheral (CB_2) cannabinoid receptors. Mol Pharmacol 1995;48:288–292.
41. Pan X, Ikeda SR, Lewis DL. Rat brain cannabinoid receptor modulates N-type Ca^{2+} channels in a neuronal expression system. Mol Pharmacol 1996;49:707–714.
42. Hampson AJ, Bornheim LM, Scanziani M et al. Dual effects of anandamide on NMDA receptor-mediated responses and neurotransmission. J Neurochem 1998;70:671–676.
43. Gebremedhin D, Lange AR, Campbell WB, Hillard CJ, Harder DR. Cannabinoid CB_1 receptor of cat cerebral arterial muscle functions to inhibit L-type Ca2+ channel current. Am J Physiol 1999;276:H2085–H2093.
44. Pan X, Ikeda SR, Lewis DL. SR 141716A acts as an inverse agonist to increase neuronal voltage-dependent Ca2+ currents by reversal of tonic CB_1 cannabinoid receptor activity. Mol Pharmacol 1998;54:1064–1072.
45. Vasquez C, Lewis DL. The CB_1 cannabinoid receptor can sequester G-proteins, making them unavailable to couple to other receptors. J Neurosci 1999;19:9271–9280.
46. Guo J, Ikeda SR. Endocannabinoids modulate N-type calcium channels and G-protein-coupled inwardly rectifying potassium channels via CB_1 cannabinoid receptors heterologously expressed in mammalian neurons. Mol Pharmacol 2004;65:665–674.
47. Bouaboula M, Poinot-Chazel C, Bourrié B et al. Activation of mitogen-activated protein kinases by stimulation of the central cannabinoid receptor CB_1. Biochem J 1995;312 (Pt 2):637–641.
48. Guzmán M, Sánchez C. Effects of cannabinoids on energy metabolism. Life Sci 1999;65:657–664.
49. Sánchez C, Galve-Roperh I, Rueda D, Guzmán M. Involvement of sphingomyelin hydrolysis and the mitogen-activated protein kinase cascade in the Delta9-tetrahydrocannabinol-induced stimulation of glucose metabolism in primary astrocytes. Mol Pharmacol 1998;54:834–843.
50. Wartmann M, Campbell D, Subramanian A, Burstein SH, Davis RJ. The MAP kinase signal transduction pathway is activated by the endogenous cannabinoid anandamide. FEBS Lett 1995;359:133–136.

51. Galve-Roperh I, Rueda D, Gómez del Pulgar T, Velasco G, Guzmán M. Mechanism of extracellular signal-regulated kinase activation by the CB(1) cannabinoid receptor. Mol Pharmacol 2002;62:1385–1392.
52. Gómez del Pulgar T, Velasco G, Guzmán M. The CB_1 cannabinoid receptor is coupled to the activation of protein kinase B/Akt. Biochem J 2000;347:369–373.
53. Davis MI, Ronesi J, Lovinger DM. A predominant role for inhibition of the adenylate cyclase/protein kinase A pathway in ERK activation by cannabinoid receptor 1 in N1E-115 neuroblastoma cells. J Biol Chem 2003;278:48973–48980.
54. Rueda D, Galve-Roperh I, Haro A, Guzmán M. The CB(1) cannabinoid receptor is coupled to the activation of c-Jun N-terminal kinase. Mol Pharmacol 2000;58:814–820.
55. Liu J, Gao B, Mirshahi F et al. Functional CB_1 cannabinoid receptors in human vascular endothelial cells. Biochem J 2000;346 (Pt 3):835–840.
56. Derkinderen P, Ledent C, Parmentier M, Girault JA. Cannabinoids activate p38 mitogen-activated protein kinases through CB_1 receptors in hippocampus. J Neurochem 2001;77:957–960.
57. Bouaboula M, Perrachon S, Milligan L et al. A selective inverse agonist for central cannabinoid receptor inhibits mitogen-activated protein kinase activation stimulated by insulin or insulin-like growth factor 1. Evidence for a new model of receptor/ligand interactions. J Biol Chem 1997;272:22330–22339.
58. Bouaboula M, Desnoyer N, Carayon P, Combes T, Casellas P. Gi protein modulation induced by a selective inverse agonist for the peripheral cannabinoid receptor CB_2: implication for intracellular signalization cross-regulation. Mol Pharmacol 1999;55:473–480.
59. Hart S, Fischer OM, Ullrich A. Cannabinoids induce cancer cell proliferation via tumor necrosis factor alpha-converting enzyme (TACE/ADAM17)-mediated transactivation of the epidermal growth factor receptor. Cancer Res 2004;64:1943–1950.
60. Berghuis P, Dobszay MB, Wang X et al. Endocannabinoids regulate interneuron migration and morphogenesis by transactivating the TrkB receptor. Proc Natl Acad Sci USA 2005;102:19115–19120.
61. Rueda D, Navarro B, Martinez-Serrano A, Guzmán M, Galve-Roperh I. The endocannabinoid anandamide inhibits neuronal progenitor cell differentiation through attenuation of the Rap1/B-Raf/ERK pathway. J Biol Chem 2002;277:46645–46650.
62. Rubovitch V, Gafni M, Sarne Y. The involvement of VEGF receptors and MAPK in the cannabinoid potentiation of Ca^{2+} flux into N18TG2 neuroblastoma cells. Brain Res Mol Brain Res 2004;120:138–144.
63. Korzh A, Keren O, Gafni M, Bar-Josef H, Sarne Y. Modulation of extracellular signal-regulated kinase (ERK) by opioid and cannabinoid receptors that are expressed in the same cell. Brain Res 2008;1189:23–32.
64. Nabemoto M, Mashimo M, Someya A et al. Release of arachidonic acid by 2-arachidonoyl glycerol and HU210 in PC12 cells; roles of Src, phospholipase C and cytosolic phospholipase A(2)alpha. Eur J Pharmacol 2008.
65. He JC, Neves SR, Jordan JD, Iyengar R. Role of the Go/i signaling network in the regulation of neurite outgrowth. Can J Physiol Pharmacol 2006;84:687–694.
66. Jordan JD, He JC, Eungdamrong NJ et al. Cannabinoid receptor-induced neurite outgrowth is mediated by Rap1 activation through G(alpha)o/i-triggered proteasomal degradation of Rap1GAPII. J Biol Chem 2005;280:11413–11421.
67. He JC, Gomes I, Nguyen T et al. The G alpha(o/i)-coupled cannabinoid receptor-mediated neurite outgrowth involves Rap regulation of Src and Stat3. J Biol Chem 2005;280:33426–33434.
68. Stefano GB, Salzet M, Magazine HI, Bilfinger TV. Antagonism of LPS and IFN-gamma induction of iNOS in human saphenous vein endothelium by morphine and anandamide by nitric oxide inhibition of adenylate cyclase. J Cardiovasc Pharmacol 1998;31:813–820.
69. Fimiani C, Mattocks D, Cavani F, Salzet M, Deutsch DG, Pryor S et al. Morphine and anandamide stimulate intracellular calcium transients in human arterial endothelial cells: coupling to nitric oxide release. Cell Signal 1999;11:189–193.
70. Mombouli JV, Schaeffer G, Holzmann S, Kostner GM, Graier WF. Anandamide-induced mobilization of cytosolic Ca^{2+} in endothelial cells. Br J Pharmacol 1999;126:1593–1600.

71. Maccarrone M, Bari M, Lorenzon T, Bisogno T, Di MV, Finazzi-Agro A. Anandamide uptake by human endothelial cells and its regulation by nitric oxide. J Biol Chem 2000; 275:13484–13492.
72. Stefano GB, Liu Y, Goligorsky MS. Cannabinoid receptors are coupled to nitric oxide release in invertebrate immunocytes, microglia, and human monocytes. J Biol Chem 1996;271: 19238–19242.
73. Prevot V, Rialas CM, Croix D et al. Morphine and anandamide coupling to nitric oxide stimulates GnRH and CRF release from rat median eminence: neurovascular regulation. Brain Res 1998;790:236–244.
74. Mukhopadhyay S, Shim JY, Assi AA, Norford D, Howlett AC. CB(1) cannabinoid receptor-G protein association: a possible mechanism for differential signaling. Chem Phys Lipids 2002;121:91–109.
75. Bisogno T, Maccarrone M, De Petrocellis L et al. The uptake by cells of 2-arachidonoylglycerol, an endogenous agonist of cannabinoid receptors. Eur J Biochem 2001;268:1982–1989.
76. De Petrocellis L, Bisogno T, Maccarrone M, Davis JB, Finazzi-Agró A, Di Marzo V. The activity of anandamide at vanilloid VR1 receptors requires facilitated transport across the cell membrane and is limited by intracellular metabolism. J Biol Chem 2001;276:12856–12863.
77. Jones JD, Carney ST, Vrana KE, Norford DC, Howlett AC. Cannabinoid receptor-mediated translocation of NO-sensitive guanylyl cyclase and production of cyclic GMP in neuronal cells. Neuropharmacology 2008;54:23–30.
78. Simmons ML, Murphy S. Induction of nitric oxide synthase in glial cells. J Neurochem 1992;59:897–905.
79. Hillard CJ, Muthian S, Kearn CS. Effects of CB(1) cannabinoid receptor activation on cerebellar granule cell nitric oxide synthase activity. FEBS Lett 1999;459:277–281.
80. Kim SH, Won SJ, Mao XO, Jin K, Greenberg DA. Molecular mechanisms of cannabinoid protection from neuronal excitotoxicity. Mol Pharmacol 2006;69:691–696.
81. Rameau GA, Chiu LY, Ziff EB. Bidirectional regulation of neuronal nitric-oxide synthase phosphorylation at serine 847 by the N-methyl-d-aspartate receptor. J Biol Chem 2004;279:14307–14314.
82. Rameau GA, Tukey DS, Garcin-Hosfield ED et al. Biphasic coupling of neuronal nitric oxide synthase phosphorylation to the NMDA receptor regulates AMPA receptor trafficking and neuronal cell death. J Neurosci 2007;27:3445–3455.
83. Jeon YJ, Yang KH, Pulaski JT, Kaminski NE. Attenuation of inducible nitric oxide synthase gene expression by delta 9-tetrahydrocannabinol is mediated through the inhibition of nuclear factor- kappa B/Rel activation. Mol Pharmacol 1996;50:334–341.
84. Cabral GA, Harmon KN, Carlisle SJ. Cannabinoid-mediated inhibition of inducible nitric oxide production by rat microglial cells: evidence for CB_1 receptor participation. Adv Exp Med Biol 2001;493:207–214.
85. Molina-Holgado F, Lledó A, Guaza C. Anandamide suppresses nitric oxide and TNF-alpha responses to Theiler's virus or endotoxin in astrocytes. Neuroreport 1997;8:1929–1933.
86. Molina-Holgado F, Molina-Holgado E, Guaza C, Rothwell NJ. Role of CB_1 and CB_2 receptors in the inhibitory effects of cannabinoids on lipopolysaccharide-induced nitric oxide release in astrocyte cultures. J Neurosci Res 2002;67:829–836.
87. Esposito G, Ligresti A, Izzo AA et al. The endocannabinoid system protects rat glioma cells against HIV-1 Tat protein-induced cytotoxicity. Mechanism and regulation. J Biol Chem 2002;277:50348–50354.
88. Molina-Holgado F, Pinteaux E, Moore JD et al. Endogenous interleukin-1 receptor antagonist mediates anti-inflammatory and neuroprotective actions of cannabinoids in neurons and glia. J Neurosci 2003;23:6470–6474.
89. Glass M, Northup JK. Agonist selective regulation of G proteins by cannabinoid CB(1) and CB(2) receptors. Mol Pharmacol 1999;56:1362–1369.
90. Prather PL, Martin NA, Breivogel CS, Childers SR. Activation of cannabinoid receptors in rat brain by WIN 55212-2 produces coupling to multiple G protein alpha-subunits with different potencies. Mol Pharmacol 2000;57:1000–1010.

91. Mukhopadhyay S, Howlett AC. Chemically distinct ligands promote differential CB_1 cannabinoid receptor-Gi protein interactions. Mol Pharmacol 2005;67:2016–2024.
92. Houston DB, Howlett AC. Differential receptor-G-protein coupling evoked by dissimilar cannabinoid receptor agonists. Cell Signal 1998;10:667–674.
93. Houston DB, Howlett AC. Solubilization of the cannabinoid receptor from rat brain and its functional interaction with guanine nucleotide-binding proteins. Mol Pharmacol 1993;43:17–22.
94. Howlett AC, Mukhopadhyay S, Shim JY, Welsh WJ. Signal transduction of eicosanoid CB_1 receptor ligands. Life Sci 1999;65:617–625.
95. Mukhopadhyay S, McIntosh HH, Houston DB, Howlett AC. The CB(1) cannabinoid receptor juxtamembrane C-terminal peptide confers activation to specific G proteins in brain. Mol Pharmacol 2000;57:162–170.
96. McAllister SD, Rizvi G, Anavi-Goffer S et al. An aromatic microdomain at the cannabinoid CB(1) receptor constitutes an agonist/inverse agonist binding region. J Med Chem 2003;46: 5139–5152.
97. Savinainen JR, Saario SM, Niemi R, Jarvinen T, Laitinen JT. An optimized approach to study endocannabinoid signaling: evidence against constitutive activity of rat brain adenosine A_1 and cannabinoid CB_1 receptors. Br J Pharmacol 2003;140:1451–1459.
98. Landsman RS, Burkey TH, Consroe P, Roeske WR, Yamamura HI. SR141716A is an inverse agonist at the human cannabinoid CB1 receptor. Eur J Pharmacol 1997;334:R1–R2.
99. MacLennan SJ, Reynen PH, Kwan J, Bonhaus DW. Evidence for inverse agonism of SR141716A at human recombinant cannabinoid CB_1 and CB_2 receptors. Br J Pharmacol 1998;124:619–622.
100. Sim-Selley LJ, Brunk LK, Selley DE. Inhibitory effects of SR141716A on G-protein activation in rat brain. Eur J Pharmacol 2001;414:135–143.
101. Meschler JP, Kraichely DM, Wilken GH, Howlett AC. Inverse agonist properties of N-(piperidin-1-yl)-5-(4-chlorophenyl)-1-(2, 4-dichlorophenyl)-4-methyl-1H-pyrazole-3-carboxamide HCl (SR141716A) and 1-(2-chlorophenyl)-4-cyano-5-(4-methoxyphenyl)-1H-pyrazole-3-carboxyl ic acid phenylamide (CP-272871) for the CB(1) cannabinoid receptor. Biochem Pharmacol 2000;60:1315–1323.
102. Price MR, Baillie GL, Thomas A et al. Allosteric modulation of the cannabinoid CB_1 receptor. Mol Pharmacol 2005;68:1484–1495.
103. Horswill JG, Bali U, Shaaban S et al. PSNCBAM-1, a novel allosteric antagonist at cannabinoid CB_1 receptors with hypophagic effects in rats. Br J Pharmacol 2007;152:805–814.
104. Ross RA. Allosterism and cannabinoid CB(1) receptors: the shape of things to come. Trends Pharmacol Sci 2007;28:567–572.
105. Howlett AC, Song C, Berglund BA, Wilken GH, Pigg JJ. Characterization of CB_1 cannabinoid receptors using receptor peptide fragments and site-directed antibodies. Mol Pharmacol 1998;53:504–510.
106. Mukhopadhyay S, Howlett AC. CB_1 receptor-G protein association. Subtype selectivity is determined by distinct intracellular domains. Eur J Biochem 2001;268:499–505.
107. Mukhopadhyay S, Cowsik SM, Lynn AM, Welsh WJ, Howlett AC. Regulation of Gi by the CB1 cannabinoid receptor C-terminal juxtamembrane region: structural requirements determined by peptide analysis. Biochemistry 1999;38:3447–3455.
108. Ulfers AL, McMurry JL, Miller A, Wang L, Kendall DA, Mierke DF. Cannabinoid receptor-G protein interactions: G(alphai1)-bound structures of IC3 and a mutant with altered G protein specificity. Protein Sci 2002;11:2526–2531.
109. Nie J, Lewis DL. The proximal and distal C-terminal tail domains of the CB_1 cannabinoid receptor mediate G protein coupling. Neuroscience 2001;107:161–167.
110. Bramblett RD, Panu AM, Ballesteros JA, Reggio PH. Construction of a 3D model of the cannabinoid CB_1 receptor: determination of helix ends and helix orientation. Life Sci 1995;56:1971–1982.
111. Choi G, Guo J, Makriyannis A. The conformation of the cytoplasmic helix 8 of the CB_1 cannabinoid receptor using NMR and circular dichroism. Biochim Biophys Acta 2005;1668:1–9.

112. Xie XQ, Chen JZ. NMR structural comparison of the cytoplasmic juxtamembrane domains of G-protein-coupled CB_1 and CB_2 receptors in membrane mimetic dodecylphosphocholine micelles. J Biol Chem 2005;280:3605–3612.
113. Grace CR, Cowsik SM, Shim JY, Welsh WJ, Howlett AC. Unique helical conformation of the fourth cytoplasmic loop of the CB_1 cannabinoid receptor in a negatively charged environment. J Struct Biol 2007;159:359–368.
114. White SH, Ladokhin AS, Jayasinghe S, Hristova K. How membranes shape protein structure. J Biol Chem 2001;276:32395–32398.
115. Fritze O, Filipek S, Kuksa V, Palczewski K, Hofmann KP, Ernst OP. Role of the conserved NPxxY(x)5,6F motif in the rhodopsin ground state and during activation. Proc Natl Acad Sci USA 2003;100:2290–2295.
116. Anavi-Goffer S, Fleischer D, Hurst DP et al. Helix 8 Leu in the CB_1 cannabinoid receptor contributes to selective signal transduction mechanisms. J Biol Chem 2007;282:25100–25113.
117. Ulfers AL, McMurry JL, Kendall DA, Mierke DF. Structure of the third intracellular loop of the human cannabinoid 1 receptor. Biochemistry 2002;41:11344–11350.
118. Liu J, Conklin BR, Blin N, Yun J, Wess J. Identification of a receptor/G-protein contact site critical for signaling specificity and G-protein activation. Proc Natl Acad Sci USA 1995;92:11642–11646.
119. Abadji V, Lucas-Lenard JM, Chin C, Kendall DA. Involvement of the carboxyl terminus of the third intracellular loop of the cannabinoid CB_1 receptor in constitutive activation of Gs. J Neurochem 1999;72:2032–2038.
120. Samama P, Cotecchia S, Costa T, Lefkowitz RJ. A mutation-induced activated state of the beta 2-adrenergic receptor. Extending the ternary complex model. J Biol Chem 1993;268:4625–4636.
121. Singh R, Hurst DP, Barnett-Norris J, Lynch DL, Reggio PH, Guarnieri F. Activation of the cannabinoid CB_1 receptor may involve a W6.48/F3.36 rotamer toggle switch. J Pept Res 2002;60:357–370.
122. Shim JY, Welsh WJ, Howlett AC. Homology model of the CB_1 cannabinoid receptor: sites critical for non-classical cannabinoid agonist interaction. Biopolymers 2003; 71:169–89.
123. Kapur A, Hurst DP, Fleischer D et al. Mutation studies of Ser7.39 and Ser2.60 in the human CB_1 cannabinoid receptor: evidence for a serine-induced bend in CB_1 transmembrane helix 7. Mol Pharmacol 2007;71:1512–1524.
124. Hurst D, Umejiego U, Lynch D et al. Biarylpyrazole inverse agonists at the cannabinoid CB_1 receptor: importance of the C-3 carboxamide oxygen/lysine3.28(192) interaction. J Med Chem 2006;49:5969–5987.
125. Shim JY, Howlett AC. Steric trigger as a mechanism for CB_1 cannabinoid receptor activation. J Chem Inf Comput Sci 2004;44:1466–1476.
126. Barnett-Norris J, Hurst DP, Lynch DL, Guarnieri F, Makriyannis A, Reggio PH. Conformational memories and the endocannabinoid binding site at the cannabinoid CB_1 receptor. J Med Chem 2002;45:3649–3659.
127. Song ZH, Bonner TI. A lysine residue of the cannabinoid receptor is critical for receptor recognition by several agonists but not WIN55212-2. Mol Pharmacol 1996;49:891–896.
128. Tong W, Collantes ER, Welsh WJ, Berglund BA, Howlett AC. Derivation of a pharmacophore model for anandamide using constrained conformational searching and comparative molecular field analysis. J Med Chem 1998;41:4207–4215.
129. Thomas BF, Adams IB, Mascarella SW, Martin BR, Razdan RK. Structure-activity analysis of anandamide analogs: relationship to a cannabinoid pharmacophore. J Med Chem 1996;39:471–479.
130. Berglund BA, Fleming PR, Rice KC, Shim JY, Welsh WJ, Howlett AC. Development of a novel class of monocyclic and bicyclic alkyl amides that exhibit CB_1 and CB_2 cannabinoid receptor affinity and receptor activation. Drug Des Discov 2000;16:281–294.
131. Padgett LW, Howlett AC, Shim JY. Binding mode prediction of conformationally restricted anandamide analogs within the CB_1 receptor. J Mol Signal 2008;3:5.
132. Lynch DL, Reggio PH. Cannabinoid CB_1 receptor recognition of endocannabinoids via the lipid bilayer: molecular dynamics simulations of CB_1 transmembrane helix 6 and anandamide in a phospholipid bilayer. J Comput Aided Mol Des 2006;20:495–509.

133. Song ZH, Slowey CA, Hurst DP, Reggio PH. The difference between the CB(1) and CB(2) cannabinoid receptors at position 5.46 is crucial for the selectivity of WIN55212-2 for CB(2). Mol Pharmacol 1999;56:834–840.
134. Shim JY, Howlett AC. WIN55212-2 docking to the CB_1 receptor and a mechanism for conformational induction. J Chem Inform Model 2006; 46:1286–1300.
135. Sealfon SC, Chi L, Ebersole BJ et al. Related contribution of specific helix 2 and 7 residues to conformational activation of the serotonin 5-HT_{2A} receptor. J Biol Chem 1995;270:16683–16688.
136. Nie J, Lewis DL. Structural domains of the CB_1 cannabinoid receptor that contribute to constitutive activity and G-protein sequestration. J Neurosci 2001;21:8758–8764.
137. Niehaus JL, Liu Y, Wallis KT et al. CB_1 cannabinoid receptor activity is modulated by the cannabinoid receptor interacting protein CRIP 1a. Mol Pharmacol 2007;72:1557–1566.
138. Sánchez C, Rueda D, Segui B, Galve-Roperh I, Levade T, Guzmán M. The CB(1) cannabinoid receptor of astrocytes is coupled to sphingomyelin hydrolysis through the adaptor protein fan. Mol Pharmacol 2001;59:955–959.
139. Jin W, Brown S, Roche JP et al. Distinct domains of the CB_1 cannabinoid receptor mediate desensitization and internalization. J Neurosci 1999;19:3773–3780.
140. Daigle TL, Kearn CS, Mackie K. Rapid CB_1 cannabinoid receptor desensitization defines the time course of ERK1/2 MAP kinase signaling. Neuropharmacology 2008;54:36–44.
141. Bakshi K, Mercier RW, Pavlopoulos S. Interaction of a fragment of the cannabinoid CB_1 receptor C-terminus with arrestin2. FEBS Lett 2007;581:5009–5016.
142. Martini L, Waldhoer M, Pusch M et al. Ligand-induced down-regulation of the cannabinoid 1 receptor is mediated by the G-protein-coupled receptor-associated sorting protein GASP1. FASEB J 2007;21:802–811.
143. Tappe-Theodor A, Agarwal N et al. A molecular basis of analgesic tolerance to cannabinoids. J Neurosci 2007;27:4165–4177.
144. Leterrier C, Laine J, Darmon M, Boudin H, Rossier J, Lenkei Z. Constitutive activation drives compartment-selective endocytosis and axonal targeting of type 1 cannabinoid receptors. J Neurosci 2006;26:3141–3153.
145. Simonin F, Karcher P, Boeuf JJ, Matifas A, Kieffer BL. Identification of a novel family of G protein-coupled receptor associated sorting proteins. J Neurochem 2004;89:766–775.
146. Jarrahian A, Watts VJ, Barker EL. D_2 dopamine receptors modulate Galpha-subunit coupling of the CB_1 cannabinoid receptor. J Pharmacol Exp Ther 2004;308:880–886.
147. Kearn CS, Blake-Palmer K, Daniel E, Mackie K, Glass M. Concurrent stimulation of cannabinoid CB_1 and dopamine D_2 receptors enhances heterodimer formation: a mechanism for receptor cross-talk? Mol Pharmacol 2005;67:1697–1704.
148. Marcellino D, Carriba P, Filip M et al. Antagonistic cannabinoid CB_1/dopamine D_2 receptor interactions in striatal CB_1/D_2 heteromers. A combined neurochemical and behavioral analysis. Neuropharmacology 2008;54:815–823.
149. Shapira M, Gafni M, Sarne Y. Independence of, and interactions between, cannabinoid and opioid signal transduction pathways in N18TG2 cells. Brain Res 1998;806:26–35.
150. Shapira M, Vogel Z, Sarne Y. Opioid and cannabinoid receptors share a common pool of GTP-binding proteins in cotransfected cells, but not in cells which endogenously coexpress the receptors. Cell Mol Neurobiol 2000;20:291–304.
151. Canals M, Milligan G. Constitutive activity of the cannabinoid CB_1 receptor regulates the function of co-expressed Mu opioid receptors. J Biol Chem 2008;283:11424–11434.
152. Rios C, Gomes I, Devi LA. mu opioid and CB_1 cannabinoid receptor interactions: reciprocal inhibition of receptor signaling and neuritogenesis. Br J Pharmacol 2006;148:387–395.
153. Hilairet S, Bouaboula M, Carriere D, Le Fur G, Casellas P. Hypersensitization of the Orexin 1 receptor by the CB_1 receptor: evidence for cross-talk blocked by the specific CB_1 antagonist, SR141716. J Biol Chem 2003;278:23731–23737.
154. Ellis J, Pediani JD, Canals M, Milasta S, Milligan G. Orexin-1 receptor-cannabinoid CB_1 receptor heterodimerization results in both ligand-dependent and -independent coordinated alterations of receptor localization and function. J Biol Chem 2006;281:38812–38824.

155. Childers SR, Pacheco MA, Bennett BA et al. Cannabinoid receptors: G-protein-mediated signal transduction mechanisms. Biochem Soc Symp 1993;59:27–50.
156. Pacheco MA, Ward SJ, Childers SR. Identification of cannabinoid receptors in cultures of rat cerebellar granule cells. Brain Res 1993;603:102–110.
157. Cinar R, Freund TF, Katona I, Mackie K, Szucs M. Reciprocal inhibition of G-protein signaling is induced by CB_1 cannabinoid and $GABA_B$ receptor interactions in rat hippocampal membranes. Neurochem Int 2008;52:1402–1409.
158. Ross RA, Gibson TM, Brockie HC et al. Structure-activity relationship for the endogenous cannabinoid, anandamide, and certain of its analogues at vanilloid receptors in transfected cells and vas deferens. Br J Pharmacol 2001;132:631–640.
159. Zygmunt PM, Petersson J, Andersson DA et al. Vanilloid receptors on sensory nerves mediate the vasodilator action of anandamide. Nature 1999;400:452–457.
160. Duan Y, Zheng J, Nicholson RA. Inhibition of [^{3}H]batrachotoxinin A-20alpha-benzoate binding to sodium channels and sodium channel function by endocannabinoids. Neurochem Int 2008;52:438–446.
161. Baranowska U, Gothert M, Rudz R, Malinowska B. Methanandamide allosterically inhibits in vivo the function of peripheral nicotinic acetylcholine receptors containing the α7-subunit. J Pharmacol Exp Ther 2008.
162. Hejazi N, Zhou C, Oz M, Sun H, Ye JH, Zhang L. Delta9-tetrahydrocannabinol and endogenous cannabinoid anandamide directly potentiate the function of glycine receptors. Mol Pharmacol 2006;69:991–997.

Chapter 12
Functional Selectivity at Opioid Receptors

Graciela Piñeyro

Abstract Opiate drugs are among the most effective analgesics available but their clinical use is restricted by tolerance, physical dependence, respiratory depression, nausea, and constipation. As a class, opioid ligands produce their effects by acting upon G protein coupled receptors (GPCRs). In this class of membrane receptors, agonist binding induces a series of conformational changes, which propagate to intracellular signaling partners as GPCRs switch from a resting to an active conformation. This active state had been classically considered unique and responsible for regulation of all signaling pathways controlled by any given receptor. However, recent studies have challenged this classical notion, calling for an alternative paradigm where receptors would exist in more than one active conformation with distinct signaling properties. Ligand ability to stablize different active states of the same receptors is currently referred to as *functional selectivity*. In this review, we summarize evidence supporting the existence of ligand-selective conformations for μ and δ-opioid receptors and analyze how functional selectivity may contribute to the production of longer lasting, better tolerated opiate analgesics.

Keywords Opioid, MOR, DOR, Tolerance, Analgesia, Receptor states, Biased agonism, Stimulus trafficking, Ligand-specific conformations

---

G. Piñeyro
Chercheure Agrégée, Département de Psychiatrie, Université de Montréal,
7331 Rue Hochelaga, Montréal, Qc, H1N3V2
e-mail: graciela.pineyro.filpo@.umontreal.ca

K. Neve (ed.), *Functional Selectivity of G Protein-Coupled Receptor Ligands*,
DOI: 10.1007/978-1-60327-335-0_12,

12.1 Introduction

Opiates are among the most effective analgesics known and elective in the treatment of severe pain. However, their therapeutic use is restricted by undesired effects such as nausea, persistent constipation, respiratory depression *(1–3)*, development of analgesic tolerance, and their potential for abuse *(4, 5)*. Not surprisingly then, separating beneficial from detrimental actions of opioid receptor ligands remains a major goal in the field of opioid research. It is well established that therapeutically available opiates produce their analgesic actions via μ-opioid receptors (MORs), but unfortunately beneficial actions produced by the activation of these receptors may not be dissociated from the production of adverse side effects *(6)*. Initial efforts to overcome these undesired actions took advantage of the existence of different opioid receptor subtypes, by selectively targeting δ-opioid receptors (DORs) whose activation produces antinociceptive actions *(7, 8)* with reduced respiratory depression *(9)*, less gastrointestinal side effects *(10)*, and minimal potential for physical dependence *(11)*. Unfortunately, DOR agonists are not free of tolerance *(12,13)* and have problems of their own, including lower analgesic efficacy than MOR ligands and production of seizures *(14)*. Consequently, alternative approaches for the development of novel analgesics are constantly being sought, and "functional selectivity" is now receiving considerable attention.

12.1.1 Conformational Diversity is the Basis of "Functional Selectivity"

Classical pharmacological theory holds that drug efficacy is a system-independent quantitative property descriptive of a drug's ability to induce the accumulation of the ligand-bound, single active conformation of the receptor *(15)*. Although this principle has proven sufficient to explain signaling and therapeutic properties of G protein coupled receptor (GPCR) ligands for almost half a century, there is now considerable evidence that a single quantitative efficacy parameter is insufficient to describe drug action in multieffector systems. In particular, the classical idea that drug efficacy is entirely determined by system-independent parameters inherent to each drug-receptor pair has been challenged by an overwhelming number of reports showing that pharmacological responses of many ligands depend on the effector pathway in which drug activity is evaluated *(16–20)*. These observations called for a modification of the classical model and prompted the idea that a single receptor may exist in multiple active conformations *(18, 21)*, some of which would have the capacity to trigger only a subset of responses within the ensemble of activities regulated by the receptor *(22)*. This new conceptual framework raised the possibility of pharmacologically delineating the type of signal elicited by any given receptor by producing a ligand that would induce/recognize the conformation responsible for the desired set of actions. In this chapter, ligand ability to stabilize/recognize a receptor conformation that triggers a distinct subset of all possible receptor responses will be referred to as "*functional selectivity*" *(22)* ("*stimulus trafficking*" or "*biased agonism*" are commonly used synonyms) *(18, 22, 23)*.

12.1.2 Purified Receptors Undergo Ligand-Specific Conformational Changes

Biophysical and biochemical studies on rhodopsin *(24)*, β2 adrenergic *(25,26)* and κ-opioid receptors (KORs) *(27)* have made it possible to localize conformational changes undergone by GPCRs upon ligand binding. A common feature to all three receptors is a structural reorganization of transmembrane domains III and VI, displacement of intracellular ends of these helices, and separation of cytosolic loops two and three to open a cleft, which allows receptor interaction with cytosolic signaling partners *(28–30)*. Since different ligands induce distinct structural rearrengements within the receptor *(31)*, the cleft defined between the two intracellular loops is also expected to be ligand specific. Such specificity raises the possibility that cytosolic signaling molecules that interact with the receptor may distinguish among conformations stabilized by different drugs. This possibility has been confirmed for DORs in plasmon-waveguide resonance (PWR) spectroscopy assays. Indeed, purified DORs stabilized by different ligands were able to distinctively recognize different Gα subunits. Three different observations led to this conclusion: (a) agonist affinity for the purified receptor was influenced by the Gα subtype with which the receptor was resuspended, (b) the profile of DOR interaction with different α subunits was dictated by the type of ligand bound to the receptor, and (c) the ability of ternary complexes to promote GTPγS binding was dependent on complex composition (see Table 12.1) *(33)*. These observations obtained using purified receptors in reconstitution systems clearly indicate that DORs bound to different agonists distinctively interact with different downstream signaling partners. These studies, however, give little information about the physiological relevance of ligand-dependent conformational diversity of opioid receptors.

12.1.3 Opioid Receptors Undergo Ligand-Specific Conformational Changes in Living Cells

There is now considerable evidence indicating that ligand-specific interaction between opioid receptors and downstream signaling partners may also occur in vivo *(46)*. However, unlike reconstitution systems, GPCRs in living cells form part of multimeric arrays that also contain Gαβγ subunits *(47, 48)*, other receptors *(49, 50)*, effectors *(51–53)*, and signaling regulators *(54)*. Several coimmunopurification assays from brain membranes or from heterologous expression systems indicate that this is certainly the case for opioid receptors *(55–58)*, Gα subunits *(32, 59–61)*, and regulators of G protein signaling (RGS) *(59, 60, 62)*, which can all be recovered as a single complex. Association between complex components usually takes place during GPCR export to the membrane or even during protein synthesis *(63, 64)*, and complex composition is influenced by levels of expression of interacting proteins, the availability of chaperones necessary for maturation *(63)* or the presence of scaffolding proteins that "glue" the complex together *(65)*. These variables are cell-specific and therefore the same receptor may reach the cell surface as part of very different multimeric arrays, depending on the cell in which the complex is expressed. Given their allosteric nature, receptors included into signaling complexes of different composition will undergo specific stereic restrictions that may distinctively influence the way they respond to ligand-promoted conformational changes. Such nuances cannot be possibly evaluated by techniques that rely on purified receptors, and this is one of the reasons why ligand-induced conformational diversity is being increasingly assessed in vivo *(46, 66)*.

Experimental approaches that monitor ligand-induced conformational changes within multimeric protein arrays are frequently based on the nonradiative transfer of energy between a fluorescent (FRET) or bioluminescent (BRET) donor and a fluorescent acceptor *(67)*. We have recently used this approach to explore ligand-induced changes in DOR interaction with heterotrimeric G proteins and among $G\alpha\beta\gamma$ subunits themselves *(46)*. Our results indicate that different ligands induced specific conformational changes at the receptor level that were conveyed downstream to G protein components. In other words, different agonists produced distinct types of interactions between DORs and the $G\alpha$ subunit and between DORs and the $G\beta\gamma$ complex, such that ligand-specific conformational changes imposed upon the receptor were transmitted to the way $G\alpha\beta\gamma$ subunits organized themselves within the G protein heterotrimer *(46)*.

In vivo conformational changes undergone by MORs and DORs have also been assessed using antibodies directed against the N-terminal domain of the receptor. This approach has allowed to demonstrate that ligands that behaved as agonists or antagonists in the GTPγS binding assay differed in their ability to expose extracellular residues recognized by the different antibodies *(68,69)*. Moreover, ligands displayed the same rank order of efficacy to induce GTPγS binding and to modify recognition by the antibody *(69)*, indicating that the antibodies are indeed state sensitive. However, these observations do not allow to conclude whether conformational changes unveiled by antibody binding reveal the existence of different active conformations of the receptor or if they represent the accumulation of increasing amounts of a single active state by progressively efficacious ligands.

12.2 Functional Selectivity and Opioid Receptor Signal Transduction

12.2.1 Ligand-Specific Activation of Gα Subunits by DORs and MORs

Studies in brain membranes and heterologous expression systems indicate that opioid receptors are pleiotropic, capable of simultaneously activating more than one G-protein subtype. However, on the one hand, since not all reports have systematically ruled out whether agonist-related differences in Gα activation are linked to the stimulation of different receptor subtypes or have not always verified whether differences among agonists are not simply related to strength of signal (higher efficacy agonists activating more Gα subtypes than weaker ones), some of the data available remain merely suggestive (Tables 12.1 and 12.2). On the other hand, clear reversal in ligand rank order of potency/efficacy to produce activation

Table 12.1 Physical and functional evidence supporting pleiotropic G protein signaling by DORs

Pleiotropic DOR-α interactions			
Physical interactions (transfected receptor/endogenous G protein)			
CHO cells coimmuno-precipitation	DPDPE	$\alpha_o = \alpha_{i2} > \alpha_{i3}$	*(32)*
Physical interactions (affinity of solubilized DOR for α)			
Solubilized DORs *PWR (plasmon-waveguide resonance)*[a]	*Deltorphin II*	*αo= αi2> αi1> αi3*	*(33)*
	DPDPE	*αo= αi2> αi3> αi1*	
	Morphine	*αi3> αo= αi> αi2*	
	SNC	*αo> αi2= αi3> αi1*	
	TAN67	*αi3> αo> αi1> αi2*	
Pleiotropic α activation			
Endogenous receptor/endogenous G protein			
Brain membranes (PAG)	Deltorphin II	$\alpha_{i2} >>> \alpha_Z$	*(34)*
	DPDPE	$\alpha_{i2} > \alpha_Z$	
Neuroblastoma NG108-15	DADLE	α_{i2}	*(35)*
	DADLE	$\alpha_o > \alpha_{i2} >> \alpha_{i3}$	*(36)*
	DADLE	$\alpha_{o2} > \alpha_{i2} > \alpha_{o1} >> \alpha_{i3}$	*(37)*
SH-SY5Y	DPDPE	$\alpha_{i1} > \alpha_{i2};\alpha_{i3};\ \alpha_{o1/2}$	*(38)*
SK-N-BE[a]	*Deltorphin I*	*αo2> αi2*	*(39)*
	DPDPE	*αi2= αPTX insensitive> αo2*	
	Etorphine	*αi2= αPTX insensitive>> αi3> αi1*	
Transfected receptor/endogenous G protein			
CHO cells	DADLE	$\alpha_{i2} \geq \alpha_{o2} > \alpha_{i3}$	*(40)*
GH3 cells	DPDPE	$\alpha_{o1} >> \alpha_{i2} = \alpha_o > \alpha_{i3}$	*(41)*
Physical methods (ligand-mediated GTPγS binding to DOR-Gα complex)			
Solubilized DORs *PWR* (plasmon-waveguide resonance)[a]	*Deltorphin II*	*αi2= αo> αi3> αi1*	*(33)*
	DPDPE	*αi1> αi2> αi3> αo*	
	Morphine	*αi1>> αo= αi3> αi2*	
	SNC	*αi1> αo> αi3> αi2*	
	TAN67	*αi3> αo> αi2> αi1*	
Ligand affinity for fusion constructs containing different Gα subtypes			
HEK 293 cells hDOR-α_{i1} vs hDOR-α_{o1}	DADLE	$\alpha_{i1} > \alpha_{o1}$	*(42)*
Physiological evidence: analgesia			
Supraspinal analgesia[b] (PAG)	β-endorphine	$\alpha_{i2};\ \alpha_{i3};\ \alpha_{x/z}$	*(43)*
	Deltorphin II DPDPE	$\alpha_{i2};\ \alpha_{i3}$	
	Deltorphin II	$\alpha_{i2};\ \alpha_{i3};\ \alpha_{o2};\ a_{11}$ and α_q	*(44)*
	DPDPE	$\alpha_{i2};\ \alpha_{i3};\ \alpha_{o2};\ a_{11}$and α_{o1}	
Spinal analgesia	DPDPE	$\alpha_{i1} = \alpha_o = \alpha_q \geq \alpha_3 > \alpha_{i2} > \alpha_z$	*(45)*

[a]These studies provide evidence of ligand-specific receptor states
[b]These studies are suggestive of ligand-specific receptor states

of different Gα subunits constitutes one of the most reliable ways to unveil agonist-specific receptor states in G protein activation assays *(17, 79, 80)*. Results fulfilling such conditions have been obtained for DORs endogenously expressed in the SK-N-BE neuroblastoma cell line. In particular, the rank order of G protein activation by D-Pen-2,5-enkephalin (DPDPE) was $\alpha_{i2} = \alpha_{PTX\text{-insensitive}} > \alpha_{o2}$ while the Gα activation profile for deltorphin I was $\alpha_{o2} >> \alpha_{i2}$ *(39)* (Table 12.1). Moreover, unlike these two selective DOR agonists, etorphine, a nonselective MOR/DOR agonist, did not stimulate α_{o2} but activated $\alpha_{i2} = \alpha_{PTX\text{-insensitive}} >> \alpha_{i3.}$ Since the SK-N-BE neuroblastoma cell line used in this study only expressed DORs *(81)*, the specific profile of Gα activation by etorphine could not be attributed to concomitant activation of other opioid receptor subtypes, further supporting the conclusion derived by comparing DPDPE and deltorphin I that endogenous levels of DOR expression support functional selectivity. DOR mutations replacing tryptophan 284 by leucine also point to agonist-specific active conformations, as indicated by a reversal in agonist rank order of efficacy to promote GTPγS binding in mutants (SNC80 > DPDPE = TAN67 = SB219825) when compared with wild type DORs (DPDPE > TAN67 = SNC 80 > SB219825) *(82)*.

The ability of different MOR agonists to activate different G$\alpha_{i/o}$ subunits has been assessed in glioma, HEK, and CHO cell lines expressing endogenous or transfected wild type MORs (Table 12.2). Although maximal agonist efficacy to stimulate different α subtypes varied from one cell line to the other, when compared within the same expression system ligands displayed the same rank order of efficacy to activate each Gα subunit tested, ruling out ligand-specific conformations as revealed by this approach (Table 12.2). This is an example of the key principle that ligand differences to activate different Gα subunits in different cell lines should not be considered proof of "functional selectivity," unless it has been determined that the α subunits tested were equally expressed across all cell lines. In contrast, the differences in rank order of affinity and distinct susceptibility of MOR-α_{o1} and MOR-α_{i2} fusion constructs to undergo activation by different agonists *(75,76)* clearly support functional selectivity at this receptor (Table 12.2). MOR mutagenesis studies are also consistent with the idea that different ligands rely upon different intracellular residues to induce G protein activation. Indeed, removal of different groups of amino acids within the third intracellular loop of MORs produced agonist-specific changes in affinity, potency and Gα stimulation, demonstrating that this putative G protein coupling region distinctively participates in MOR activation by DAMGO or morphine *(83)*. Truncation of C-terminal residues yielded similar results, producing differential modification in signaling by peptidic ligands like DAMGO and alkaloids like morphine *(84)*.

DAMGO and morphine differences in promoting GTPγS binding by different Gα subunits has also been assessed in brain membranes. Intracerebroventricular (i.c.v.) administration of these two ligands induced the activation of different Gα subtypes *(34)*, but since DOR implication in morphine-mediated actions was not ruled out, the possibility that differences could be related to morphine and DAMGO binding to different receptor subtypes cannot be excluded. In another set of studies, different Gα subunits were selectively inactivated within the periaqueductal gray (PAG) and analgesic actions of selective MOR ligands assessed. Despite being quite indirect,

Table 12.2 Physical and functional evidence supporting pleiotropic G protein signaling by MORs

Pleiotropic MOR-α interactions – coimmunoprecipitation studies			
Endogenous receptor/endogenous G protein			
Brain membranes	Non activated receptor	α_{i1}; α_{i2}; α_{i3}; α_o	*(61)*
PAG membranes	Non activated receptor	α_{i1}; α_{i2}; α_o; α_z; $\alpha_{q/11}$	*(59,60)*
Pleiotropic α activation			
Endogenous receptor/endogenous G protein			
Brain membranes (PAG)[b]	β-endorphin-(1–31)	$\alpha_z >> \alpha_{i2}$	*(59)*
	DAMGO	$\alpha_{i2} = \alpha_o > \alpha_z$	
	Morphine	$\alpha_z = \alpha_o > \alpha_{i2}$	
Smooth muscle	DAMGO	α_{i2} and α_o	*(70)*
Neuroblastoma SH-SY5Y	DAMGO	$\alpha_{i3} >> \alpha_{i1} = \alpha_{i2} = \alpha_{o1/2}$	*(38)*
Endogenous receptor/exogenous G protein			
Glioma C6	DAMGO	$\alpha_{i3} > \alpha_{oA} > \alpha_{i1} > \alpha_{i2}$	*(71)*
	Endomorphin 1/2		
	Etorphine		
	Fentanyl		
	Meperidine		
	Morphine		
Transfected receptor/endogenous G protein			
HEK293s cells	DAMGO	$\alpha_{i3} >>> \alpha_{i1/2} > \alpha_o$	*(72,73)*
	Endomorphin 1		
	Morphine		
	Naloxone	$\alpha_{i3} >> \alpha_{i1/2} = \alpha_o$	
CHO cells	DADLE	$\alpha_{i2} \geq \alpha_{o2} > \alpha_{i3}$	*(74)*
	DAMGO		
	Morphine		
Fusion constructs			
Agonist affinity for different MOR-α constructs			
HEK293s *MOR-αi1vs MOR-αi2*[a]	*Dynorphin A*	*αi2>αi1*	*(75)*
	Endomorphin 1		
	Morphine		
	Endomorphine 2	*αi1>αi2*	
E. coli MOR-α_{o1} vs MOR-α_{i2}[b]	DAMGO	$\alpha_{i2} > \alpha_{o1}$	*(76)*
	Endomorphin 1/2		
	Morphine	$\alpha_{i2} = \alpha_{o1}$	
G protein activation within different complexes			
HEK293s *MOR-αi1vs MOR-αi2*[a]	*DAMGO*	*αi1>αi2*	*(75)*
	Dermorphin		
	Endomorphin 2		
	Morphine	*αi1= αi2*	
	β-endorphin	*αi1; not αi2*	
	Dynorphin A		
	Endomorphin 1		
	Met-enkephalin		
Agonist potency			
E. coli MOR-α_{o1} vs MOR-α_{i2}[b]	DAMGO	$\alpha_{o1} \geq \alpha_{i2}$	*(76)*
	Endomorphin 1	$\alpha_{i2} \geq \alpha_{o1}$	
	Endomorphin 2	$\alpha_{o1} \geq \alpha_{i2}$	
	Morphine	$\alpha_{i2} > \alpha_{o1}$	

(continued)

Table 12.2 (continued)

Pleiotropic MOR-α interactions – coimmunoprecipitation studies			
Functional evidence: analgesia			
Supraspinal analgesia[b] (PAG)	Buprenorphine	$\alpha_z = \alpha_{i3} = \alpha_{i2} > \alpha_{o2} = \alpha_q$	*(77,78)*
	DAMGO	$\alpha_{i2} = \alpha_z > \alpha_q$	
	Endomorphine-1	$\alpha_z \geq \alpha_{i1} > \alpha_{i3}$	
	Endomorphine-2	$\alpha_{i1} \geq \alpha_{i2} = \alpha_{i3} = \alpha_z$	
	Heroin	$\alpha_z = \alpha_{o1} = \alpha_{o2} > \alpha_q = \alpha_{11} > \alpha_{i3} = \alpha_{i2} = \alpha_{i1}$	
	Methadone	$\alpha_z = \alpha_{i3} = \alpha_{i2} = \alpha_{i1} = \alpha_{o1} = \alpha_{11}$	
	Morphine	$\alpha_z > \alpha_{i2}$	
Supraspinal analgesia[b]	Morphine	$\alpha_{i2} \geq \alpha_o$	*(45)*
	Morphine-6β-glucuronide	$\alpha_{i1} = \alpha_z$	
Spinal analgesia[b]	Morphine	$\alpha_{i2} = \alpha_z$	
	Morphine-6β-glucuronide	$\alpha_{i1} = \alpha_z = \alpha_o > \alpha_{i3}$	

[a]These studies provide *evidence* of ligand-specific receptor states
[b]These studies are suggestive of ligand-specific receptor states

results from these studies support the idea that opioid signaling within the living brain may be ligand-specific. In particular, DAMGO ($\alpha_{i2} = \alpha_z > \alpha_q$), endomorphin 1 ($\alpha_z \geq \alpha_{i1} > \alpha_{i3}$), and endomorphin 2 ($\alpha_{i1} \geq \alpha_{i2} = \alpha_{i3} = \alpha_z$) clearly relied upon different Gα subtypes to induce analgesia *(34, 45, 77, 78)*. However, the limitation in these reports was the inability to determine whether α subunits involved in analgesia induced by DAMGO, endomorphin 1, and endomorphin 2 were directly activated by agonist-bound MORs or if each agonist recruited different neuronal networks, which in turn relied upon different signaling cascades to produce antinociception. Unfortunately, even if specificity in α activation might seem a logical means for dissociating analgesic actions from unwanted side effects of opioid ligands *(85)*, the problem with this rationale lies upon the lack of clear evidence indicating that analgesia and secondary effects are produced by different α subtypes.

12.2.2 Ligand-Specific Regulation of Intracellular Signaling Cascades by DORs and MORs

Comparison of ligand efficacy to modulate cyclic-adenosine monophosphate (cAMP) production, protein kinase C (PKC) activation, or extracellular regulated kinase (ERK) stimulation has provided evidence that functional selectivity of opioid ligands extends beyond their ability to distinctively modulate G protein activation. For example, etorphine and morphine behave as full and partial agonists in the regulation of cAMP production by MORs, respectively, but only morphine induces PKC-dependent phosphorylation of cytoplasmic substrates *(86)*. Since etorphine is a full agonist in the cyclase pathway, its failure to promote PKC activity that is successfully evoked by the partial agonist morphine excludes low strength of signal, and points to ligand-specific signaling conformations as a possible explanation for the differential activation of PKC by both ligands. This idea is further reinforced by the observation that ERK activation by etorphine and morphine relies upon different mechanisms and induces distinct distribution of the activated kinase pERK. In particular, on the one hand, ERK activation by morphine is PKC-dependent and once activated remains confined to the cytoplasm. On the other hand, ERK activation by etorphine relies on β-arrestin2 and the activated kinase accumulates in the nucleus. Differences in ERK activation by DAMGO and morphine are not confined to overexpression systems since they were also observed in primary neuronal cultures *(86)*. Interestingly, ERK stimulation by etorphine but not morphine resulted in increased GRK2 and β-arrestin2 transcription *(86)*. As discussed in the following section, agonist ability to induce MOR phosphorylation and β-arrestin recruitment has been inversely correlated with ligand potential for tolerance. From this perspective, it would be tempting to speculate that ligands activating signaling pathways that enhance GRK2/β-arrestin transcription would result in less analgesic tolerance. However, despite its intrinsic logic this reasoning could prove overly simplistic given that tolerance is a multifactorial phenomenon, which remains poorly understood (see later).

DOR ligands are also capable of distinguishing among different receptor conformations with distinct signaling properties. Binding studies performed following short-term exposure to an inverse agonist revealed that this type of treatment reduced the amount of binding sites recognized by [^{3}H]DPDPE by the same amount that it increased sites recognized by [^{3}H]TICP (TICP: Tyr- [CH(2)-NH]Cha-Phe-OH). Since total receptor protein and the amount of sites labeled by a neutral antagonist remained unmodified, the opposing changes in DPDPE and TICP binding capacities were interpreted as an indication that DORs exist in different conformations, which discriminate between the two ligands *(87)*. This interpretation was later corroborated by two other findings: (a) BRET assays demonstrated that each ligand induced distinct conformational rearrangement of the constitutive DOR-αi1β1γ2 complex *(46)* and (b) functional assays showed that DPDPE inhibited adenylate cyclase activity and promoted ERK activation, while TICP behaved as an "inverse agonist" in the adenylate cyclase cascade but as an agonist in the MAPK pathway *(17,88)*. Like DPDPE, ERK activation by TICP was sensitive to pertussis toxin (PTX) *(17)* indicating that the conformation stabilized by this "inverse agonist" is not completely inactive with respect to G-protein-mediated signals. Thus, TICP should be better considered a *dual efficacy* ligand capable of stabilizing an active receptor state that, unlike the conformation stabilized by the classical agonist DPDPE, it is capable of dissociating adenylate cyclase inhibition from ERK phosphorylation. If one takes into account that sustained adenylate cyclase inhibition by opiates is associated with superactivation of the cAMP signaling cascade, development of cellular tolerance *(89,90)*, and precipitation of withdrawal *(91)*, ligands that activate DOR signaling while avoiding a decrease in cAMP production could potentially induce analgesia with lower incidence of withdrawal symptoms. In this sense, it is worth noting that adenylate cyclase inhibition does not seem obligatory for analgesic actions of opioids, and inhibition of cAMP production may actually attenuate acute analgesic effects of morphine *(92,93)*. But then, βγ-mediated regulation of G-protein-gated inwardly rectifying K + channels (GIRKs) is an essential component of the analgesic response produced by opioid receptor ligands *(94)*. Thus, agents capable of dissociating GIRK modulation from adenylate cyclase regulation could potentially yield opioid analgesics with reduced potential for tolerance and physical dependence.

12.2.3 Multimeric Signaling Complexes and Functional Selectivity of Opioid Receptor Ligands

Functional selectivity implies that ligands that activate the same receptor may not necessarily induce the same pattern of postreceptor signals. The way in which this selectivity is achieved may be differently conceived depending on how receptors are thought to associate with their signaling partners. According to the more traditional model where receptor interaction with G proteins, effectors, and regulatory proteins is believed to be triggered by receptor activation *(95, 96)*, conformational restrictions imposed upon ligand binding (*conformational induction*) would determine preferential recruitment and activation of one signaling partner over another (Fig. 12.1a). However, a growing number of reports indicate that receptors reach the membrane as part of preassembled signaling complexes *(46, 58, 63, 88, 97)*. This type of organization raises the possibility of developing ligands capable of distinguishing signaling complexes of different composition. Indeed, "complex-specific" ligands may prove an efficacious way to produce functionally selective opioid ligands provided that steric restrictions imposed by different complex components force the receptor into distinct conformations. In this alternative but not necessarily exclusive model, functional selectivity would not depend on ligands' ability to recruit different proteins but on their capacity to recognize and activate the receptor associated with the specific type of signaling partner whose activation is being sought (*conformational selection,* Fig. 12.1b). As detailed later, the list of signaling molecules constitutively interacting with opioid receptors is increasingly growing, making conformational selection a promising avenue for development of functionally selective ligands at diverse effectors.

12.2.3.1 Non-G Protein Signaling Partners Directly Interacting with Opioid Receptors

Apart from G proteins *(46)*, MORs and DORs constitutively interact with transducers and activators of transcription STAT5A and STAT5B *(98)*, phospholipase D2 (PLD2) *(99)*, and calmodulin *(100)*. Each of these signaling molecules directly contacts the receptor, either at the C-terminus in the case of STATs and PLD2 *(98,99)* or at the third intracellular loop in the case of calmodulin *(100)*. The latter is released from the complex upon agonist binding, as a consequence of G protein activation *(100)*. Agonist-mediated STAT release is determined by activation of

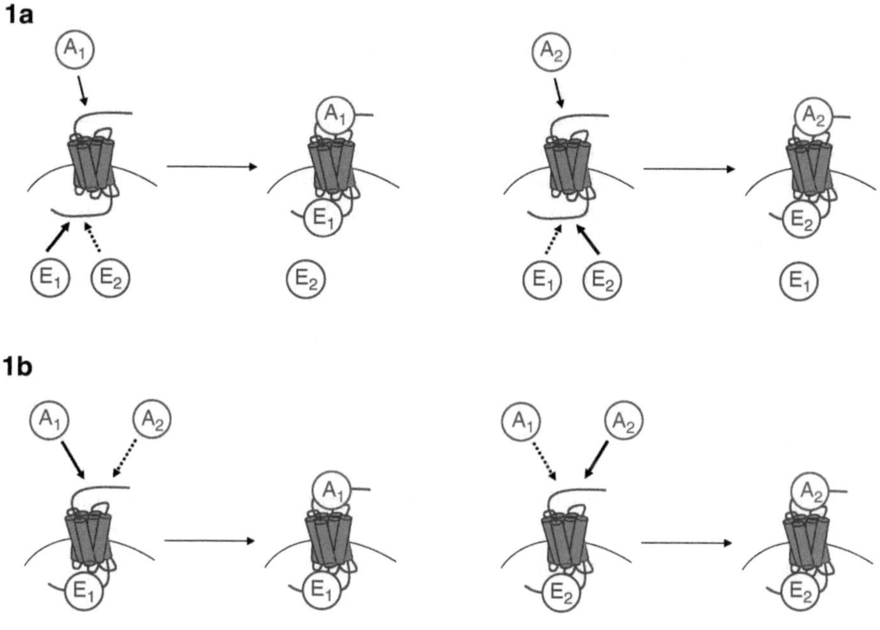

Fig. 12.1 Putative mechanisms of functional selectivity. (**a**) Conformational induction and ligand-specific recruitment of different intracellular proteins. In this model, functional selectivity would be determined by conformational restrictions imposed on the receptor upon ligand binding: ligands A1 and A2 stabilize two different conformations of the same receptor which have different affinity for signaling proteins E1 and E2. (**b**) Conformational selection and ligand specific binding to preformed signaling complexes. In this model, functional selectivity would depend on ligand ability to recognize one signaling complex over the other, depending on their distinct composition: ligands A1 and B1 display differential affinity for complexes containing signaling proteins E1 and E2 (*See Color Plates*)

Src *(98)*, which itself constitutively associates with opioid receptors *(88)*. Finally, PLD2 separation from MORs requires ligand-dependent recruitment and activation of the ADP ribosylation factor (ARF) *(99)*. Although it is theoretically possible that selective activation of any of these effectors might lead to the development of better tolerated, longer acting opioid analgesics, at present there is no evidence supporting this assumption. At the most, given the active role of calmodulin-dependent adenylate cyclases (ACI/ACVIII) in development of analgesic tolerance, it is possible to speculate that a ligand capable of activating GIRKs while avoiding calmodulin activation could provide a means of prolonging analgesic efficacy of MOR and DOR ligands.

12.2.3.2 Opioid Receptor Oligomerization

Opioid receptors undergo oligomerization among themselves *(50)* or with other GPCRs *(55)* that control the immune response or are involved in pain regulation, including chemokine receptors *(101, 102)*, α_2-adrenoceptors *(97, 103, 104)*, neurokinin-1 receptors *(105)*, and cannabinoid-1 receptors *(106)*. Pharmacological and signaling properties of opioid receptors that form part of heterodimers are frequently different from those of the contributing protomers *(50)*, suggesting that protein–protein interactions among different receptor subtypes may impose structural constraints that modify both protomer interaction with its signaling partners as well as the binding pocket for different ligands. For example, MOR/DOR heterodimers have significantly reduced affinity for morphine, DAMGO, and DPDPE when compared with MORs or DORs expressed by themselves *(56)*. In addition, heterodimeric MORs and DORs preferentially couple to $G\alpha_z$ instead of $G\alpha_i$ *(107)*, and ERK activation by MORs included in a heterodimer switches from a G protein to a βarr2-mediated mechanism. When coexpressed, MOR/DOR association among themselves and the $G\alpha_z$ subunit takes place in the endoplasmic reticulum *(58)*, such that pharmacological and signaling properties of heterodimerized opioid receptors are determined before they reach the cell membrane. Compounds that preferentially recognize receptors contained within such complexes would be modulating a very specific subpopulation of MORs and DORs that signal preferentially through $G\alpha_z$ rather that $G\alpha_i$. Whether the resulting switch in signaling might be beneficial in the development of better tolerated, longer acting analgesics remains to be determined.

12.3 Functional Selectivity and Regulation of Opioid Receptor Signaling

Together with receptor activation, opioid ligands trigger a constellation of regulatory events that lead to a progressive decrease in receptor signaling capacity. These include changes that take place at the receptor, G protein, and effector levels, all of which have been implicated in the development of analgesic tolerance. Given that loss of analgesic efficacy is one of the major drawbacks in the therapeutic use of opioid analgesics the possibility of using "functional selectivity" to dissociate signaling from regulatory phenomena is of great interest. In the following section, we will analyze evidence favoring the idea that opioid agonists may stabilize receptor conformations with distinct regulatory properties and discuss the extent to which these differences could be exploited to produce opioid analgesics with reduced potential for tolerance.

12.3.1 Ligand-Specific Regulation of Opioid Receptor Desensitization

12.3.1.1 Receptor Phosphorylation

Phosphorylation is classically considered the initial step in the process of desensitization, and numerous kinases including G protein receptor kinases

(GRKs) *(108, 109)*, Ca^{+2}-/calmodulin-dependent kinases *(110, 111)*, PKCs *(110, 112, 113)*, mitogen-activated protein kinase (MAPKs) *(114, 115)*, and tyrosine kinases *(17, 88, 116)* participate in MOR/DOR desensitization. Given the assumption that drug efficacy is determined by the accumulation of a single active state, classical receptor theory predicts that signaling and desensitization are directly correlated, i.e., the more efficacious the ligand the greater the number of receptors stabilized in a conformation that will serve as GRK substrate or that is capable of stimulating second messenger-dependent kinases that could negatively regulate signaling. Consistent with this model, several studies have shown direct proportionality between signaling efficacy and desensitization *(117–119)*. However, there are also several exceptions which support the notion that opioid receptor ligands may be distinctively recognized by desensitizing kinases. For example, the full agonist DAMGO fails to stimulate PKC but activates GRK2, which is essential for MOR desensitization by this agonist. Conversely, the partial agonist morphine does not significantly activate GRK2 but induces PKC activity that causes MOR desensitization in a ligand-dependent manner *(113)*. In the same line of evidence, although the classical agonist DPDPE and the double efficacy ligand TICP produce similar Src activation, only the conformation stabilized by DPDPE is a substrate for this non-receptor tyrosine kinase *(17, 88)*. Truncation studies further support the idea that conformations stabilized by different opioid ligands may be distinctively recognized by cytosolic kinases. For example, removal of Ser/Thr residues within the C-terminus of DORs results in loss of DPDPE but not SNC-80 ability to induce phosphorylation *(120, 121)*, indicating that each of these agonists exposes different Ser/Thr residues to cytosolic kinases.

12.3.1.2 Receptor Internalization

It is also well accepted that opioid agonists vary in their ability to induce opioid receptor internalization, but there has been a long standing debate whether these differences are related to drug efficacy *(117, 122)* or if they actually reveal ligand-specific regulatory properties *(89, 123)*. Central to the discussion is morphine, which unlike DAMGO or etorphine, weakly recruits β-arrestin *(123–125)* and fails to consistently induce internalization *(89, 123, 126, 127)*. However, since morphine is a partial agonist in most signaling pathways in which it has been tested, inefficient β-arrestin recruitment and poor internalization are inconclusive whether the conformation stabilized by this ligand has distinct internalization properties, or if deficient endocytosis is simply due to morphine's limited ability to stabilize the active state of the receptor. More recently, the possibility that internalization and signaling efficacy may be dissociated from one another has been conclusively addressed using other opioid ligands. Namely, herkinorin, a synthetic derivative of Salvinorin A that selectively binds MORs *(128, 129)*, produced similar adenylate cyclase inhibition and ERK activation as the full agonist DAMGO but, unlike the latter, herkinorin failed to recruit β-arrestin2 and induce internalization *(129)*.

Similar dissociation between signaling and endocytosis was observed when comparing the UFP-512 and morphine. Although both ligands displayed similar efficacy in the adenylate cyclase pathway, only UFP-512 induced sequestration of membrane receptors *(130)*. These observations clearly contradict the prediction of the classical model and cannot be adequately explained unless invoking the existence of ligand-specific conformations that are distinctively recognized by signaling effectors and cytosolic proteins responsible for GPCR internalization.

In addition to β-arrestin, MOR internalization seems to require PLD2 *(99, 131, 132)*. PLD2 activation is thought to contribute to receptor endocytosis by recruiting the adaptor protein AP2, which facilitates sequestration by promoting β-arrestin interaction with the clathrin cage *(133, 134)*. PLD2 is activated by internalizing opioids like DAMGO but not by morphine *(99)*, and this observation has been interpreted as ligand-specific activation of the enzyme *(135)*. However, as stated earlier, given the partial efficacy of morphine in generating other signals it is not possible to rule out that its failure to activate PLD2 is simply related to differences in strength of signaling.

12.3.1.3 Post-Endocytic Sorting of Opioid Receptors

Following sequestration, receptors may either recycle to the cell surface or be targeted for lysosomal degradation, processes that contribute to resensitization or long-lasting desensitization, respectively *(135)*. The decision whether opioid receptors will be sorted toward recycling or degradation relies upon posttranslational modifications of the receptor as well as on its noncovalent interaction with cytoplasmic "sorting" proteins *(136, 137)*. The role of posttranslational modifications is exemplified by the fact that MOR isoforms lacking Thr at position 394 (MOR1B) have quicker recycling and resensitization kinetics than isoforms containing Thr at the same position *(138)*. Agonist-dependent ubiquitination of a di-leucine motif within the third intracellular loop may also direct internalized opioid receptors toward degradation *(139)*, although this mechanism seems to rely on proteasomal rather than lysosomal targeting *(140)*. But then, lysosomal degradation of DORs may take place independent of ubiquitination *(141, 142)* via receptor interaction with the cytoplasmic sorting protein GASP (G protein coupled receptor-associated sorting protein) *(136, 143, 144)*. Although MORs also possess the string of residues necessary for GASP interaction *(144)*, lower affinity for this lysosomal targeting protein *(136, 143)* and presence of the C-terminal sequence LENLEAE *(137)* could facilitate active targeting of MORs toward the recycling pathway. LENLEAE is not the only determinant for MOR recycling since certain of the different MOR isoforms generated by alternative splicing are rapidly resensitized and sent back to the membrane despite lacking this sequence *(138, 145)*.

Taken together these observations indicate that post-endocytic sorting plays a crucial role in determining whether receptor internalization will serve a resensitization or desensitization function. Since mutagenesis studies show that access to resensitization steps is closely related to receptor affinity for sorting proteins *(146)*, it is reasonable to think that different ligands could preferentially direct opioid receptors toward recycling/resensitization or to lysosomal degradation, depending on their ability to stabilize conformations with differential affinity for proteins that target receptors to these two alternative pathways. In fact, ligand-directed post-endocytic trafficking of DORs has been reported for DPDPE and etorphine, which favor receptor degradation or recycling, respectively *(147)*. A reversal in agonist rank order of efficacy for G protein activation (Deltorphin II ≈ DPDPE ≥ SNC80) *(148)* and capacity to induce DOR down-regulation (SNC80 > DPDPE ≥ Deltorphin II) *(121)* also supports the idea that receptor degradation may be ligand specific, as do mutagenesis studies showing that C-terminal truncation blocked DOR down-regulation by DPDPE but not SNC80 *(120, 121)*.

12.3.2 Functional Selectivity and Opioid Tolerance

Analgesic tolerance is a complex phenomenon that involves adaptive changes at different organizational levels within the central nervous system. These changes start as molecular and cellular adaptations and progressively develop to interfere with normal functioning of pain perception circuits *(91)*. The exact molecular bases of tolerance remain unclear, but signaling regulation is thought to play a central role in the process *(12, 135)*. Among the most commonly invoked contributing mechanisms are a rapid decrease in receptor signaling activity (desensitization) *(12, 125, 149)* and adaptive changes in intracellular signaling pathways *(89, 150)*. Both of these adaptive changes seem greatly determined by receptor trafficking, and given that endocytic and sorting properties of opioid receptors are subject to functional selectivity by different ligands, it is possible to speculate that developing ligands with the appropriate trafficking properties should allow the production of opioid ligands with longer lasting analgesic actions. The question that remains is what are the "optimal trafficking properties"?

According to the hypothesis that favors desensitization as a molecular determinant of analgesic tolerance, internalization would play an important role in ensuring that desensitized receptors are rapidly removed from the membrane and recycled back to the cell surface in an active form *(12, 125)*. Hence, according to this model, noninternalizing agonists such as morphine or herkinorin would promote membrane accumulation of desensitized receptors causing greater tolerance than ligands that promote sequestration. This reasoning is consistent with the observation that analgesic tolerance in animals is inversely correlated with ligand capacity to induce internalization *(151)*, and is consistent with the assumption that opioid agonists that stabilize conformations easily recognized by the endocytic machinery should produce longer lasting analgesic actions. Furthermore, if preventing analgesic tolerance is indeed dependent on receptor recycling, a long-lasting analgesic agent would require the stabilization of a receptor conformation that is not only effectively internalized but that is also poorly recognized by lysosomal targeting proteins.

The second hypothesis views tolerance mainly as the consequence of cellular mechanisms that counteract sustained opioid receptor activation *(89, 150, 152)*. In particular, a counteregulatory increase in adenylate cyclase activity is considered a major contributing factor to changes in gene expression and neuronal excitability underlying tolerance and dependence *(153)*. Interestingly, in contrast to the previous hypothesis, this model proposes that the stimulus responsible for tolerance development is sustained activation of membrane receptors *(150, 152, 154)*, such that noninternalizing ligands would produce the type of prolonged signal that is necessary for compensatory adaptations in intracellular signaling cascades. On the basis of this assumption, and in opposition to the previous hypothesis, one could speculate that ligands that stabilize internalizing conformations with poor recycling capacity should produce longer acting analgesic actions.

12.4 Conclusion

Molecular evidence analyzed in this chapter clearly points to the existence of multiple active conformations for MORs and DORs, and provides solid support to the idea that it is possible to develop functionally selective ligands for these receptors. However, until there is a better understanding of the mechanisms involved in the production of undesired side effects and the development of tolerance, it is difficult to predict the type of signaling and regulatory properties that should be promoted (or avoided) to achieve longer lasting analgesic actions with fewer side effects.

References

1. Goldstein FJ. Adjuncts to opioid therapy. J Am Osteopath Assoc 2002;102(9 Suppl 3):S15–21.
2. Grunkemeier DM, Cassara JE, Dalton CB, Drossman DA. The narcotic bowel syndrome: clinical features, pathophysiology, and management. Clin Gastroenterol Hepatol 2007;5(10):1126–39; quiz 1–2.
3. Walker JM, Farney RJ, Rhondeau SM, et al. Chronic opioid use is a risk factor for the development of central sleep apnea and ataxic breathing. J Clin Sleep Med 2007;3(5):455–61.
4. Ballantyne JC, LaForge KS. Opioid dependence and addiction during opioid treatment of chronic pain. Pain 2007;129(3):235–55.
5. Noble M, Schoelles K. Opioid treatment for chronic back pain and its association with addiction. Ann Intern Med 2007;147(5):348–9; author reply 9–50.
6. Kieffer BL, Gaveriaux-Ruff C. Exploring the opioid system by gene knockout. Prog Neurobiol 2002;66(5):285–306.
7. Rapaka RS, Porreca F. Development of delta opioid peptides as nonaddicting analgesics. Pharm Res 1991;8(1):1–8.
8. Dickenson AH. Plasticity: implications for opioid and other pharmacological interventions in specific pain states. Behav Brain Sci 1997;20(3):392–403; discussion 35–513.
9. Cheng PY, Wu D, Soong Y, McCabe S, Decena JA, Szeto HH. Role of mu 1- and delta-opioid receptors in modulation of fetal EEG and respiratory activity. Am J Physiol 1993;265(2 Pt 2): R433–8.

10. Coop A, Rice KC. Role of delta-opioid receptors in biological processes. Drug News Perspect 2000;13(8):481–7.
11. Cowan A, Zhu XZ, Mosberg HI, Omnaas JR, Porreca F. Direct dependence studies in rats with agents selective for different types of opioid receptor. J Pharmacol Exp Ther 1988;246(3):950–5.
12. Gainetdinov RR, Premont RT, Bohn LM, Lefkowitz RJ, Caron MG. Desensitization of G protein-coupled receptors and neuronal functions. Annu Rev Neurosci 2004;27:107–44.
13. von Zastrow M. A cell biologist's perspective on physiological adaptation to opiate drugs. Neuropharmacology 2004;47 Suppl 1:286–92.
14. Jutkiewicz EM, Baladi MG, Folk JE, Rice KC, Woods JH. The convulsive and electroencephalographic changes produced by nonpeptidic delta-opioid agonists in rats: comparison with pentylenetetrazol. J Pharmacol Exp Ther 2006;317:1337–48.
15. Furchgott RF. Metabolic factors that influence contractility of vascular smooth muscle. Bull N Y Acad Med 1966;42(11):996–1006.
16. Azzi M, Charest PG, Angers S, et al. Beta-arrestin-mediated activation of MAPK by inverse agonists reveals distinct active conformations for G protein-coupled receptors. Proc Natl Acad Sci USA 2003;100(20):11406–11.
17. Audet N, Paquin-Gobeil M, Landry-Paquet O, Schiller PW, Pineyro G. Internalization and Src activity regulate the time course of ERK activation by delta opioid receptor ligands. J Biol Chem 2005;280(9):7808–16.
18. Urban JD, Clarke WP, von Zastrow M, et al. Functional selectivity and classical concepts of quantitative pharmacology. J Pharmacol Exp Ther 2007;320(1):1–13.
19. Mailman RB. GPCR functional selectivity has therapeutic impact. Trends Pharmacol Sci 2007;28(8):390–6.
20. Gilchrist A, Blackmer T. G-protein-coupled receptor pharmacology: examining the edges between theory and proof. Curr Opin Drug Discov Devel 2007;10(4):446–51.
21. Kenakin T. Principles: receptor theory in pharmacology. Trends Pharmacol Sci 2004;25(4):186–92.
22. Kenakin T. Functional selectivity through protean and biased agonism: who steers the ship? Mol Pharmacol 2007;72(6):1393–401.
23. Bosier B, Hermans E. Versatility of GPCR recognition by drugs: from biological implications to therapeutic relevance. Trends Pharmacol Sci 2007;28(8):438–46.
24. Hubbell WL, Altenbach C, Hubbell CM, Khorana HG. Rhodopsin structure, dynamics, and activation: a perspective from crystallography, site-directed spin labeling, sulfhydryl reactivity, and disulfide cross-linking. Adv Protein Chem 2003;63:243–90.
25. Gether U, Lin S, Kobilka BK. Fluorescent labeling of purified beta 2 adrenergic receptor. Evidence for ligand-specific conformational changes. J Biol Chem 1995;270(47):28268–75.
26. Gether U, Lin S, Ghanouni P, Ballesteros JA, Weinstein H, Kobilka BK. Agonists induce conformational changes in transmembrane domains III and VI of the beta2 adrenoceptor. Embo J 1997;16(22):6737–47.
27. Schwartz TW, Frimurer TM, Holst B, Rosenkilde MM, Elling CE. Molecular mechanism of 7TM receptor activation–a global toggle switch model. Annu Rev Pharmacol Toxicol 2006;46:481–519.
28. Swaminath G, Steenhuis J, Kobilka B, Lee TW. Allosteric modulation of beta2-adrenergic receptor by Zn(2+). Mol Pharmacol 2002;61(1):65–72.
29. Yao X, Parnot C, Deupi X, et al. Coupling ligand structure to specific conformational switches in the beta2-adrenoceptor. Nat Chem Biol 2006;2(8):417–22.
30. Li JH, Han SJ, Hamdan FF, et al. Distinct structural changes in a G protein-coupled receptor caused by different classes of agonist ligands. J Biol Chem 2007;282(36):26284–93.
31. Hoffmann C, Zurn A, Bunemann M, Lohse MJ. Conformational changes in G-protein-coupled receptors-the quest for functionally selective conformations is open. Br J Pharmacol 2008;153 Suppl 1:S358–66.
32. Law SF, Reisine T. Changes in the association of G protein subunits with the cloned mouse delta opioid receptor on agonist stimulation. J Pharmacol Exp Ther 1997;281(3):1476–86.

33. Alves ID, Ciano KA, Boguslavski V, et al. Selectivity, cooperativity and reciprocity in the interactions between the delta opioid receptor, its ligands and G-proteins. J Biol Chem 2004;17:17.
34. Garzon J, Garcia-Espana A, Sanchez-Blazquez P. Opioids binding mu and delta receptors exhibit diverse efficacy in the activation of Gi2 and G(x/z) transducer proteins in mouse periaqueductal gray matter. J Pharmacol Exp Ther 1997b;281(1):549–57.
35. McKenzie FR, Milligan G. Delta-opioid-receptor-mediated inhibition of adenylate cyclase is transduced specifically by the guanine-nucleotide-binding protein Gi2. Biochem J 1990;267(2):391–8.
36. Offermanns S, Schultz G, Rosenthal W. Evidence for opioid receptor-mediated activation of the G-proteins, Go and Gi2, in membranes of neuroblastoma x glioma (NG108-15) hybrid cells. J Biol Chem 1991;266(6):3365–8.
37. Prather PL, Loh HH, Law PY. Interaction of delta-opioid receptors with multiple G proteins: a non-relationship between agonist potency to inhibit adenylyl cyclase and to activate G proteins. Mol Pharmacol 1994;45(5):997–1003.
38. Laugwitz KL, Offermanns S, Spicher K, Schultz G. mu and delta opioid receptors differentially couple to G protein subtypes in membranes of human neuroblastoma SH-SY5Y cells. Neuron 1993;10(2):233–42.
39. Allouche S, Polastron J, Hasbi A, Homburger V, Jauzac P. Differential G-protein activation by alkaloid and peptide opioid agonists in the human neuroblastoma cell line SK-N-BE. Biochem J 1999;342 (Pt 1):71–8.
40. Prather PL, McGinn TM, Erickson LJ, Evans CJ, Loh HH, Law PY. Ability of delta-opioid receptors to interact with multiple G-proteins is independent of receptor density. J Biol Chem 1994;269(33):21293–302.
41. Prather PL, Song L, Piros ET, Law PY, Hales TG. delta-Opioid receptors are more efficiently coupled to adenylyl cyclase than to L-type Ca(2+) channels in transfected rat pituitary cells. J Pharmacol Exp Ther 2000;295(2):552–62.
42. Moon HE, Cavalli A, Bahia DS, Hoffmann M, Massotte D, Milligan G. The human delta opioid receptor activates G(i1)alpha more efficiently than G(o1)alpha. J Neurochem 2001;76(6):1805–13.
43. Sanchez-Blazquez P, Garcia-Espana A, Garzon J. In vivo injection of antisense oligodeoxynucleotides to G alpha subunits and supraspinal analgesia evoked by mu and delta opioid agonists. J Pharmacol Exp Ther 1995;275(3):1590–6.
44. Sanchez-Blazquez P, Garzon J. delta Opioid receptor subtypes activate inositol-signaling pathways in the production of antinociception. J Pharmacol Exp Ther 1998;285(2):820–7.
45. Standifer KM, Rossi GC, Pasternak GW. Differential blockade of opioid analgesia by antisense oligodeoxynucleotides directed against various G protein alpha subunits. Mol Pharmacol 1996;50(2):293–8.
46. Audet N, Gales C, Archer-Lahlou E, et al. BRET assays reveal ligand-specific conformational changes within preformed signalling complexes containing delta opioid receptor (DOR) and heterotrimeric G proteins. J Biol Chem 2008;283:15078–88.
47. Bunemann M, Frank M, Lohse MJ. Gi protein activation in intact cells involves subunit rearrangement rather than dissociation. Proc Natl Acad Sci USA 2003;100(26):16077–82.
48. Gales C, Rebois RV, Hogue M, et al. Real-time monitoring of receptor and G-protein interactions in living cells. Nat Methods 2005;2(3):177–84.
49. George SR, O'Dowd BF, Lee SP. G-protein-coupled receptor oligomerization and its potential for drug discovery. Nat Rev Drug Discov 2002;1(10):808–20.
50. Gupta A, Decaillot FM, Devi LA. Targeting opioid receptor heterodimers: strategies for screening and drug development. Aaps J 2006;8(1):E153–9.
51. Davare MA, Avdonin V, Hall DD, et al. A beta2 adrenergic receptor signaling complex assembled with the Ca2 + channel Cav1.2. Science 2001;293(5527):98–101.
52. Lavine N, Ethier N, Oak JN, et al. G protein-coupled receptors form stable complexes with inwardly rectifying potassium channels and adenylyl cyclase. J Biol Chem 2002;277(48):46010–9.
53. Rebois RV, Robitaille M, Gales C, et al. Heterotrimeric G proteins form stable complexes with adenylyl cyclase and Kir3.1 channels in living cells. J Cell Sci 2006;119(Pt 13):2807–18.

54. Abramow-Newerly M, Roy AA, Nunn C, Chidiac P. RGS proteins have a signalling complex: interactions between RGS proteins and GPCRs, effectors, and auxiliary proteins. Cell Signal 2006;18(5):579–91.
55. Jordan BA, Devi LA. G-protein-coupled receptor heterodimerization modulates receptor function. Nature 1999;399(6737):697–700.
56. George SR, Fan T, Xie Z, et al. Oligomerization of mu- and delta-opioid receptors. Generation of novel functional properties. J Biol Chem 2000;275(34):26128–35.
57. Gomes I, Gupta A, Filipovska J, Szeto HH, Pintar JE, Devi LA. A role for heterodimerization of mu and delta opiate receptors in enhancing morphine analgesia. Proc Natl Acad Sci USA 2004;101(14):5135–9.
58. Hasbi A, Nguyen T, Fan T, et al. Trafficking of preassembled opioid mu-delta heterooligomer-Gz signaling complexes to the plasma membrane: coregulation by agonists. Biochemistry 2007;46(45):12997–3009.
59. Garzon J, Rodriguez-Munoz M, Sanchez-Blazquez P. Morphine alters the selective association between mu-opioid receptors and specific RGS proteins in mouse periaqueductal gray matter. Neuropharmacology 2005a;48(6):853–68.
60. Garzon J, Rodriguez-Munoz M, Lopez-Fando A, Sanchez-Blazquez P. Activation of mu-opioid receptors transfers control of Galpha subunits to the regulator of G-protein signaling RGS9-2: role in receptor desensitization. J Biol Chem 2005b;280(10):8951–60.
61. Chalecka-Franaszek E, Weems HB, Crowder AT, Cox BM, Cote TE. Immunoprecipitation of high-affinity, guanine nucleotide-sensitive, solubilized mu-opioid receptors from rat brain: coimmunoprecipitation of the G proteins G(alpha o), G(alpha i1), and G(alpha i3). J Neurochem 2000;74(3):1068–78.
62. Georgoussi Z, Leontiadis L, Mazarakou G, Merkouris M, Hyde K, Hamm H. Selective interactions between G protein subunits and RGS4 with the C-terminal domains of the mu- and delta-opioid receptors regulate opioid receptor signaling. Cell Signal 2006;18(6):771–82.
63. Dupre DJ, Hebert TE. Biosynthesis and trafficking of seven transmembrane receptor signalling complexes. Cell Signal 2006;18(10):1549–59.
64. Dupre DJ, Robitaille M, Ethier N, Villeneuve LR, Mamarbachi AM, Hebert TE. Seven transmembrane receptor core signaling complexes are assembled prior to plasma membrane trafficking. J Biol Chem 2006;281(45):34561–73.
65. Kreienkamp HJ. Organisation of G-protein-coupled receptor signalling complexes by scaffolding proteins. Curr Opin Pharmacol 2002;2(5):581–6.
66. Nikolaev VO, Hoffmann C, Bunemann M, Lohse MJ, Vilardaga JP. Molecular basis of partial agonism at the neurotransmitter alpha2A-adrenergic receptor and Gi-protein heterotrimer. J Biol Chem 2006;281(34):24506–11.
67. Milligan G, Bouvier M. Methods to monitor the quaternary structure of G protein-coupled receptors. Febs J 2005;272(12):2914–25.
68. Gupta A, Decaillot FM, Gomes I, et al. Conformation state-sensitive antibodies to G-protein-coupled receptors. J Biol Chem 2007;282(8):5116–24.
69. Gupta A, Rozenfeld R, Gomes I, et al. Post-activation-mediated changes in opioid receptors detected by N-terminal antibodies. J Biol Chem 2008;283(16):10735–44.
70. Murthy KS, Makhlouf GM. Opioid mu, delta, and kappa receptor-induced activation of phospholipase C-beta 3 and inhibition of adenylyl cyclase is mediated by Gi2 and G(o) in smooth muscle. Mol Pharmacol 1996;50(4):870–7.
71. Clark MJ, Furman CA, Gilson TD, Traynor JR. Comparison of the relative efficacy and potency of mu-opioid agonists to activate G{alpha}i/o proteins containing a pertussis toxin-insensitive mutation. J Pharmacol Exp Ther 2006;317:858–64.
72. Burford NT, Tolbert LM, Sadee W. Specific G protein activation and mu-opioid receptor internalization caused by morphine, DAMGO and endomorphin I. Eur J Pharmacol 1998;342(1):123–6.
73. Burford NT, Wang D, Sadee W. G-protein coupling of mu-opioid receptors (OP3): elevated basal signalling activity. Biochem J 2000;348 Pt 3:531–7.
74. Chakrabarti S, Prather PL, Yu L, Law PY, Loh HH. Expression of the mu-opioid receptor in CHO cells: ability of mu-opioid ligands to promote alpha-azidoanilido[32P]GTP labeling of multiple G protein alpha subunits. J Neurochem 1995;64(6):2534–43.

75. Massotte D, Brillet K, Kieffer B, Milligan G. Agonists activate Gi1 alpha or Gi2 alpha fused to the human mu opioid receptor differently. J Neurochem 2002;81(6):1372–82.
76. Stanasila L, Lim WK, Neubig RR, Pattus F. Coupling efficacy and selectivity of the human mu-opioid receptor expressed as receptor-Galpha fusion proteins in Escherichia coli. J Neurochem 2000;75(3):1190–9.
77. Sanchez-Blazquez P, Rodriguez-Diaz M, DeAntonio I, Garzon J. Endomorphin-1 and endomorphin-2 show differences in their activation of mu opioid receptor-regulated G proteins in supraspinal antinociception in mice. J Pharmacol Exp Ther 1999;291(1):12–8.
78. Sanchez-Blazquez P, Gomez-Serranillos P, Garzon J. Agonists determine the pattern of G-protein activation in mu-opioid receptor-mediated supraspinal analgesia. Brain Res Bull 2001;54(2):229–35.
79. Berg KA, Stout BD, Cropper JD, Maayani S, Clarke WP. Novel actions of inverse agonists on 5-HT2C receptor systems. Mol Pharmacol 1999;55(5):863–72.
80. Kenakin T. Drug efficacy at G protein-coupled receptors. Annu Rev Pharmacol Toxicol 2002;42:349–79.
81. Polastron J, Mur M, Mazarguil H, Puget A, Meunier JC, Jauzac P. SK-N-BE: a human neuroblastoma cell line containing two subtypes of delta-opioid receptors. J Neurochem 1994;62(3):898–906.
82. Hosohata Y, Varga EV, Stropova D, et al. Mutation W284L of the human delta opioid receptor reveals agonist specific receptor conformations for G protein activation. Life Sci 2001;68(19–20):2233–42.
83. Chaipatikul V, Loh HH, Law PY. Ligand-selective activation of mu-oid receptor: demonstrated with deletion and single amino acid mutations of third intracellular loop domain. J Pharmacol Exp Ther 2003;305(3):909–18.
84. Surratt CK, Johnson PS, Moriwaki A, et al. -mu opiate receptor. Charged transmembrane domain amino acids are critical for agonist recognition and intrinsic activity. J Biol Chem 1994;269(32):20548–53.
85. Prather PL. Inverse agonists: tools to reveal ligand-specific conformations of G protein-coupled receptors. Sci STKE 2004;2004(215):pe1.
86. Zheng H, Loh HH, Law PY. Beta-arrestin-dependent mu-opioid receptor-activated extracellular signal-regulated kinases (ERKs) translocate to nucleus in contrast to G protein-dependent ERK activation. Mol Pharmacol 2008;73(1):178–90.
87. Pineyro G, Azzi M, De Lean A, Schiller P, Bouvier M. Short-term inverse-agonist treatment induces reciprocal changes in delta-opioid agonist and inverse-agonist binding capacity. Mol Pharmacol 2001;60(4):816–27.
88. Archer-Lahlou E, Audet N, Amraei MG, Huard K, Paquin-Gobeil M, Pineyro G. Src promotes delta opioid receptor (DOR) desensitization by interfering with receptor recyling. J Cell Mol Med 2008.
89. Whistler JL, Chuang HH, Chu P, Jan LY, von Zastrow M. Functional dissociation of mu opioid receptor signaling and endocytosis: implications for the biology of opiate tolerance and addiction. Neuron 1999;23(4):737–46.
90. Williams JT, Christie MJ, Manzoni O. Cellular and synaptic adaptations mediating opioid dependence. Physiol Rev 2001;81(1):299–343.
91. Nestler EJ, Alreja M, Aghajanian GK. Molecular control of locus coeruleus neurotransmission. Biol Psychiatry 1999;46(9):1131–9.
92. Nicholson D, Reid A, Sawynok J. Effects of forskolin and phosphodiesterase inhibitors on spinal antinociception by morphine. Pharmacol Biochem Behav 1991;38(4):753–8.
93. Bradaia A, Berton F, Ferrari S, Luscher C. beta-Arrestin2, interacting with phosphodiesterase 4, regulates synaptic release probability and presynaptic inhibition by opioids. Proc Natl Acad Sci USA 2005;102(8):3034–9.
94. Mitrovic I, Margeta-Mitrovic M, Bader S, Stoffel M, Jan LY, Basbaum AI. Contribution of GIRK2-mediated postsynaptic signaling to opiate and alpha 2-adrenergic analgesia and analgesic sex differences. Proc Natl Acad Sci USA 2003;100(1):271–6.

95. Gilman AG. G proteins: transducers of receptor-generated signals. Annu Rev Biochem 1987;56:615–49.
96. Bourne HR. How receptors talk to trimeric G proteins. Curr Opin Cell Biol 1997;9(2):134–42.
97. Vilardaga JP, Nikolaev VO, Lorenz K, Ferrandon S, Zhuang Z, Lohse MJ. Conformational cross-talk between alpha2A-adrenergic and mu-opioid receptors controls cell signaling. Nat Chem Biol 2008;4(2):126–31.
98. Mazarakou G, Georgoussi Z. STAT5A interacts with and is phosphorylated upon activation of the mu-opioid receptor. J Neurochem 2005;93(4):918–31.
99. Koch T, Brandenburg LO, Schulz S, Liang Y, Klein J, Hollt V. ADP-ribosylation factor-dependent phospholipase D2 activation is required for agonist-induced mu-opioid receptor endocytosis. J Biol Chem 2003;278(11):9979–85.
100. Wang D, Sadee W, Quillan JM. Calmodulin binding to G protein-coupling domain of opioid receptors. J Biol Chem 1999;274(31):22081–8.
101. Parenty G, Appelbe S, Milligan G. CXCR2 chemokine receptor antagonism enhances DOP opioid receptor function via allosteric regulation of the CXCR2-DOP receptor hetero-dimer. Biochem J 2008;412:245–56.
102. Pello OM, Martinez-Munoz L, Parrillas V, et al. Ligand stabilization of CXCR4/delta-opioid receptor heterodimers reveals a mechanism for immune response regulation. Eur J Immunol 2008;38(2):537–49.
103. Jordan BA, Gomes I, Rios C, Filipovska J, Devi LA. Functional interactions between mu opioid and alpha 2A-adrenergic receptors. Mol Pharmacol 2003;64(6):1317–24.
104. Rios C, Gomes I, Devi LA. Interactions between delta opioid receptors and alpha-adrenoceptors. Clin Exp Pharmacol Physiol 2004;31(11):833–6.
105. Pfeiffer M, Kirscht S, Stumm R, et al. Heterodimerization of substance P and mu-opioid receptors regulates receptor trafficking and resensitization. J Biol Chem 2003;278(51):51630–7.
106. Mackie K. Cannabinoid receptor homo- and heterodimerization. Life Sci 2005;77(14):1667–73.
107. Fan T, Varghese G, Nguyen T, Tse R, O'Dowd BF, George SR. A role for the distal carboxyl tails in generating the novel pharmacology and G protein activation profile of mu and delta opioid receptor hetero-oligomers. J Biol Chem 2005;280(46):38478–88.
108. Marie N, Aguila B, Allouche S. Tracking the opioid receptors on the way of desensitization. Cell Signal 2006;18(11):1815–33.
109. Pineyro G, Archer-Lahlou E. Ligand-specific receptor states: implications for opiate receptor signalling and regulation. Cell Signal 2007;19(1):8–19.
110. Mestek A, Hurley JH, Bye LS, et al. The human mu opioid receptor: modulation of functional desensitization by calcium/calmodulin-dependent protein kinase and protein kinase C. J Neurosci 1995;15(3 Pt 2):2396–406.
111. Koch T, Kroslak T, Mayer P, Raulf E, Hollt V. Site mutation in the rat mu-opioid receptor demonstrates the involvement of calcium/calmodulin-dependent protein kinase II in agonist-mediated desensitization. J Neurochem 1997;69(4):1767–70.
112. Bailey CP, Smith FL, Kelly E, Dewey WL, Henderson G How important is protein kinase C in mu-opioid receptor desensitization and morphine tolerance? Trends Pharmacol Sci 2006;27(11):558–65.
113. Johnson EA, Oldfield S, Braksator E, et al. Agonist-selective mechanisms of mu-opioid receptor desensitization in human embryonic kidney 293 cells. Mol Pharmacol 2006;70(2):676–85.
114. Schmidt H, Schulz S, Klutzny M, Koch T, Handel M, Hollt V. Involvement of mitogen-activated protein kinase in agonist-induced phosphorylation of the mu-opioid receptor in HEK 293 cells. J Neurochem 2000;74(1):414–22.
115. Polakiewicz RD, Schieferl SM, Dorner LF, Kansra V, Comb MJ. A mitogen-activated protein kinase pathway is required for mu-opioid receptor desensitization. J Biol Chem 1998;273(20):12402–6.
116. Pak Y, O'Dowd BF, Wang JB, George SR. Agonist-induced, G protein-dependent and -independent down-regulation of the mu opioid receptor. The receptor is a direct substrate for protein-tyrosine kinase. J Biol Chem 1999;274(39):27610–6.

117. Kovoor A, Celver JP, Wu A, Chavkin C. Agonist induced homologous desensitization of mu-opioid receptors mediated by G protein-coupled receptor kinases is dependent on agonist efficacy. Mol Pharmacol 1998;54(4):704–11.
118. Borgland SL, Connor M, Osborne PB, Furness JB, Christie MJ. Opioid agonists have different efficacy profiles for G protein activation, rapid desensitization, and endocytosis of mu-opioid receptors. J Biol Chem 2003;278(21):18776–84.
119. Woolf PJ, Linderman JJ. Untangling ligand induced activation and desensitization of G-protein-coupled receptors. Biophys J 2003;84(1):3–13.
120. Okura T, Cowell SM, Varga E, et al. Differential down-regulation of the human delta-opioid receptor by SNC80 and [D-Pen(2),D-Pen(5)]enkephalin. Eur J Pharmacol 2000;387(2):R11–3.
121. Okura T, Varga EV, Hosohata Y, et al. Agonist-specific down-regulation of the human delta-opioid receptor. Eur J Pharmacol 2003;459(1):9–16.
122. Celver J, Xu M, Jin W, Lowe J, Chavkin C. Distinct domains of the mu-opioid receptor control uncoupling and internalization. Mol Pharmacol 2004;65(3):528–37.
123. Whistler JL, von Zastrow M. Morphine-activated opioid receptors elude desensitization by beta-arrestin. Proc Natl Acad Sci USA 1998;95(17):9914–9.
124. Zhang J, Ferguson SS, Barak LS, et al. Role for G protein-coupled receptor kinase in agonist-specific regulation of mu-opioid receptor responsiveness. Proc Natl Acad Sci U S A 1998;95(12):7157–62.
125. Bohn LM, Dykstra LA, Lefkowitz RJ, Caron MG, Barak LS. Relative opioid efficacy is determined by the complements of the G protein-coupled receptor desensitization machinery. Mol Pharmacol 2004;66(1):106–12.
126. Keith DE, Murray SR, Zaki PA, et al. Morphine activates opioid receptors without causing their rapid internalization. J Biol Chem 1996;271(32):19021–4.
127. Keith DE, Anton B, Murray SR, et al. mu-Opioid receptor internalization: opiate drugs have differential effects on a conserved endocytic mechanism in vitro and in the mammalian brain. Mol Pharmacol 1998;53(3):377–84.
128. Harding WW, Tidgewell K, Schmidt M, et al. Salvinicins A and B, new neoclerodane diterpenes from Salvia divinorum. Org Lett 2005;7(14):3017–20.
129. Groer CE, Tidgewell K, Moyer RA, et al. An opioid agonist that does not induce micro-opioid receptor--arrestin interactions or receptor internalization. Mol Pharmacol 2007;71(2):549–57.
130. Aguila B, Coulbault L, Boulouard M, et al. In vitro and in vivo pharmacological profile of UFP-512, a novel selective delta-opioid receptor agonist; correlations between desensitization and tolerance. Br J Pharmacol 2007;152(8):1312–24.
131. Koch T, Brandenburg LO, Liang Y, et al. Phospholipase D2 modulates agonist-induced mu-opioid receptor desensitization and resensitization. J Neurochem 2004;88(3):680–8.
132. Koch T, Wu DF, Yang LQ, Brandenburg LO, Hollt V. Role of phospholipase D2 in the agonist-induced and constitutive endocytosis of G-protein coupled receptors. J Neurochem 2006;97(2):365–72.
133. Liscovitch M, Cantley LC. Signal transduction and membrane traffic: the PITP/phosphoinositide connection. Cell 1995;81(5):659–62.
134. De Camilli P, Emr SD, McPherson PS, Novick P. Phosphoinositides as regulators in membrane traffic. Science 1996;271(5255):1533–9.
135. Koch T, Hollt V. Role of receptor internalization in opioid tolerance and dependence. Pharmacol Ther 2008;117(2):199–206.
136. Whistler JL, Enquist J, Marley A, et al. Modulation of postendocytic sorting of G protein-coupled receptors. Science 2002;297(5581):615–20.
137. Tanowitz M, von Zastrow M. A novel endocytic recycling signal that distinguishes the membrane trafficking of naturally occurring opioid receptors. J Biol Chem 2003;278(46):45978–86.
138. Koch T, Schulz S, Schroder H, Wolf R, Raulf E, Hollt V. Carboxyl-terminal splicing of the rat mu opioid receptor modulates agonist-mediated internalization and receptor resensitization. J Biol Chem 1998;273(22):13652–7.
139. Wang W, Loh HH, Law PY. The intracellular trafficking of opioid receptors directed by carboxyl tail and a di-leucine motif in Neuro2A cells. J Biol Chem 2003;278(38):36848–58.

140. Chaturvedi K, Bandari P, Chinen N, Howells RD. Proteasome involvement in agonist-induced down-regulation of mu and delta opioid receptors. J Biol Chem 2001;276(15):12345–55.
141. Tanowitz M, Von Zastrow M. Ubiquitination-independent trafficking of G protein-coupled receptors to lysosomes. J Biol Chem 2002;277(52):50219–22.
142. Hislop JN, Marley A, Von Zastrow M. Role of mammalian vacuolar protein-sorting proteins in endocytic trafficking of a non-ubiquitinated G protein-coupled receptor to lysosomes. J Biol Chem 2004;279(21):22522–31.
143. Heydorn A, Sondergaard BP, Ersboll B, et al. A library of 7TM receptor C-terminal tails. Interactions with the proposed post-endocytic sorting proteins ERM-binding phosphoprotein 50 (EBP50), N-ethylmaleimide-sensitive factor (NSF), sorting nexin 1 (SNX1), and G protein-coupled receptor-associated sorting protein (GASP). J Biol Chem 2004;279(52):54291–303.
144. Simonin F, Karcher P, Boeuf JJ, Matifas A, Kieffer BL. Identification of a novel family of G protein-coupled receptor associated sorting proteins. J Neurochem 2004;89(3):766–75.
145. Koch T, Schulz S, Pfeiffer M, et al. C-terminal splice variants of the mouse mu-opioid receptor differ in morphine-induced internalization and receptor resensitization. J Biol Chem 2001;276(33):31408–14.
146. Thompson D, Pusch M, Whistler JL. Changes in G protein-coupled receptor sorting protein affinity regulate postendocytic targeting of G protein-coupled receptors. J Biol Chem 2007;282(40):29178–85.
147. Marie N, Lecoq I, Jauzac P, Allouche S. Differential sorting of human delta-opioid receptors after internalization by peptide and alkaloid agonists. J Biol Chem 2003;278(25):22795–804.
148. Quock RM, Burkey TH, Varga E, et al. The delta-opioid receptor: molecular pharmacology, signal transduction, and the determination of drug efficacy. Pharmacol Rev 1999;51(3):503–32.
149. Bohn LM, Lefkowitz RJ, Gainetdinov RR, Peppel K, Caron MG, Lin FT. Enhanced morphine analgesia in mice lacking beta-arrestin2. Science 1999;286(5449):2495–8.
150. Finn AK, Whistler JL. Endocytosis of the mu opioid receptor reduces tolerance and a cellular hallmark of opiate withdrawal. Neuron 2001;32(5):829–39.
151. Grecksch G, Bartzsch K, Widera A, Becker A, Hollt V, Koch T. Development of tolerance and sensitization to different opioid agonists in rats. Psychopharmacology (Berl) 2006;186(2):177–84.
152. Martini L, Whistler JL. The role of mu opioid receptor desensitization and endocytosis in morphine tolerance and dependence. Curr Opin Neurobiol 2007;17(5):556–64.
153. Bailey CP, Connor M. Opioids: cellular mechanisms of tolerance and physical dependence. Curr Opin Pharmacol 2005;5(1):60–8.
154. He L, Fong J, von Zastrow M, Whistler JL. Regulation of opioid receptor trafficking and morphine tolerance by receptor oligomerization. Cell 2002;108(2):271–82.

Chapter 13
Functional Selectivity at Non-Opioid Peptide Receptors

Anushree Bhatnagar and Sadashiva Karnik

Abstract Pleiotropic signaling pathways activated by peptide hormone G protein-coupled receptors (GPCRs) are thought to mediate the physiological and pathogenic actions of these hormones. The mechanism by which the same ligand–receptor system mediates multiple effects is unclear. In general, cell-based discrepancy in signaling is thought to be responsible and accordingly drug development efforts have been refocused on modulating molecules downstream of receptors. Recent developments suggest that ligand–receptor systems are binary interaction units. A change in receptor or ligand will result in a specific change in functional outcome such as selective activation of pathways or skipping certain steps of a pathway. This phenomenon of ligand-induced functional selectivity is a new paradigm that could potentially lead to design of "smart" drugs that can enhance therapeutic potential of receptors and minimize potential adverse effects.

Keywords Functional selectivity, Selective activation, Ligand-induced conformation, Surrogate ligands, GPCR mechanisms, Peptide hormone receptors

13.1 Introduction

Peptide hormone G protein-coupled receptors (GPCRs) are probably the largest heterogeneous group representing all the five main human GPCR families – *Rhodopsin*, *Secretin*, *Adhesion*, *Glutamate*, and *Frizzled* – within the GPCR super family. These receptors are predicted to share a common structure–function relationship *(1)*, – the seven transmembrane alpha helical bundle switches conformation(s) that initiate intracellular signal upon binding a specific extracellular ligand – despite

A. Bhatnagar and S. Karnik (✉)
Department of Molecular Cardiology, NB50 Lerner Research Institute,
Cleveland Clinic Foundation, Cleveland, OH 44195
e-mail: karniks@ccf.org

K. Neve (ed.), *Functional Selectivity of G Protein-Coupled Receptor Ligands*,
DOI: 10.1007/978-1-60327-335-0_13,

their enormous evolutionary success to elicit biological response to heterogeneous ligands. The prevailing view for long has been that the GPCRs upon activation interact with heterotrimeric G proteins stimulating GDP/GTP exchange and dissociation of α and βγ-subunits. Each of these subunits acts on a large variety of effectors such as enzymes or ion channels that modulate cellular concentration of second messengers or alter transmembrane electrical potential. Phosphorylation and the binding of β-arrestin to phosphorylated GPCRs terminate the G protein activation state of the receptor. The β-arrestin bound phospho-GPCR-agonist complex enters the endocytic cycle, and ultimately replenishes at least a fraction of plasma membrane receptor in a resensitized state. Until recently, this type of agonist-triggered sequential cascade has been the standard paradigm of GPCR signal transduction.

Increasingly, this view is challenged by discoveries of novel signaling phenomena in peptide hormone GPCRs but also in various other GPCRs leading to a more complex signal transduction scenario. Not only a prolonged G protein-independent phase of signaling by the β-arrestin bound to internalized, phosphorylated receptor but also the recruitment of novel signal transducers, scaffolding proteins, and adaptor molecules are documented for many GPCRs. These new observations suggest a cell-based selectivity in signaling response from a GPCR, which relies on the stoichiometry of the cellular signaling components driving the response. The net effect of receptor activation, therefore, is controlled by the differential affinities of the agonist-bound receptor for mediators of various pathways. Many excellent reviews have dealt with these issues previously *(2–6)*.

Several proteins that directly interact with GPCRs engage the receptor quite independent of G protein activation suggesting perhaps the production of receptor active conformations that favor their recruitment selectively *(7)*. The parsimonious view prevalent in the GPCR pharmacology is that agonists produce a single active state favorable to G protein recruitment and activation. However, the versatility of the chemical structure of peptides, in principle, can stabilize a plethora of receptor active states. Heterogeneity of active states of a receptor in a cell can lead to selectivity bias for various interacting proteins and consequently to ligand-induced selectivity in signaling, a phenomenon of functional selectivity. Unlike the cell based-selectivity, we envisage the peptide ligand (or any other ligand) "setting the trajectory" for functional selectivity during activation transition and a specific bias in the downstream functional outcome. Thus, the receptor activated by a specific ligand may bypass some signaling cascades that are known to be primarily coupled to the receptor or even certain steps and preferentially activate others in a linear cascade. For over a decade, either natural or laboratory generated mutations in receptors were known to cause such bias, often deemed an artifact of mutation. Rather, the evolutionarily endowed conformational repertories of GPCRs account for enormous variety of downstream functional outcome. Recently functional selectivity is gaining considerable attention *(6)*. In this chapter, we provide an overview of functional selectivity in peptide GPCRs and delve into our own experience with functional selectivity in angiotensin receptor. Developing drugs with receptor-based functional selectivity is a novel approach and should be differentiated from a conventional drug development paradigm *(2)*.

13.2 Brief Survey of the Phenomenon

The generic mechanism for the peptide hormone GPCR functions involves G protein activation followed by the β-arrestin signaling cascade. Classical view is that agonists bind and induce conformation in receptors that activate heterotrimeric G proteins and lead to canonical second-messenger signaling. The activated receptors are substrates for G protein coupled receptor kinases, some of which in turn are activated by the dissociated G protein subunits. Phosphorylation by GRKs of the agonist-bound receptors causes β-arrestin recruitment, inhibition of G protein signaling and receptor desensitization. β-arrestins scaffold the receptors to membrane-trafficking machinery and cause receptor internalization from the cell surface and sequestration from G proteins *(8)*. Numerous signaling molecules have been found to be recruited to the GPCRs in the endocytic compartment and initiate a G protein independent phase of signals *(9)*. That the second phase of signaling could be activated directly without the activation of G proteins when some ligands bind to peptide hormone receptors and that a large variety of novel signaling molecules can bind to the activated receptor without G proteins was not common knowledge until recently.

Assays currently in use are inadequate to fully harness the potential of functional selectivity for almost all receptor–ligand systems. Cell signaling assays, ERK1/2 activation, or various transcription reporter systems cannot reveal hidden and/or novel receptor conformation-driven signal transduction pathways. Suggestive evidence for functional selectivity, at least as a theoretical possibility or virtual phenomenon, has come from pharmacological analysis of deviation observed in canonical behavior of the G protein initiated signaling cascade. The various steps involved in an agonist-mediated G protein activation, desensitization, down regulation, and the cytoplasmic signal transduction cascade are thought to be sequential. Therefore, receptor internalization and subsequent β-arrestin signaling is positively correlated with G protein activation by the receptor. By inference, a ligand incapable of inducing G protein activation by the receptor is unlikely to produce almost all other responses including receptor internalization *(10, 11)*. However, studies from several receptor types indicate that there is divergence in what is commonly thought of as a classical agonist-receptor behavior.

13.2.1 Selectivity in G Protein Coupling

The ability of a GPCR to bind and activate a specific G protein heterotrimer is one of the most impressive examples of ligand-induced molecular recognition specificity. Conventionally, the type of G protein it activates describes specific GPCR functionality. However, for several peptide hormone GPCRs, this is a conditional property i.e., different ligands can induce association with different G protein signaling pathways in an analog-specific manner.

Thiazolidinones, a novel class of positive allosteric modulators of the follicle-stimulating hormone receptor (FSHR), actually induce association of the FSHR with either $G_{\alpha s}$ or $G_{\alpha i}$ signaling pathways depending on the stereo-isomer bound to FSHR. These are novel therapeutics developed to improving reproductive success mediated by FSHR. Discrete modifications in the chemical structure of the thiazolidinone agonists produced compounds with different pharmacological properties. Select thiazolidinone analogs act as positive allosteric modulators, which activated adenylate cyclase signaling ($G_{\alpha s}$) and others induced the activation of the $G_{\alpha i}$ signaling pathway. Some analogs of this class indeed activated both pathways as would be expected, based on structure–function consideration *(12)*.

A single functional gonadotropin releasing hormone (GnRH) receptor type is found in humans, which has been proposed to assume different conformations to display different selectivity for GnRH analogs and intracellular signaling proteins complexes. GnRH and its analogs are widely employed in modulating gonadotropin and sex steroid secretion to treat infertility, precocious puberty, and many hormone-dependent diseases including endometriosis, uterine fibroids, and prostatic cancer. The GnRH analog effects on reproductive behaviors is exclusively mediated by the $G_{\alpha q}$ signaling pathway coupled to the pituitary GnRH receptor and the GnRH analog effects in many extra pituitary cells in the nervous system and periphery includes in addition the $G_{\alpha q}$-independent ERK pathways. The GnRH analogs mediate both effects and different signaling pathways are utilized with distinctly different relative efficacies. Coupling of the GnRH receptor to the ERK pathway induced by several selective GnRH analogs involves the assembly of a multiprotein signaling complex in association with specialized microdomains of the plasma membrane while by-passing $G_{\alpha q}$ signaling *(13)*.

Thrombosis depends on activation of platelets, shape change, aggregation, secretion, and calcium mobilization, which requires cooperative interaction of P2Y and the protease activated receptors, PAR1 and/or PAR4. The activation of platelets through the PAR-1 receptor induced by the C-terminal peptide derived from human platelet P2Y*(1)* receptor (TFRRRLSRATR), or its analogs TFRRR, YFLLRNP-peptides, and the physiological agonist thrombin involves selective activation of $G_{\alpha 12/13}$ pathways at low concentrations and the activation of both $G_{\alpha q}$ and $G_{\alpha 12/13}$ pathways at higher concentrations. The $G_{\alpha 12/13}$ coupling at low concentration is responsible for platelet shape changes but does not cause aggregation. The events at higher concentrations cause platelet aggregation. Thus, regulation of platelet aggregation depends on the spatio-temporal selectivity of the activation of specific G protein coupled pathways induced by the native ligands *(14)*.

13.2.2 Bypassing the G Protein Activation Step

In case of the cholecystokinin receptor, the antagonist D-Tyr-Gly-[(Nle28, 31,D-Trp30) cholecystokinin-26-32]-phenethyl ester does not produce receptor activation but produces profound internalization of the receptor *(15)*. This is an instance of ligand-induced receptor internalization in absence of complete lack of cholecystokinin-mediated tissue responses. This phenomenon is not unique to cholecystokinin receptor ligands because SR121463B, an inverse agonist at the V2 vasopressin receptor for G protein activation induces recruitment of β-arrestin and stimulation of ERK1/2 *(16)*. In the C5a complement receptor, mutations that dissociate ligand-dependent endocytosis from G protein signaling have been found in a large-scale genetic screen *(17, 18)*.

Such deviation from canonical behavior is the basis for thwarting spread of HIV. A potent inhibitor of HIV-1 infection, aminooxypentane-RANTES (AOP-RANTES), is an analogue of the endogenous CCR5 agonist RANTES. AOP-RANTES promotes rapid internalization of the CCR5 chemokine receptor and, inhibits CCR5 receptor recycling back to the cell surface but does *not* produce CCR5-mediated chemotaxis of T-cells *(19–21)*. The lack of activation of chemotaxis, the canonical function of CCR5, by AOP-RANTES is a deviation but in this instance a desired effect. Thus APO-RANTES protection against HIV-1 infection, a very relevant therapeutic activity, should be considered as functional selectivity. The HIV-1 mediated infection of healthy cells that leads to AIDS is known to occur through the interactions of the viral coat protein gp120, the T-cell membrane proteins CD4, and CCR5. The therapeutically relevant approach to blocking this process is the removal of CCR5 from the cell surface through CCR5 receptor internalization. The chemokine peptide, RANTES normally produces chemotaxis but is not very efficacious in blocking HIV infection. AOP-RANTES can promote rapid internalization of CCR5 receptors bypassing steps in active chemotaxis, the cascade that normally leads to internalization.

13.2.3 Altered Ligand-Dependence of Different Steps

Mutations that dissociate ligand-dependent endocytosis from G protein activation were reported for the V2 vasopressin receptor (V2R) in which this distinction is the basis of a human genetic disease, "nephrogenic syndrome of inappropriate antidiuresis" (NSIAD). The patients, presumptively diagnosed with the syndrome of inappropriate antidiuretic hormone secretion, express gain-of-function point mutations of Arg-137 of the V2R *(22)*. Arg-137 is located in the conserved "DRY" motif of the receptor, and these mutations cause constitutive activation of the receptor leading to NSIAD *(23)*. The NSIAD mutant V2R exhibits constitutive G protein signaling; yet its internalization and arrestin association are strongly ligand-dependent. Thus the mutation affects requirement of ligand for G protein activation but not internalization, suggesting existence of distinct ligand-induced conformations of the V2R, differing in signaling and internalization in this human disease state. These results point to a model of receptor activation in which ligand-specific conformations are capable of differentially activating distinct signaling partners and receptor-based conformational mechanisms that can account for the observed functional selectivity.

13.3 Mechanistic Prelude

Pharmacological evidence accumulated over several past years actually contradicts the classical view that ligand binding to the peptide receptors stimulated or inhibited all receptor functions to an equal extent, suggesting a new paradigm. The concept of "signaling selectivity," the ability of a particular ligand to selectively induce a unique receptor conformation that is capable of stimulating or inhibiting subsets of receptor activities has to be considered. The examples described above suggest that design of such ligands might provide opportunities for the development of novel therapies.

Discovery of various new ways for initiating β-arrestin recruitment without canonical GPCR signaling is an example. The seminal studies attributed the function of β-arrestins as a "brake" on G protein activation by a GPCR. The arrestin family of proteins is now considered universal regulators of GPCR signaling that promote receptor internalization and sequestration of desensitized receptors and recycle the receptors to cell surface, and that also modulate additional signaling pathways, in some instances pathways that are independent of G proteins. Under the new paradigm of functional selectivity, one would predict that it is possible to design ligands that will independently modulate a specific state of the GPCR–arrestin complex. From evaluation of many peptide ligands and receptors, it is now clear that the efficacies for β-arrestin recruitment can be specifically modulated. Although the mechanistic details are rather well described for GPCR-mediated β-arrestin signaling selectivity, the functional selectivity principle should be applicable to other equally important examples, such as Janus kinase coupling in the case of angiotensin receptors *(24)*, CD4-Gp120 assembly with CCR5 receptors *(25)*, EGF receptor signalplex assembly *(24)*, and externalization of misfolded receptors by molecular chaperones.

We prefer to use the term "functional selectivity" in place of many acronyms in use to describe this phenomenon. Members of the Nomenclature Committee of IUPHAR have echoed the importance of an appropriate mechanistic classification of functional selectivity. In fact, the NC-IUPHAR urges those who are interested to actively participate in the ongoing dialogue on the issue of terminology by submitting their opinions directly to http://www.iuphar.org/nciuphar.html

13.4 Ang II Receptor-Specific Observations

The type I receptor for the octapeptide hormone angiotensin II (AT_1) regulates blood pressure and electrolyte homeostasis and is the target of antihypertensive drugs such as candisartan and losartan, which are angiotensin-receptor blockers (ARBs). The AT_1 receptor is a GPCR, mediates various Ang II-dependent physiological responses such as cardiac contraction and growth, smooth muscle constriction, cell proliferation and motility, and aldosterone secretion. Upon Ang II binding, the AT_1 receptor activates G protein dependent formation of inositol triphosphate (IP_3) and diacylglycerol leading to release of stored Ca^{2+} from the endoplasmic reticulum and Ca^{2+}-dependent vasoconstriction. The AT_1 receptor is rapidly phosphorylated by

GPCR kinases (GRK) and β-arrestin is recruited to turn-off G protein activation. Ang II-stimulation of AT_1 receptor has been shown to lead to slow but persistent β-arrestin-mediated activation of ERK1/2 in the cytosol following cessation of $G\alpha_q$ activation *(26)*. This sustained β-arrestin-dependent ERK1/2 signaling is thought to be an important chronic phase of Ang II signaling that is important for novel physiological end points linked to AT_1. Activated receptor has been shown to induce conformational changes in β-arrestin leading to interaction with other molecules or posttranslation modifications, such as phosphorylation and ubiquitination *(27)*.

13.4.1 Discovering Functional Selectivity at AT_1 Receptors

The AT_1 receptor is a prototypical model for the study of functional selectivity. AngII signaling involves canonical relationship between steps of $G_{\alpha q}$ activation, phosphorylation of receptor, β-arrestin recruitment, internalization, and β-arrestin signaling *(23, 25, 26, 27)*. All of the above-mentioned activities for the AT_1 receptor were inhibited by the ARBs losartan and candesartan. Catt and coworkers demonstrated early that the Ang II-antagonist Sarile (Sar^1, Ile^8-AngII) did not display a classical antagonism; that is, Sarile binding did not inhibit all receptor functions to an equal extent *(26)*. Specifically, AT_1 receptor failed to accumulate inositol phosphate upon Sarile binding; however, the internalization of the receptor was not blocked. Lack of correlated inhibition of $G_{\alpha q}$ activation, receptor phosphorylation, β-arrestin recruitment by the receptor, and internalization is now documented for several mutant receptors *(28–30)*. For instance, the AT_1 receptor mutant where the conserved DRY motif is replaced with AAA does not activate $G_{\alpha q}$, but induces β-arrestin recruitment and downstream signaling. A systematic study using specifically designed analogs of Ang II indicated that some analogs of AngII do bypass G protein activation and GRK recruitment steps and preferentially mediate ERK1/2 activation, which is thought to be β-arrestin dependent *(31)*.

A proof-of-principle ligand for functional selectivity, [Sar^1, Ile^4, Ile^8] AngII (SII), was discovered. SII stimulates β-arrestin-dependent ERK1/2 activation, in the absence of measurable G protein agonism, from the native AT_1 receptor *(32)*. The AT_1 receptor mutants, D74A, H256A, Y292F and DRY→AAA when activated by Ang II display β-arrestin mediated ERK1/2 activation without $G_{\alpha q}$-PLC activation. Therefore, it is very important to be aware that both the ligand and the receptor display critical selectivity in producing functional selectivity (Fig. 13.1.). For instance, SII does not bypass the G protein-activating conformation of the constitutively activated N111G mutant AT_1 receptor. This mutant displays strong bias against phosphorylation, β-arrestin recruitment, and internalization *(32, 33)*.

The studies using the N111G mutant demonstrated that G protein activation does not always stimulate receptor phosphorylation and β-arrestin recruitment and contradicted the prevailing belief in correlated efficacies for G protein stimulation and phosphorylation by GRKs. In the case of the wild-type AT_1 receptor, it was soon discovered that G protein activation is not necessary for actuating β-arrestin-dependent functions. Several ligands that recruit β-arrestin and/or induce receptor internalization

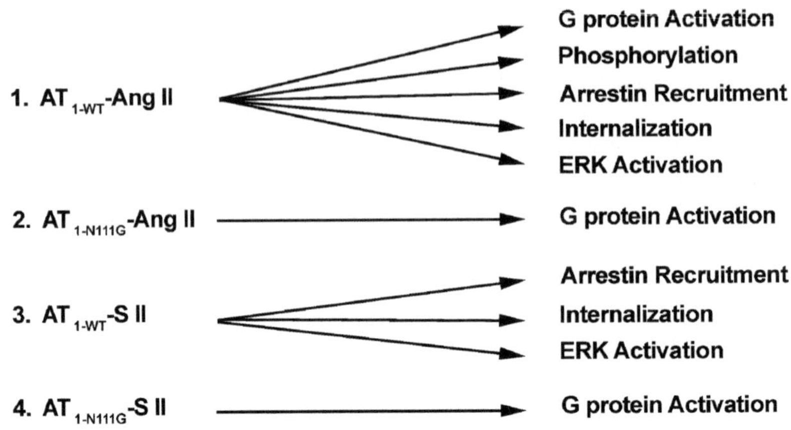

Fig. 13.1 Binary ligand–receptor interaction leading to functional selectivity. AT_{1N111G} is a constitutively active mutant of the AT_1 receptor in which residue Asn111 is mutated to Gly. Resposes elicited by the binding of Ang II or the functionally selective ligand [Star^1, Ile^4, Ile^8] Ang II (S II) to wild type or mutant receptors are indicated at the right (see ref. *49, 50*).

Table 13.1 Approach for development of analogs that display functional selectivity

Analog	IP	PO	Int	ERK1/2
Ang II	100	100	100	100
[Sar^1]Ang II	115	—	—	—
[Ala^1]Ang II	—	80	—	80
[Sar^{1}Gln^2]Ang II	—	—	—	—
[Sar^{1}Ala^3]Ang II	115	—	—	—
[Sar^{1}Ala^4]Ang II	50	80	—	70
[Sar^{1}Ala^5]Ang II	—	—	—	—
[Sar^{1}Ala^6]Ang II	—	—	—	—
[Sar^{1}Ala^7]Ang II	—	—	—	—
[Sar^{1}Ala^8]Ang II	18	75	—	35
[Sar^1]Ang II-amide	85	—	—	80
[Sar^{1}Cha^4]Ang II	48	90	—	75
[Sar^{1}Cha^8]Ang II	40	50	—	65
[Sar^{1}Ile^8]Ang II	20	75	—	50
[Sar^{1}DiP^8]Ang II	—	75	—	—
[Sar^{1}Ile^{4}Ile^8]Ang II	0	80	—	50
[Sar^{1}Gly^4 Gly^8]Ang II	0	—	—	—

All values represent percentage of Ang II control at saturation
—, no change from control; IP, inositol phosphate; PO, phosphorylation; Int, internalization, ERK1/2, activated extracellular signal regulated kinases 1 and 2
Ang II: NH2-Asp-Arg-Val-Try-Ile-His-Pro-Phe-COOH

without stimulating any detectable G protein signaling have been designed and tested (Table 13.1). Similarly, mutant receptors that do not couple to G proteins bind Ang II leading to β-arrestin signaling as shown by siRNA silencing of β-arrestin *(34)*.

13.4.2 In Vivo Relevance of Functional Selectivity at AT1 Receptors

That functional selectivity is relevant in vivo when ligands, such as SII, act on endogenous AT_1 receptors is gaining considerable attention at present in laboratory research. Daniel and coworkers *(35)* used in vitro and in vivo approaches to clarify the signaling and salt-water drinking responses to SII in rats. Hypothalamic cortical injections of Ang II increased both water and brine intake, a response due to AT_1 receptor stimulation of phospholipase C activation and the subsequent formation of diacylglycerol and IP_3, which then activated protein kinase C (PKC) and increased intracellular calcium. SII mimicked Ang II stimulation of MAPK in AT_1 receptor-transfected cells and rat brain tissue but blocked the calcium mobilizing effects of Ang II in both settings: Moreover, central injection of SII increased intake of 1.5% NaCl, despite blocking the IP_3 formation and water intake. Distinct conclusions that can be drawn from these studies are (a) SII antagonizes the water-drinking response that requires the G protein coupled signaling, and (b) SII stimulated the brine-drinking response, revealed for the first time to be a G protein independent behavioral response. In physiological studies, the salt-intake response is known to be more delayed, which appears to be consistent with the delayed time course of β-arrestin dependent ERK1/2 activation stimulated by the SII-bound AT_1 receptor in cells. These studies are important in laying the foundation for the hypothesis that functional selectivity by the AT_1 receptor has behavioral relevance as shown and may be mechanistically more important than currently appreciated particularly in considering therapeutic intervention.

Recently, SII has been shown to induce contractility via β-arrestin signaling in isolated cardiac myocytes *(36)*, suggesting that functional selectivity through β-arrestin will have a significant role in cardiac function. Aplin and coworkers *(37, 38)* examined spatiotemporal consequences of SII-mediated activation of ERK compared with Ang II activation. To examine the biological effects of ERK1/2 activity resulting from differential activation of the AT_1 receptor in the heart, SII and Ang II were used in native preparations of cardiac myocytes and Langendorff-perfused beating hearts. SII did not activate inotropic or chronotropic responses, which are G-protein dependent, yet stimulated the β-arrestin-dependent ERK1/2 activation. The Ang II activated pool of ERK1/2 rapidly translocated to the nucleus, while the β-arrestin-scaffold ERK1/2 pool remained in the cytosol. In a different study, Hunton et al. demonstrated β-arrestin-mediated chemotaxis of HEK-293 cells with stable expression of the AT_1 receptor in response to Ang II *(39)*. The Ang II peptide analog SII induces chemotaxis in these cells that is unaffected by pertussis toxin-mediated suppression of $G_{\alpha i}$. Suppression of β-arrestin2 expression using small interfering RNA (siRNA) essentially eliminated AT_1 receptor-mediated chemotaxis induced by either Ang II or the SII-Ang II peptide but had no effect on epidermal growth factor (EGF) induced chemotaxis. Inhibiting p38 mitogen-activated protein kinase decreased AT_1 receptor-mediated chemotaxis and eliminated EGF-mediated chemotaxis, suggesting that p38 plays a role in chemotaxis. These data suggest that SII-Ang II can mediate chemotaxis through mechanisms that are G protein independent.

The signaling pathways shown to be activated by SII until now include Src, ERK1/2, Akt, and PI3K which are mostly β-arrestin mediated and are considered to be cytoprotective and proliferative *(27, 40)*.

13.5 Knowledge-Based Ligand Design Strategy

13.5.1 Designing Functionally Selective Ligands

In general, drug design strategies should consider the thermodynamic process of receptor conformational isomerization, which leads to spontaneous display of active states capable of coupling to G protein signaling or β-arrestin signaling or any other signaling. How the isomerized receptor state evolves into a specific signaling entity is elegantly described elsewhere in this book, namely, the extended ternary complex and the cubic ternary complex models.

The characteristics of both the ligand and the receptor should be considered when thinking of a model for functional selectivity. For practical reasons, we will consider a formation of a binary complex as the simplest model scheme for potentially generating distinct active conformations, which can account for functional selectivity. The wild-type receptor complexed (a ligand with receptor molecule) with native ligand, such as Ang II, would generate the full spectrum of active conformations that can account for all different signals that directly arise from the receptor. The binary interaction unit could be systematically altered, such as when a mutant receptor complexes with native ligand, which could generate a subset of active conformations and a distinctly different spectrum of functions. Alternatively, an altered ligand bound to the wild-type receptor could also generate a subset of the active conformations of the WT receptor rather than smaller amounts of a single active state. The ligand modification approach could be refined to achieve a functionally selective ligand-receptor binary unit. We have successfully applied this approach to Ang II-AT_1 receptor binary unit and discovered a ligand (Ang II-analog) that complexes with the wild-type AT_1 receptor and produces functional selectivity.

Classical approaches in Ang II analog development used aortic smooth muscle contraction assays, which lead to the identification of agonists, partial agonists, and antagonists. Deviations from canonical physiological behavior were described for many Ang II analogs, in term of "delayed action" (chronic effects) and "tachyphylaxis" (desensitization), well before the biochemistry and molecular pharmacology of the AT_1 receptor was described. With the availability of transfected receptors in pure cell culture and discrete assays for receptor binding, G protein signaling, receptor phosphorylation, endocytosis, and G protein independent ERK1/2 activation studies, a new look at Ang II analogs was feasible.

The Tyr^4 and Phe^8 side chains were known to be the major feature of Ang II that determines its biological activity in the aortic contraction assay. However, the specific role of other Ang II residues in the regulation of receptor activity was undetermined.

Miura and Karnik investigated the effect of various side-chain substitutions of Ang II (including single or combined substitution of Tyr^4 and Phe^8) in interaction with the AT_1 receptor and its activation *(28)*. The aromatic Tyr^4 and Phe^8 side chains were found to be switches for G protein activation. The Tyr^4 side chain is necessary for high affinity binding and activation, whereas the aromaticity of Phe^8 side chain is necessary for G protein activation and plays no role in receptor affinity. Subsequently, Holloway et al. investigated specific roles of the various Ang II side chains in promoting internalization, phosphorylation, and ERK1/2 signaling of the AT_1 receptor *(31)*. Surprisingly, AT_1 receptor internalization was unaffected by substitution of Ang II side chains at any position. Assuming that AT_1 receptor phosphorylation is associated with receptor internalization, would the same analogs affect receptor phosphorylation? Unexpectedly, several substitutions at Phe^8 were much less effective than Ang II at stimulating receptor phosphorylation but not others suggesting that these two steps are easily dissociated when position-eight side chain is not aromatic. Was β-arrestin-dependent ERK1/2 signaling activated by the AT_1 receptor differentially affected by substitutions in Ang II? Substitutions at position eight that affect side-chain aromaticity and decrease AT_1 receptor phosphorylation did not inhibit arrestin-dependent (~50%) ERK1/2 activation. This indicated that Phe^8 is required for early phase of ERK1/2 activation via the G protein-dependent pathway. However, the late phase of ERK1/2 activation, which is β-arrestin dependent, is preserved in most analogs including those lacking both Tyr^4 and Phe^8 side-chains. A series of analogs with aliphatic substitutions at both position 4 and 8 were found to display full agonism for β-arrestin-dependent ERK1/2 activation while completely lacking G protein agonism. This series of Ang II analogs is novel and unique in generating nearly perfect functional selectivity, and several other analogs that produce functional selectivity retain a weaker broad spectrum signaling activity.

These studies provided four important conclusions: *(1)* specific substitutions on Ang II did not equally affect G protein activation, receptor phosphorylation, and ERK1/2 activation; *(2)* there is a key role for aromaticity of Phe^8 in receptor phosphorylation and early phase activation of ERK1/2, with important differences in tolerance to specific substitutions; *(3)* receptor internalization was unaffected by substitution at any position in Ang II, suggesting that the requirements for AT_1 receptor internalization seem to be less stringent than receptor activation and signaling, and *(4)* aliphatic substitutions at both position 4 and 8 produce ligands that display functional selectivity when complexes with the wild-type AT_1 receptor.

Furthermore, this series of studies demonstrated that the type of assay used to characterize ligands would greatly define the efficacy of molecules in eliciting varied responses. Thus, if the traditional radioligand-binding assay is employed, then the ligands discovered by that screen will be specific binding site inhibitors. Functional screening will yield ligands that can modulate the particular biological function without exerting apparent effects on other activities. By exploiting this approach, it should be possible to elicit novel biological effects from well-studied GPCRs such as AT_1 receptor. It should be possible to design ligands, which can selectively restore aberrant functions displayed by disease-causing mutations.

13.5.2 Pharmacological Rescue of Disease Causing Mutant GPCRs

Loss of selective aspects of overall receptor functionality is a common cause of disease due to mutation, and pharmacologically augmenting the lost functionality is emerging as a promising therapeutic strategy. An important aspect of drug design came from studies of nephrogenic diabetes isipidus (NDI) wherein over 150 different mutations within the coding region of the vasopressin type 2 receptor (*V2R*) gene were identified. Most of the mutations lead to misfolding of the receptor such that the ER targets these proteins to proteosomal degradation, with the resulting loss of cell surface receptors leading to the loss in signaling that causes this disease condition. Bouvier and colleagues *(41)* and Hawtin *(42)* have shown that selective V2-receptor antagonists (for example, SR121463A, VPA985, and SR49059) can lead to 8- to 15-fold increase in expression of functional V2 vasopressin receptor at the cell surface, rescuing these receptors from proteasome degradation. Robben et al. *(43)* have recently reported in their in vitro studies that constitutively internalized receptors are the major cause for congenital nephrogenic diabetes insipidus. Their studies also show that high affinity antagonists OPC31260 and OPC41061 induced functional rescue, and proposed as clinically safe, thus, promising candidates to relieve NDI. Studies by Bernier et al. *(44)* characterized the mutation R137H in the *V2R* gene, which was reported to lead to constitutive endocytosis. Using various pharmacological techniques, the author demonstrated that SR49059 mediates its action on R137H V2R by acting as a chaperone to target the misfolded protein to the cell surface and was not due to inhibition of β-arrestin-mediated constitutive endocytosis or the stabilization of the receptor at the cell surface. These studies alluded to the requirement of specific pharmacological agents for either transport to cell surface or retention in the plasma membrane of the mutant receptor, hence rescue of the phenotype.

13.6 Conclusion and Speculation

The functional selectivity phenomenon is widespread within the GPCR superfamily, and its importance in pathophysiology and drug development is greater than currently recognized. With regards to drug development, uses of screening strategies based on specific functional assays are essential. For example, cleavage of receptor-engaged β-arrestin *(45, 49)* could be a novel functional selectivity for which we lack appropriate assays and the knowledge of physiological endpoints. For those GPCRs that serve not only as a simple on/off switches but also as a multiple signaling path regulators, utilization of sensitive assays for differential coupling properties have indeed elucidated functional selectivity of ligands *(46)*. In synaptic communication, the C terminus of DFrizzled2 (post-synaptic GPCR for Wingless) is cleaved and translocated into the nucleus for modulating gene regulation *(47)*. The translocation of the cleaved C-terminus is dependent on Wingless signaling, but the proteolysis of internalized DFrizzled is ligand-independent. Development of functionally selective ligands for each of the different steps in the process would require refined assays. When discovered, such ligands bear enormous promise for better understanding the role of different events in synaptic communication. Indeed, significant advances have been made toward discovering GPCR dimer-specific ligands that display novel analgesic properties. The opioid agonist ligand 6α-guanidinonaltrindole (6α-GNTI) selectively activates only opioid receptor heterodimers but not homodimers. The opioid receptor heterodimers are functionally relevant in vivo, responsible for analgesia in the spinal cord but not in the brain *(48)*. Identification of functional selectivity and its exploitation in drug development will result in pharmacological tools that will have enormous potential in therapy and reducing unwanted side effects.

Acknowledgments Supported by NRSA grant # F32 HL088893 to Anushree Bhatnagar and RO1 grant # HL57470 to Sadashiva Karnik

References

1. Karnik SS, Gogonea C, Patil S, Saad Y, Takezako T. Activation of G-protein-coupled receptors: a common molecular mechanism. Trends in Endocrinol Metab. 2003; 14 (9): 431–7.
2. Kenakin T. Ligand-selective receptor conformations revisited: the promise and the problem. Trends Pharmacol Sci. 2003; 24(7): 346–54.
3. Kenakin T. Predicting therapeutic value in the lead optimization phase of drug discovery. Nat Rev Drug Discov. 2003; 2(6): 429–38.
4. Kenakin T. Functional selectivity through protean and biased agonism: who steers the ship? Mol Pharmacol. 2007; 72(6):1393–401.
5. Perez DM, Karnik SS. Multiple signaling states of G-protein-coupled receptors. Pharmacol Rev. 2005; 57(2): 147–61.
6. Urban JD, Clarke WP, von Zastrow M, et al. Functional selectivity and classical concepts of quantitative pharmacology. J Pharmacol Exp Ther. 2007; 320(1): 1–13.
7. Sun Y, McGarrigle D, Huang XY. When a G protein-coupled receptor does not couple to a G protein. Mol Biosyst. 2007; 3(12):849–54.
8. Luttrell LM, Lefkowitz RJ. The role of beta-arrestins in the termination and transduction of G-protein-coupled receptor signals. J Cell Sci. 2002; 115:455–65.
9. Bhatnagar A, Sheffler DJ, Kroeze WK, Compton-Toth B, Roth BL. Caveolin-1 interacts with 5-HT2A serotonin receptors and profoundly modulates the signaling of selected Galpha-coupled protein receptors. J Biol Chem. 2004; 279(33): 34614–23.
10. Oppermann M, Mack M, Amanda E. Proudfoot I, Olbrich H. Differential effects of CC chemokines on CC chemokine receptor 5 (CCR5): phosphorylation and identification of phosphorylation sites on the CCR5 carboxyl terminus J Biol Chem. 1999; 274: 8875–85.
11. Koenig JA and Edwardson JM. Endocytosis and recycling of G protein-coupled receptors. Trends Pharmacol Sci. 1997; 18: 276–87.
12. Arey BJ, Yanofsky SD, Claudia Pérez M, et al. Differing pharmacological activities of thiazolidinone analogs at the FSH receptor. Biochem Biophys Res Commun. 2008; 368(3): 723–8.
13. Bliss SP, Navratil AM, Breed M, Skinner DC, Clay CM, Roberson MS. Signaling complexes associated with the type I gonadotropin-releasing hormone (GnRH) receptor: colocalization of extracellularly regulated kinase 2 and GnRH receptor within membrane rafts. Mol Endocrinol. 2007; 21(2):538–49.

14. Mao Y, Jin J, Kunapuli SP. Characterization of a new peptide agonist of the protease-activated receptor-1 Biochem Pharmacol. 2008; 75(2): 438–47.
15. Roettger BF, Ghanekar D, Rao R, et al. Antagonist-stimulated internalization of the G protein-coupled cholecystokinin receptor, Mol Pharmacol. 1997; 51: 357–62.
16. Azzi M, Pascale G. Charest, SA, et al. β–Arrestin-mediated activation of MAPK by inverse agonists reveals distinct active conformations for G protein-coupled receptors, Proc Natl Acad Sci USA 2003; 100: 11406–11.
17. Baranski TJ, Herzmark P, Lichtarge O, et al. C5a receptor activation. Genetic identification of critical residues in four transmembrane helices. J Biol Chem. 1999; 274(22): 15757–65.
18. Whistler JL, Gerber BO, Meng EC, Baranski TJ, von Zastrow M, Bourne HR. Constitutive activation and endocytosis of the complement factor 5a receptor: evidence for multiple activated conformations of a G protein-coupled receptor. Traffic. 2002; 3(12):866–77.
19. Simmons G, Clapham PR, Picard L, et al. Potent inhibition of HIV-1 infectivity in macrophages and lymphocytes by a novel CCR5 antagonist. Science, 1997; 276: 276–79.
20. Amara A, Le Gall S, Schwartz O, et al. HIV coreceptor downregulation as antiviral principle: SDF-1 alpha -dependent internalization of the chemokine receptor CXCR4 contributes to inhibition of HIV replication J Exp Med. 1997; 186: 139–46.
21. Rodriguez-Frade JM. Similarities and Differences in RANTES- and (AOP)-RANTES-triggered signals: implications for chemotaxis. J Cell Biol. 1999; 144: 755.
22. Feldman BJ, Rosenthal SM, Vargas GA, et al. Nephrogenic syndrome of inappropriate antidiuresis. N Engl J Med. 2005; 352(18): 1884–90.
23. Rosenthal SM, Feldman BJ, Vargas GA, Gitelman SE. Nephrogenic syndrome of inappropriate antidiuresis (NSIAD): a paradigm for activating mutations causing endocrine dysfunction. Pediatr Endocrinol Rev. 2006; 4Suppl 1:66–70.
24. Yin G, Yan C, Berk BC. Angiotensin II signaling pathways mediated by tyrosine kinases. Int J Biochem Cell Biol. 2003; 35(6): 780–3.
25. Biorn AC, Cocklin S, Madani N, et al. Mode of action for linear peptide inhibitors of HIV-1 gp120 interactions. Biochemistry 2004; 43(7):1928–38.
26. Hunyady L and Catt KJ. Pleiotropic AT1 receptor signaling pathways mediating physiological and pathogenic actions of angiotensin II. Mol Endocrinol. 2006; 20(5): 953–70.
27. Lefkowitz RJ and Shenoy SK. Transduction of receptor signals by β-arrestins, Science 2005; 308: 512–17.
28. Miura S, Feng YH, Husain A, Karnik SS. Role of aromaticity of agonist switches of angiotensin II in the activation of the AT_1 receptor. J Biol Chem. 1999; 274 (11): 7103–10.
29. Feng YH, Karnik SS. Role of transmembrane helix IV in G-protein specificity of the angiotensin II type 1 receptor. J Biol Chem. 1999; 274(50): 35546–52.
30. Feng YH, Miura S, Husain A, Karnik SS. Mechanism of constitutive activation of the AT_1 receptor: influence of the size of the agonist switch-binding residue Asn (111). Biochemistry 1998; 37(45): 15791–8.
31. Holloway AC, Qian H, Pipolo L, et al. Side-chain substitutions within angiotensin II reveal different requirements for signaling, internalization, and phosphorylation of type 1A angiotensin receptors. Mol Pharmacol. 2002; 61(4): 768–77.
32. Wei H, Ahn S, Shenoy SK, et al. Independent beta-arrestin2 and G protein-mediated pathways for angiotensin II activation of extracellular signal-regulated kinases 1 and 2. Proc Natl Acad Sci USA 2003; 100(19):10782–7.
33. Thomas WG, Qian H, Chang CS, Karnik S. Agonist-induced phosphorylation of the angiotensin II (AT(1A)) receptor requires generation of a conformation that is distinct from the inositol phosphate-signaling state. J Biol Chem. 2000; 275(4): 2893–900.
34. Wei H, Ahn S, Barnes WG, Lefkowitz RJ. Stable interaction between beta-arrestin2 and angiotensin type 1A receptor is required for beta-arrestin2-mediated activation of extracellular signal-regulated kinases 1 and 2. J Biol Chem. 2004; 279(46): 48255–61.
35. Daniels D, Yee DK, Faulconbridge LF, and Fluharty SJ. Divergent behavioral roles of angiotensin receptor intracellular signaling cascades. Endocrinology, 2005; 146: 5552–60.

36. Rajagopal K, Whalen EJ, Violin JD, Stiber JA, Rosenberg PB, Premont RT. Beta-arrestin2-mediated inotropic effects of the angiotensin II type 1A receptor in isolated cardiac myocytes. Coffman TM, Rockman HA, Lefkowitz RJ. Proc Natl Acad Sci USA 2006; 103(44): 16284–9.
37. Aplin M, Christensen GL, Schneider M, et al. Differential extracellular signal-regulated kinases 1 and 2 activation by the angiotensin type 1 receptor supports distinct phenotypes of cardiac myocytes. Basic Clin Pharmacol Toxicol. 2007; 100 (5): 296–301.
38. Aplin M, Christensen GL, Schneider M, et al. The angiotensin type 1 receptor activates extracellular signal-regulated kinases 1 and 2 by G protein-dependent and -independent pathways in cardiac myocytes and langendorff-perfused hearts. Basic Clin Pharmacol Toxicol. 2007; 100(5): 289–95.
39. Hunton DL, Barnes WG, Kim J, et al. Beta-arrestin2-dependent angiotensin II type 1A receptor-mediated pathway of chemotaxis. Mol Pharmacol. 2005; 67(4): 1229–36.
40. DeWire SM. Beta-arrestins and cell signaling. Annu Rev Physiol. 2007; 69, 483–510.
41. Morello JP, Salahpour A, Laperriere A, et al. Pharmacological chaperones rescue cell-surface expression and function of misfolded V2 vasopressin receptor mutants. J Clin Invest. 2000; 105:887–95.
42. Hawtin SR. Charged residues of the conserved DRY triplet of the vasopressin V1a receptor provide molecular determinants for cell surface delivery and internalization. Mol Pharmacol. 2005; 68(4): 1172–82.
43. Robben JH, Sze M, Knoers, NVAM, Deen PMT. Functional rescue of vasopressin V2 receptor mutants in MDCK cells by pharmacochaperones: relevance to therapy of nephrogenic diabetes insipidus, Am J Physiol Renal Physiol. 2007; 292(1): F253–F260.
44. Bernier V, Lagacé M, Lonergan M, Arthus MF, Bichet DG, Bouvier M. Functional rescue of the constitutively internalized V2 vasopressin receptor mutant R137H by the pharmacological chaperone action of SR49059. Mol Endocrinol. 2004; 18(8): 2074–84.
45. Lee C, Bhatt S, Shukla S, et al. Site-specific cleavage of GPCR-engaged Beta-arrestin: influence of the AT1 receptor conformation on scissile site selection. J Biol Chem. 2008; M803062200
46. Tateyama M and Kubo Y. Dual signaling is differentially activated by different active states of the metabotropic glutamate receptor 1a. Proc Natl Acad Sci USA 2006; 103 (4) 1124–8.
47. Mathew D, Ataman B, Chen J, Zhang Y, Cumberledge S, Budnik V. Wingless signaling at synapses is through cleavage and nuclear import of receptor DFrizzled2. Science 2005; 310: 1344–7.
48. Waldhoer M, Fong J, Jones RM, Lunzer MM, Sharma SK, Kostenis E, Portoghese PS, Whistler JL. A heterodimer-selective agonist shows in vivo relevance of G protein-coupled receptor dimers. Proc Natl Acad Sci USA 2005; 102 (25): 9050–5.
49. Lee C, Hwang SA, Jang SH, Chung HS, Bhat MB, Karnik SS. Manifold active-state conformations in GPCRs: agonist-activated constitutively active mutant AT_1 receptor preferentially couples to Gq compared to the wild-type AT_1 receptor. FEBS Lett. 2007;581(13): 2517–22.
50. Hunyady L. Molecular mechanisms of angiotensin II receptor internalization. J Am Soc Nephrol. 1999; 10 Suppl 11:S47–56.

Index

Color Plates

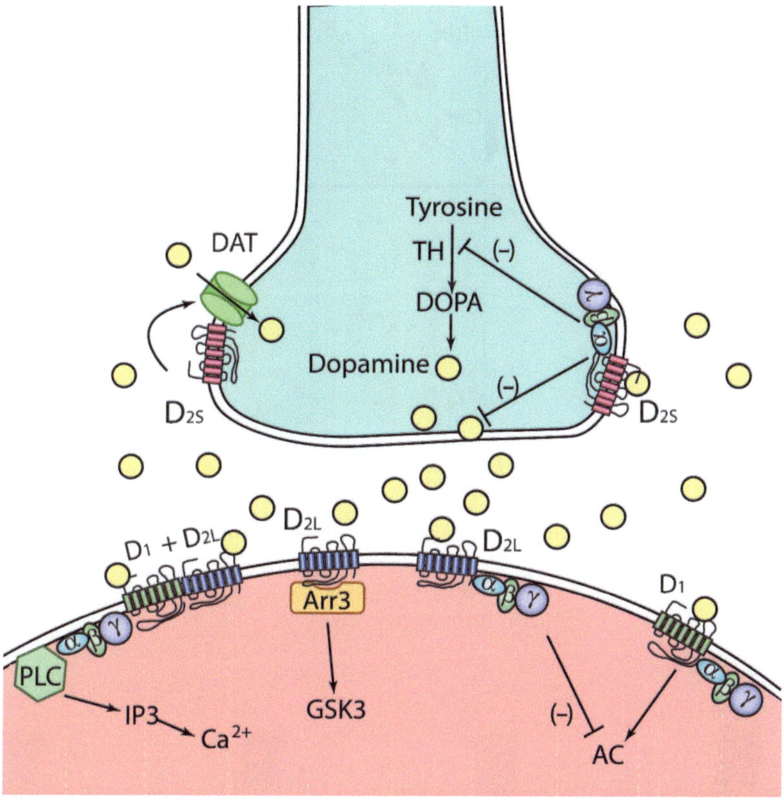

Fig. 6.2 Dopamine autoreceptors and postsynaptic receptors. D_1 and D_2 postsynaptic receptors activate G proteins to stimulate and inhibit, respectively, adenylate cyclase (AC). Activation of postsynaptic D_2 receptors also leads to arrestin-mediated signaling that activates glycogen synthase kinase 3 (GSK3). Binding of dopamine to both receptors in the D_1/D_2 heteromer, in contrast, activates phospholipase C (PLC), producing inositol trisphosphate (IP3) and leading to mobilization of Ca^{2+}. D_{2L} is the major postsynaptic D_2 receptor in the basal forebrain. In dopaminergic terminals, dopamine autoreceptors, believed to be primarily the D_{2S} receptor, inhibit the tyrosine hydroxylase- (TH-) catalyzed synthesis of dopamine, inhibit dopamine release, and enhance the reuptake of dopamine via the dopamine transporter (DAT). Dopamine autoreceptors are more sensitive than postsynaptic receptors, responding to lower concentrations of dopamine. There is abundant evidence for involvement of G protein-mediated pathways such as inhibition of adenylate cyclase and activation of K^+ channels in the regulation of dopamine synthesis and release, whereas the involvement of arrestin-mediated pathways has not been tested. SKF83959 activates the D_1/D_2 heteromer but not the D_1 receptor, and SKF83822 activates the D_1 receptor but not the heteromer. In many model systems, dihydrexidine is an agonist at D_2 autoreceptors but an antagonist at D_2 postsynaptic receptors, whereas the reverse is characteristic of aripiprazole

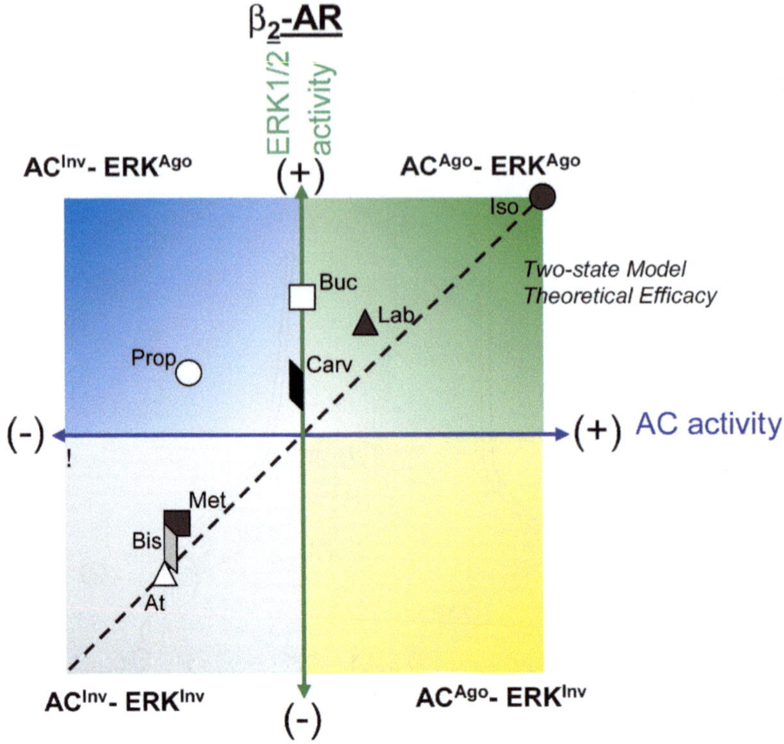

Fig. 7.1 Pluridimensional efficacy – β2-adrenergic receptor. (*For complete caption refer page 115*)

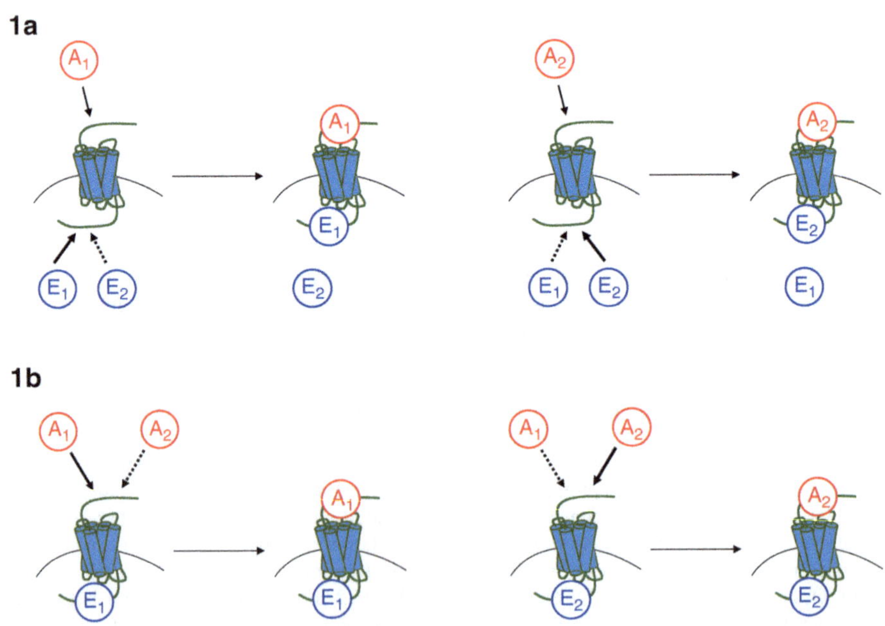

Fig. 12.1 Putative mechanisms of functional selectivity. (*For complete caption refer page 253*)

MIX
Papier aus verantwortungsvollen Quellen
Paper from responsible sources
FSC® C105338

If you have any concerns about our products,
you can contact us on
ProductSafety@springernature.com

In case Publisher is established outside the EU,
the EU authorized representative is:
Springer Nature Customer Service Center GmbH
Europaplatz 3, 69115 Heidelberg, Germany

Printed by Libri Plureos GmbH
in Hamburg, Germany